I0896269

Biostimulation Remediation Technologies for Groundwater Contaminants

Ashok K. Rathoure
Independent Researcher, India

A volume in the Advances in Environmental Engineering and Green Technologies (AEEGT) Book Series

IGI Global
DISSEMINATOR OF KNOWLEDGE

Published in the United States of America by
IGI Global
Engineering Science Reference (an imprint of IGI Global)
701 E. Chocolate Avenue
Hershey PA, USA 17033
Tel: 717-533-8845
Fax: 717-533-8661
E-mail: cust@igi-global.com
Web site: http://www.igi-global.com

Copyright © 2018 by IGI Global. All rights reserved. No part of this publication may be reproduced, stored or distributed in any form or by any means, electronic or mechanical, including photocopying, without written permission from the publisher.
Product or company names used in this set are for identification purposes only. Inclusion of the names of the products or companies does not indicate a claim of ownership by IGI Global of the trademark or registered trademark.

Library of Congress Cataloging-in-Publication Data

Names: Rathoure, Ashok K., 1983- editor.
Title: Biostimulation remediation technologies for groundwater contaminants / Ashok K. Rathoure, editor.
Description: Hershey, PA : Engineering Science Reference, 2018. | Includes bibliographical references and index.
Identifiers: LCCN 2017028944| ISBN 9781522541622 (hardcover) | ISBN 9781522541639 (ebook)
Subjects: LCSH: Bioremediation. | Groundwater--Purification. | Groundwater--Pollution. | Soil pollution.
Classification: LCC TD192.5 .B559 2018 | DDC 628.1/68--dc23 LC record available at https://lccn.loc.gov/2017028944

This book is published in the IGI Global book series Advances in Environmental Engineering and Green Technologies (AEEGT) (ISSN: 2326-9162; eISSN: 2326-9170)

British Cataloguing in Publication Data
A Cataloguing in Publication record for this book is available from the British Library.

All work contributed to this book is new, previously-unpublished material.
The views expressed in this book are those of the authors, but not necessarily of the publisher.

For electronic access to this publication, please contact: eresources@igi-global.com.

Advances in Environmental Engineering and Green Technologies (AEEGT) Book Series

ISSN:2326-9162
EISSN:2326-9170

Editor-in-Chief: Sang-Bing Tsai, University of Electronic Science and Technology of China Zhongshan Institute, China & Ming-Lang Tseng, Lunghwa University of Science and Technology, Taiwan & Yuchi Wang, University of Electronic Science and Technology of China Zhongshan Institute, China

MISSION

Growing awareness and an increased focus on environmental issues such as climate change, energy use, and loss of non-renewable resources have brought about a greater need for research that provides potential solutions to these problems. Research in environmental science and engineering continues to play a vital role in uncovering new opportunities for a "green" future.

The **Advances in Environmental Engineering and Green Technologies (AEEGT)** book series is a mouthpiece for research in all aspects of environmental science, earth science, and green initiatives. This series supports the ongoing research in this field through publishing books that discuss topics within environmental engineering or that deal with the interdisciplinary field of green technologies.

COVERAGE

- Pollution Management
- Electric Vehicles
- Waste Management
- Sustainable Communities
- Alternative Power Sources
- Industrial Waste Management and Minimization
- Green Technology
- Radioactive Waste Treatment
- Cleantech
- Contaminated Site Remediation

IGI Global is currently accepting manuscripts for publication within this series. To submit a proposal for a volume in this series, please contact our Acquisition Editors at Acquisitions@igi-global.com or visit: http://www.igi-global.com/publish/.

The Advances in Environmental Engineering and Green Technologies (AEEGT) Book Series (ISSN 2326-9162) is published by IGI Global, 701 E. Chocolate Avenue, Hershey, PA 17033-1240, USA, www.igi-global.com. This series is composed of titles available for purchase individually; each title is edited to be contextually exclusive from any other title within the series. For pricing and ordering information please visit http://www.igi-global.com/book-series/advances-environmental-engineering-green-technologies/73679. Postmaster: Send all address changes to above address. ©© 2018 IGI Global. All rights, including translation in other languages reserved by the publisher. No part of this series may be reproduced or used in any form or by any means – graphics, electronic, or mechanical, including photocopying, recording, taping, or information and retrieval systems – without written permission from the publisher, except for non commercial, educational use, including classroom teaching purposes. The views expressed in this series are those of the authors, but not necessarily of IGI Global.

Titles in this Series

For a list of additional titles in this series, please visit:
https://www.igi-global.com/book-series/advances-environmental-engineering-green-technologies/73679

Green Production Strategies for Sustainability
Sang-Binge Tsai (University of Electronic Science and Technology of China (Zhongshan Institute), China & Civil Aviation University of China, China) Bin Liu (Shanghai Maritime University, China) and Yongian Li (Nankai University,China)
Engineering Science Reference • ©2018 • 325pp • H/C (ISBN: 9781522535379) • US $235.00

Innovative Strategies and Frameworks in Climate Change Adaptation Emerging Research...
Alexander G. Flor (University of the Philippines Open University, Philippines) and Benjamina Gonzalez Flor (University of the Philippines, Philippines)
Engineering Science Reference • ©2018 • 165pp • H/C (ISBN: 9781522527671) • US $165.00

Utilizing Innovative Technologies to Address the Public Health Impact of Climate Change...
Debra Weiss-Randall (Florida Atlantic University, USA)
Engineering Science Reference • ©2018 • 295pp • H/C (ISBN: 9781522534143) • US $185.00

Economical and Technical Considerations for Solar Tracking Methodologies and Opportunities for Energy Management
S. Soulayman (Higher Institute for Applied Sciences and Technology, Syria)
Engineering Science Reference • ©2018 • 647pp • H/C (ISBN: 9781522529507) • US $245.00

Hydrology and Best Practices for Managing Water Resources in Arid and Semi-Arid Lands
Christopher Misati Ondieki (Kenyatta University, Kenya) and Johnson U. Kitheka (South Eastern Kenya University, Kenya)
Engineering Science Reference • ©2018 • 266pp • H/C (ISBN: 9781522527190) • US $205.00

For an entire list of titles in this series, please visit:
https://www.igi-global.com/book-series/advances-environmental-engineering-green-technologies/73679

IGI Global
DISSEMINATOR OF KNOWLEDGE

701 East Chocolate Avenue, Hershey, PA 17033, USA
Tel: 717-533-8845 x100 • Fax: 717-533-8661
E-Mail: cust@igi-global.com • www.igi-global.com

Editorial Advisory Board

I. O. Agbagwa, *University of Port Harcourt, Nigeria*
Idress Hamad Attitalla, *Omar Al-Mukhtar University, Libya*
Arun Bhatt, *G. B. Pant Engineering College, India*
Manoj Bhatt, *G. B. Pant Engineering College, India*
B. S. Bisht, *Hemvati Nandan Bahuguna Garhwal University, India*
S. Choi, *Pusan National University, South Korea*
Tejpal Dhewa, *Delhi University, India*
Nelson Durán, *Chemical Institute, Brazil*
Rajib Roychowdhury, *Visva-Bharati Santiniketan, India*
A.M. Saxena, *Lucknow University, India*
N. Singh, *Hemvati Nandan Bahuguna Garhwal University, India*
P. Soni, *Forest Research Institute, India*
Meena Srivastava, *Pt. Deen Dayal Upadhyay Govt. Girls P.G. College, India*
M. K. Tripathi, *Central Institute of Agricultural Engineering, India*
Sandip Tripathi, *NIMS University, India*
Silas Granato Villas-Boas, *University of Auckland, New Zealand*
G. T. P. Wong, *Academia Sinica, Taiwan*

Table of Contents

Detailed Table of Contents

The effective dissolution of metals is widely known with the help of microorganisms called bioleaching or biomining used for the extraction of metals from their ores. Usually the microorganisms involved in biomining are chemolithoautotrophic and extremophilic in nature, since they are living in highly acidic environments (pH 1-3.0) containing heavy concentrations of metals. The commonly found genera of archea are Sulfolobus, Acidianus, Metallosphaera, and Sulfurisphaera. Throughput microbial genomics and proteomics analysis provides novel insights of metabolism mechanisms of bioleaching microbes. These microbes are having significant impact on the bioremediation of acid mine drainage (AMD) resulted from many industrial operations. Using these microbes, various metals including Ni, Cd, Cu, Fe, As, Pb, Hg, Cr, Mn, Zn, etc. are removed from the environment. Biomining microorganisms are having significant applications in the biotechnological processes including extraction of gold from ores, extraction of nickel from low-grade sulfide ores, extraction of copper from chalcopyrite, etc.

Textile industries are one of the prevalent water consumers and contaminators ensuing high generation of waste water. Wastewater from dyeing and printing units is often rich in organic compounds, colours, and heavy metals containing residues of various dyes and chemicals. Among these waste water pollutants, heavy metals are of serious environmental concern in recent years. Metals are extensively used for manufacture of synthetic dyestuffs and for colouration of natural dyes on textiles to achieve different shades and hues. Heavy metals such as lead (Pb), chromium (Cr), cadmium (Cd), iron (Fe), zinc (Zn), copper (Cu), etc. are widely used for production of colour pigments of textile dyes. These heavy metals, highly toxic, get conveyed to the environment, and can bio-accumulate in the human body, aquatic life, natural water-bodies, and also possibly become trapped in the soil. The chapter deals with the utilization of heavy metals in textile wet processing, their important characteristics, various toxic aspects, and different methods for their elimination from the textile effluent liquors.

Heavy metal contamination in water is a serious concern to the environment and human health. High concentrations of heavy metals in the environment can be toxic to a variety of living species. Natural bio-absorbents are abundant and inexpensive and considered a waste if not managed properly. The role of bio-absorbents has been widely studied and has been utilized for the removal of heavy metals. The objective of the chapter is to search the database for different absorbents and their efficiency for the removal of heavy metals. Key words related to the study have been used to select different papers published by the researchers all over the world. A rigorous three-tier process has been utilized by the authors to select the papers from the database for the current study. This chapter has identified a few research gaps in the field of heavy metal removal by using different low cast absorbents that need to be taken into account in future research.

The heavy metal pollution problem is all over the world. Plant growth promoting bacteria (PGPB) has transformed heavy metals present in the soil, which removes and minimizes their toxic effects. This chapter highlights the role of PGPB for remediation of heavy metals, their mechanism of action, and their applications approach of hyperaccumulation. Further, it also highlights the role of uptake and detoxification of metals by cellular mechanisms which facilitate the bioremediation of heavy metals from contaminated areas. Bacteria may also enhance nutrient uptake, increasing plant growth and defenses while diminish heavy metals intake and their toxic effects. Therefore, this chapter focuses on the mechanisms by which microorganisms can mobilize or immobilize metals in soils and the bioremediation strategies are addressed for the improvement of phytoextraction as an innovative process for enhancement of heavy metals removal from soil.

Rising cases of environmental mercury hazards has led to a need for cost-effective mercury treatment techniques. Extensive use of mercury from ancient times has resulted in water contamination that may require remediation. Mercury contamination is tedious to treat and may pose a risk to human health and the environment. To deal with this threat of mercury contamination, industrial wastes and wastewaters containing mercury requires treatment for its removal and immobilization. This chapter provides a synopsis of the availability, performance, and technologies for management of mercury in water. It covers the innovative methods to treat the mercury contamination like biosorption. In this chapter, the technological aspects available for the mercury treatment technologies are reviewed. It describes the theory, design, and operation of the technologies; provides information on commercial availability and use; and includes data on performance, where available.

The primary functions of ecosystems, wherever located, or whatever their constituents on earth, are to sustain life. They provide vital needs, commonly described as "ecological services," for all the living things residing within that particular system, or for those on its fringes. Such services include shelter, food, maintaining soils and hydrology, and providing livelihoods for its dwellers. Water is considered an important resource for mankind and other living organisms. Therefore, maintaining the water to a high quality is crucial. Bioremediation of heavy metals and micro pollutants by microorganisms (algae, bacteria, fungi, and yeast) has been used for the last two decades. All micro-organisms and microalgae are gaining increasing attention because microalgae can easily grow and is cheap to process and able to accumulate high metal content.

This chapter deals with screening of native Pteridophytes (ferns) in the vicinity of fly ash (FA) contaminated areas of National Thermal Power Corporation (NTPC) in Kanti and Kahalgaon, Bihar, India. These ferns were used for phytoremediation of heavy metals in the aforementioned areas. The metal tolerance ability and diversity of the native ferns were examined on several diversity indices (population density, concentration of dominance, heterogeneity, equitability) and the results were compared with reference site (nearby forests). The present data revealed that 11 species of ferns (10 terrestrial and 1 climbing fern) had rich frequency distribution and better performance on FA sites. The results also revealed that all ferns had accumulated significant metal content in their tissues. However, metal content was varied in different fern species. Effect of FA on photosynthetic pigments were also analyzed and compared with the reference site. The results of the above study suggest the suitability and usefulness of these ferns for reclamation of FA contaminated wastelands.

In the chapter, the authors describe how a plant can be selected as a natural indicator of heavy metal (mercury) contamination, how one can select a plant species to monitor a particular type of heavy metal and use those plants to remove the contaminants from the area under consideration. The chapter also contains a brief idea of environmental contamination by heavy metals and how the situation can be managed by the techniques of modern plant biotechnology. The authors add some reports and data of their in-vitro studies of mercury toxicity on Ananas sp., generally known as pineapple, for better understanding of the text.

Increasing water pollution due to heavy metals is a major global concern, and favorable remediation techniques are required. Heavy metal contamination affects both flora and fauna as it enters the food web. The development of nanotechnology and novel nanomaterials production has attracted researchers worldwide. Both carbon- and metal-based nanomaterials proved great adsorbents for heavy metal remediation because of their unique properties such as thermal and chemical stability and high surface. Novel green route for nanomaterials synthesis make the nanomaterials production an environmentally friendly, low cost, and user-friendly approach. This chapter reviews the heavy metal pollution causes and utilization of different nanomaterials for the remediation process.

The functional aspect of nanotechnology (NBT) is driven either to accelerate the performance of materials and/or to reduce the quantity of materials that are used for

Foreword

Bioremediation is one of the most promising technological approaches to the problem of hazardous waste. It is a technology for removing pollution from environment, restoring contaminated site and preventing future pollution. Bioremediation can be performed in situ or ex situ. Microorganisms directly degrade contaminants rather than merely transferring them from one medium to another, employ metabolic degradation pathways and can be used in situ to minimize disturbance of the clean-up site. Hence, microorganisms can be effective, economical and non-disruptive tools for eliminating hazardous chemicals. Its advantage generally outweigh the disadvantage, therefore may be used as management tool.

In the recent past, various physical and chemical techniques have been employed for the removal of pollutants from wastewater. However most of these methods have certain drawbacks. Biological treatment is often efficient and economical. Many microorganisms are able to accumulate and degrade different pollutants.

Bioremediation is not a panacea for soil and groundwater contamination. A successful, cost-effective bioremediation process is dependent on hydrogeologic conditions, contaminant signature, microbial ecology, and other spatial/temporal factors that vary widely.

In order to write a forward to the present book focused on this technology, I would like to appreciate the sincere efforts of my colleague Dr. Ashok Kumar Rathoure, Environmental Scientist to publish this book which contains latest information on the various aspects of remediation technology to remove environmental contaminations especially groundwater pollutants, toxic metals like Arsenic, Cadmium, Nickel, Uranium, etc.

I hope this book will serve as a strong reference material and milestone to the new and current researcher and also help the scientist working in identifying and discovering the gaps in this research area. I consider this book a value addition to the scientific knowledge on remediation technologies focusing on groundwater contamination abatement and quality improvement.

Pawan Kumar Bharti
Shriram Institute for Industrial Research, India & Society for Environment, Health, Awareness of Nutrition and Toxicology (SEHAT), India

Preface

Ground waters exist as a central part of the larger hydrologic cycle of a specified region. Interactions between groundwater and surface water bodies (recharge and discharge zones) provide one of the major pathways through which groundwater contaminants interact with humans and the wider terrestrial environment. These interactions can be beneficial by diluting the contaminated groundwater. This can be a major factor in the reduction of the impact of groundwater contamination. Alternatively, contamination may become concentrated in bottom sediments through precipitation and sorption processes or may be taken up and accumulated in plants and animals. It is also possible that contaminants will be transported some distance from the point of discharge only to become deposited elsewhere. This deposition usually occurs at an interface of some kind, such as when suspended particulates are deposited when a river flows into a static water body. Changes in water chemistry can occur downstream in a river system or where two rivers meet. Information on the distribution of contaminant concentrations in the subsurface is required to evaluate the needs and the design of groundwater remediation techniques. Information on the natural quality of surface and ground waters is needed to evaluate the risks posed by the contamination and to establish remediation targets. Background water quality data should be obtained from samples collected from the aquifer that contains the contamination, and sampling should be undertaken at unaffected locations.

The presence of contaminants in groundwater at concentrations above background levels demonstrates a high potential health and ecological risk. Common contaminants include a variety of volatile hydrocarbons-such as benzene, toluene, ethylene and xylene-found in fuels; heavy paraffins and chlorinated organic compounds including polychlorinated biphenyls; inorganic compounds *e.g.* lead, chromium, cadmium, arsenic and mercury and radionuclides such as tritium etc. Oil spill accidents are a widespread problem due to the vast production, refining, storage, and distribution of petroleum-derived products. Terrestrial oil spillages are more common than aquatic ones, although they typically pollute areas only locally and receive less publicity than the dramatic marine tanker or platform incidents. While the bio-stimulation based methods are generally slow, the in situ chemical oxidation by which organic contaminants can be destroyed or converted into more biodegradable forms is considered fast.

Remediation is the only way to clean up the contaminants in soils and groundwater. Remediation refers to the process of environmental cleanup of contaminated sites and the techniques to reduce or eliminate contamination from soil or groundwater. General technical methodologies include transfer of contaminants solely by themselves, with contaminated soils or groundwater to other places for final treatment or disposal; confinement of contaminants in place and destruction of contaminants in place. Mechanistic principles are physical, chemical, and biological processes. Remediation pathways include transfer of contaminants alone or with contaminated groundwater to other place for final treatment or disposal, confinement, and destruction of contaminants in place. Technical principles for remediation can be divided into physical, chemical and biological processes.

Groundwater contamination and groundwater remediation will be the primordial, burning and vexing issue of the present day and future of human civilization.

ORGANIZATION OF THE BOOK

The book is organized into 15 chapters. A brief description of each of the chapters follows:

Chapter 1 presents the biomining of microorganisms with diverse structural and functional characteristics that are prone to perform the various commercial processes efficiently like metal extractions, removal of any mineral (desulfurization) as compared to the chemical extractions. Due to the extremophilic nature, microorganisms are used to oxidize sulfur minerals (Acidophiles), to isolate the thermostable proteins for diagnostic as well as therapeutic purposes (Hyperthermophiles), to remove metals at the metal contaminated sites by biostimulation process etc. The biomining microorganisms are used for accelerating the removal of pollutants in the environment.

Chapter 2 shows the sources of contamination of water by toxic heavy metals from textile industries and its remediation. Industrial effluents discharged from textiles contain a higher amount of metals. Heavy metals hold a specific position in textile industry, right from the synthesis of various synthetic dyes and dye intermediates to their utilization as mordents for improving the affinity of the dye towards the fiber. It is anticipated to use these metal/metallic compounds in the permissible limits endorsed by the Pollution Control Board so as to preserve the environment. The toxic behavior of some of these heavy metals makes their contamination in the effluent liquors harmful to the environment. These effluents are released on the land as well as discharged into the surface water which ultimately end up in ground water by leaching and lead to contamination of it due to accumulation of toxic metallic components. Industrial effluent is also a main source of direct and often continuous input of pollutants into aquatic ecosystems with long-term insinuations on ecosystem functioning. The toxicity level of these heavy metals can be assessed by various analytical approaches. Their removal from the effluent liquor is necessary and various primary, secondary and tertiary treatment methods are utilized for their elimination.

Chapter 3 presents heavy metal removal efficiency. The presence of different heavy metals in water is considered very dangerous not for only aquatic life but also for humans as well. In order to control toxicity of heavy metals and to recover heavy metals from water, efforts must be made to completely eliminate the risk of heavy metals. This can be achieved by creating public awareness about toxicity of heavy metals and government policies need to be enforced strictly to regulate the discharge of effluents contaminated with heavy metals. Only few studies regenerated the bio-adsorbents and this need to be researched further. However, the identification of various techniques that are cheap and sustainable may help in decreasing the burden of metal pollution in water. Various absorbents have been already researched by various workers that can remove heavy metals from water with great efficiency and are best suited to different isotherm models.

Chapter 4 shows the biosorption of highly toxic and radioactivity metal Uranium. Uranium exposure to the human take place through direct exposure or through drinking water, contaminated with Uranium which leads to health related complications such as kidney damage, neurological disorders and many more. So, it has become more important to remove the Uranium from drinking water. Many conventional methods such as chemical precipitation, electro-dialysis, ion-exchange, reverse osmosis, phytoremediation, ultra-filtration have been applied for Uranium removal from the aqueous solution but there are some demerits of using the conventional methods include production of huge amount of sludge and metal hydroxide that blockade the membrane, more expensive and take much time for removal of metals. It has become necessary to search for new technique for removal of Uranium. So, biosorption method is applied. In this technique, Uranium is removal by biosorption by either living microorganisms (algae, fungi and yeasts) or their dead biomasses through electrostatic interactions between the metal ions in solutions and cell walls of microbial cells.

Chapter 5 speaks about demethylation of arsenic and nickel from tannery effluent. The effluent discharged from the tannery industries is responsible for polluting the environment. It consists of acids, bases, or heavy metals contaminate the water bodies drastically. Arsenic is used as pesticide in the form of sodium arsenate which is further discharged along with the waste responsible for pollution. The nickel and arsenic affects the organism in several ways by causing cancer, birth defects, asthma, heart defects etc. The relative use of arsenic in tannery industry should be replaced with other pesticide in order to control arsenic pollution. Further removal of metal pollution can be achieved by means of cost effective bioremediation technology instead of the conventional physical and chemical methods. The microbial bioremediation is essentially effective due to the presence of transforming ability of metals either by oxidation, reduction, methylation and demethylation approach.

Chapter 6 demonstrates the ability of bacteria to mobilize or immobilize heavy metals in soil. The microorganisms can improve phytoextraction by increasing plant biomass or by enhancing the availability of heavy metals to the plant. However, use of microorganisms to improve plant biomass via improved N or P nutrition can have a significant positive impact on phytoextraction. Those soils, which are contaminated and possess the higher amount of toxic metal than plant tolerance, then at such sites phytoremediation through rhizobacteria is preferred and carried out. This method not only increases the plant tolerance towards metals but also stabilizes the growth of plants involving the remediation of metal polluted soil. The soil fertility which is being lost due to heavy metal accumulation has been restored through this method. As the role of plant growth-promoting rhizobacteria is known however a lot remains there to understand regarding the mechanism of phytoremediation. While on the other hand, bioremediation through rhizobacteria is an economic and ecological friendly process.

Chapter 7 illustrates the technological interventions in management of mercury contaminated water. Among the recent studies biosorption by filamentous fungi is the most frequently used technology to treat waste water contaminated with mercury. The technology has been used to meet regulatory cleanup levels, is commercially available to treat waste water, and generates a residual that typically does not require further treatment before disposal. There is always a scope of improvement and addition of new technologies or process improvement, the intent of this chapter is to call for action with the development of sound efficient and economical methods to reduce the contamination of mercury.

Chapter 8 is about microalgae which are used as promising tool for remediation of heavy metals. Bioremediation of heavy metals by algae (micro and macro algae) is an economic and environmental friendly method. Mineral nutrients and heavy metals are removed by biosorption, binding of metals on the cell surface, ion exchange, physical adsorption, chemical modification, efflux of metal ions. Many researchers reported the algae tolerance towards heavy metals. Bioremediation of heavy metals by algae depends on different factors ranging from the anatomy of the algae species to the environmental conditions or growth medium of algae.

Chapter 9 presents the phytoremediation study that conclude the ferns grow luxuriantly on fly ash without visual phytotoxicity having high population density; basal cover and concentration of dominance (Cd) on FA site in comparison to control site even in presence of high metal content and less total N and P. Besides, significant accumulation of metal contents like Fe, Si, Al, Ni, Cr, Cd, and As and consequently slight decrease in photosynthetic pigments also indicates its metal tolerant ability. These ferns are also applicable for phytoremediation of heavy metals and thus minimize the pollution rate.

In Chapter 10, the authors used the pineapple plant as a model for performing tissue culture procedures for detection and management of environmental health because the plant is easily available and can grow in diverse range of environment and requires no special maintenance. The *Ananas sp.* commonly known as pineapple, can be used as a natural indicator or bio indicator of mercury or any other heavy metal toxicity by planting the plants around the water bodies. As evident through the simple plant tissue culture experiment that presence of mercuric compounds interferes with normal physiological activities of the plant resulting in decrease in growth, loss of vitality, depigmentation and finally death of the plant species. And if the study of the plant's genetic material were performed, the degradation of the plant DNA could be observed, indicating that mercury also damages the genetic material of an organism permanently. So if the plantations around the water bodies start exhibiting these features even in the slightest form, is indicative of mercury toxicity. As evident through the experiment that if the concentration mercury compound exceeds 50 mg/L with respect to the water body under consideration, the plants will be showing the effects indicative of the mercury contamination. After that the conventional methods can be applied to measure the exact amount of mercury concentration. This procedure is useful because planting easily available plants as bio indicators are more reliable and cost effective as compared to installing a toxicity monitoring apparatus. Further toxicity studies of heavy metal under consideration could be studied in a large extent and in a faster rate. The rate of uptake of heavy metals by the plants from their surrounding environment can be studied *in vitro* which will help in deciding which plant species to be utilized for optimum removal of ground water contaminants. Finally it can be concluded that the technique of tissue culture is thus very essential for managing environmental pollution using plants. And there is a serious need for environmental engineers with expertise on Plant tissue culture and associated Recombinant DNA Technologies for creation of novel variety of plants that can be utilized for phytoremediation purposes in an environment friendly manner.

Chapter 11 indicates the interaction of heavy metal ions with nano-materials. Heavy metals are considered highly toxic to environment and thus, cost effective, environment friendly and effective remediation techniques must be developed. Still, the technology for remediation of heavy metal ions has not reached the optimum conditions. Emerging utilization of nano-materials for remediation is proved beneficial due to their special properties. High adsorption properties were reported by several researchers through use of carbon and metal nano-materials. This chapter illustrate that researchers in most cases utilize carbon nanotubes as CNM sorbents. The green synthesis approach for synthesis of nanoparticles proved favorable and in it further developments are required. This concludes the application of nanotechnology in heavy metal remediation processes.

Chapter 12 is about the research innovation in nano-bioremediation. Special emphasis on this technology depends on the potentials benefits received on the applicability in the field of medicine, electronics, agriculture and environment for social wellbeing. A far reaching sustainable process development has been achieved for tailoring the nanoparticles on size and shape control through nano-biotechnology. Adaptation of nano-remediation has the efficacy to reduce the overall cost for environmental clean-up at large scale application. The reduction in the contaminant level even could reach to near zero in in-situ site nano-remediation. Whereas, ex-situ nano-remediation are still under infant stage and are shown to have promising results in decontamination of microbes, organic and inorganic pollutants. Nano-remediation being in exciting stage of development is considered as potential tool for mitigating environmental pollution. These technological improvements are likely to be a favorable proxy for conventional treatments, but potential risks associated with environmental nano-toxicity are still unclear. Once these research gaps were disclosed, it is certain that nano-remediation will no doubt be a promising technology for achieving environmental sustainability.

Chapter 13 presents the nano-materials for bioremediation. There is no doubt of efficiency of utilization of Nano-materials in wastewater treatment; however, this technology has some serious downsides that need to be negotiated, since nanoparticles might release into the environment during preparation and treatment processes, where they can accumulate for long time and cause serious risks. In order to reduce the health risk there is need of a future research to prepare such catalysts having least toxicity to the environment. More work is required to re-evaluate the eco-toxicity potential for each new modification in catalyst and for existing materials. In addition, life cycle assessment of Nano-materials is crucially required to address their overall benefits and risks. Nanotechnology is rarely adopted to mass processes such that up until now, most of the Nano-materials have not been cost competitive when compared with conventional materials such as activated carbon, and thus future applications will focus on efficient processes where only small quantities of Nano-materials are required and larger research is being done in green synthesis of nanoparticles and its integration with bioremediation so that it causes least eco-toxicity and is an answer to sustainable development via biological ways of treatment.

Chapter 14 speaks about nano-composites used for remediation of wastewater. Adsorption plays an important role in removing heavy metal s from effluent. Nano-composites are the most popular adsorbent material and have been used widely in waste water treatment for decades. However, commercial nano-composites have drawbacks in terms of heavy metal removal efficiency and cost problems. So the overall idea is to find suitable cost-effective alternatives to replace traditional method. Under these circumstances, new adsorbent materials, especially cost-effective adsorbent materials, need to be developed. The low-cost adsorbent materials usually have the problem of low adsorption capacity; thus cheap and effective modification methods should be introduced to improve the adsorption capacity while controlling the cost of activation at the same time. Low-cost materials have been divided into four categories based on the nature of the materials. The common characteristics of these materials are abundance either naturally occurring or low-value by-products. By analyzing the composition and developing low-cost treatment methods, they become value-added materials with good adsorbent properties.

Chapter 15 is about green nanotechnology and recent trends. Carbon nanotubes, quantum dots work as sensors as well as remediators. Silver and gold nanoparticles work as catalyst for decomposition of pollutants from waste. Nano materials can be used as carrier for pesticides and chemical fertilizers delivery. These carriers deliver the chemicals in equal distribution into soil in required amount so that it can avoid soil pollution effectively. Nano-materials can limit the usage of chemicals, pesticides, fertilizers with the use of biosynthesis of nano-materials from microbial system as it will emerge as very novel technology for bioremediation of soil and protection of crop for mankind. Nanoparticles developed by using microbial system provide green nanotechnology for remediation and helps in keeping the environment clean. Microflora in soil used for synthesis of nano-materials and used as bio-remediator further. These types of nano-materials help in bio-sorption of heavy metals from contaminated areas (soil or water). Nanotechnology has the capability to increase efficiency of any process in terms of use of material and energy, with minimum amount of generation of toxic wastes. This application of nanotechnology to greener production processes is referred to as green nanotechnology. This helps for bioremediation in natural or organic ways and maintains ecological balance. This gives rise for evolution of green nanotechnology. Green nanotechnology needs to develop with speedy advancement to make cleaner environment with great ecological balance on earth.

The technological methods for remediating contaminated groundwater are described in terms of providing an overview which does not claim to be complete. Although substantial research is currently being carried out in this field, only a few innovative methods. The application of technological methods for remediating contaminated ground waters has to be considered in conjunction with management options such as diversion and development of alternative water sources. Available technologies are grouped generally into *in situ* methods aiming at contaminants in place and engineered treatment methods involving an alteration of groundwater flow, quantity and quality to achieve compliance with set goals. Groundwater remediation by natural flushing allows the natural groundwater movement and geochemical processes to decrease the contaminant concentrations to acceptable levels over a specified period of time. This method is increasingly accepted in area where the use of groundwater can be temporarily restricted or engineered cleanup methods do not offer particular advantage over the natural processes.

Ashok K. Rathoure
Independent Researcher, India

Acknowledgment

This book is the result of the dedicated effort of numerous individuals. We would like to acknowledge the help of all the people involved in this project and, more specifically, to the authors and reviewers that took part in the review process. Without their support, this book would not have become a reality.

First, we would like to thank each one of the authors for their contributions. Our sincere gratitude goes to the chapter's authors who contributed their time and expertise to this book.

Second, we wish to acknowledge the valuable contributions of the reviewers regarding the improvement of quality, coherence and content presentation of chapters. Most of the authors also served as referees; we highly appreciate their double task.

Additionally, we would like to thank our Editorial Advisory Board.

Our Production Coordinator skillfully managed the production of the textbook. We wish to thank the publisher for the sustained interest shown by him during the entire work. A sincere expression of the thanks goes to the entire staff of IGI Global.

Generous participation of faculties and students to eliminate errors in the text and to eliminate errors in the text and to refine the presentation will be greatly acknowledged.

Last and not least, we beg forgiveness of all those who have been with us over the course of the years and whose names we have failed to mention.

Ashok K. Rathoure
Independent Researcher, India

Chapter 1
Biomining Microorganisms' Molecular Aspects and Applications in Biotechnology and Bioremediation

Jyoti Srivastava
Indian Institute of Technology (Banaras Hindu University), India

Pradeep Srivastava
Indian Institute of Technology (Banaras Hindu University), India

ABSTRACT

The effective dissolution of metals is widely known with the help of microorganisms called bioleaching or biomining used for the extraction of metals from their ores. Usually the microorganisms involved in biomining are chemolithoautotrophic and extremophilic in nature, since they are living in highly acidic environments (pH 1-3.0) containing heavy concentrations of metals. The commonly found genera of archea are Sulfolobus, Acidianus, Metallosphaera, and Sulfurisphaera. Throughput microbial genomics and proteomics analysis provides novel insights of metabolism mechanisms of bioleaching microbes. These microbes are having significant impact on the bioremediation of acid mine drainage (AMD) resulted from many industrial operations. Using these microbes, various metals including Ni, Cd, Cu, Fe, As, Pb, Hg, Cr, Mn, Zn, etc. are removed from the environment. Biomining microorganisms are having significant applications in the biotechnological processes including extraction of gold from ores, extraction of nickel from low-grade sulfide ores, extraction of copper from chalcopyrite, etc.

DOI: 10.4018/978-1-5225-4162-2.ch001

Copyright © 2018, IGI Global. Copying or distributing in print or electronic forms without written permission of IGI Global is prohibited.

INTRODUCTION

An ore is a naturally occurring source from which metals can be extracted. Ores are generally present in oxide forms. Different metals are having their respective ores, e.g. bauxite (Al_2O_3) is a very common ore found on earth from which metal aluminium can be extracted. Haematite (Fe_2O_3) is an ore of iron, pyrolusite (MnO_2) is an oxide ore of manganese. Mining is defined as the approach of extraction of metal ores from deep underground (Johnson, 2014). The metal ores are naturally present in earth's crust in very large amount. Ores are commonly found in oxide forms e.g. bauxite (Al_2O_3), haematite (Fe_2O_3), rutile (TiO_2) or in sulfide forms e.g. pyrite (FeS_2), chalcopyrite ($CuFeS_2$) etc. The metals are extracted from its ores by -

1. **Ores Concentration:** To remove the unwanted rocks by purifying and concentrating ores (Wang et al., 2010).
2. **Reduction to Crude Metal:** The metals in the form of crude ones are reduced from their oxide forms for easy purifications (Lovley, 1995).
3. **Refining:** In order to obtain the pure metal from crude mixture of different metals purification can be followed (Tuncuk et al., 2012).

Electrolysis process can be applicable to obtain pure metal from metal oxides (Chen et al., 2000). In electrolysis of $Al_2O_3.2H_2O$, positively charged metal Al^{3+} can be attracted to negatively charged cathode whereas O^{2-} attracts towards positively charged anode.

BACKGROUND

Historically *in situ* leaching of copper was reported by Galen, a naturalist and physician in AD166 which were later came to know due to microbial community (Cummins &McAuliffe). Copper sulfate was used to be extracted from the rock by flow of water on its surface which later on percolates into and resultant solution contained high concentrations of copper sulfates. Further crystals of copper sulfate were recovered by evaporating the solution. However, the Romans were used to obtain the pure copper by a process of heating and melting, collectively known as smelting. The Chinese people in 1096 used to follow the cementation process for extracting the copper from copper sulfate. However, the bioleaching process was first observed and understood by Paracelsus the Great (1493 – 1541) (Cummins &McAuliffe). He observed the deposition of copper onto iron at the spring in Zifferbrunnen, Hungary. He aided in the process of bioleaching and during 1750 around 200ton per annum copper was recovered using bioleaching in the same area.

Since ancient times 3000 BC, the leaching of copper, gold, silver and other minerals were mined along the site of Rio Tinto, Spain. Rio Tinto is a river originated from Sierra Morena Mountains of Andalusia. The river is known for its red colour and highly acidic water (pH 2). The deep red colour is due to the dissolution of iron in the water. The acid mine drainage (AMD) from the river causes severe impact on the environment due to the high concentrations of metals dissolved in the river. In 1873, Rio Tinto Company was established to operate this mine and it became one of the largest worldwide International Company at the end of 20th century.

Case Study: Talvivaara Mining Company

This is a Finland based mining company that has optimized and developed its own bioleaching technology. This technology utilizes the application of bacteria that can simultaneously extract several different metals from the same mineral deposits. Talvivaara is Europe's largest nickel producer company (Riekkola-Vanhanen, 2010).

Codelco Mining Company

This is the world's largest copper mining company based in Chile (Aroca, 2001). Along with Codelco, there are some other biomining companies that utilize *Acidithiobacillusferrooxidans*and *Thiobacillusferrooxidans* bacteria for the extraction of copper from its insoluble minerals (Johnson et al., 2001).

BHP Billiton

Broken Hill Proprietary (BHP) Billiton is an Australian based multinational mining company involved in extraction of metals and petroleum. It is the world's largest mining company. The major bioleaching operations involves for the extraction of iron, potash, copper etc (Rawlings, 2005).

Case Study: In Situ Leaching ISL

The first ISL was approached in early 1960s to recover uranium in Wyoming, USA and used for commercial purpose (Seidel, 1976). Today ISL is followed in Kazakh, Uzbek, and most US uranium production mines. ISL can also be applied to recover other metals like copper, gold. Production of uranium was linearly increased in about 1990 and 70% of total Bulgarian production, around 300 t U/a (Johnson, 2014).

Case Study: Heap Leaching

Up to 95% of copper and uranium were extracted / leached from quartz chlorite ore by this method using *Thiobacillusferrooxidans* at Jaduguda district, Singhbhum Thrust Belt (STB), Jharkhand, India (Mathur et al., 2000).

Biomining or Bioleaching

With the help of microorganisms the recovery of metals from mineral ores is known as 'biomining' or 'bioleaching' (Schippers et al., 2013). Biomining is environment and eco-friendly approach of purifying metals from their ores. Biomining recently played a significant role in the biotechnology era due to the easier process of extraction of metals than the mechanical and chemical methods. Moreover the extraction through the microbial community results in to lesser damage of environment. At around 15% of copper mining industries utilizes this technology across the globe. The biological means of extraction is a significant and eco-friendly alternative to the traditional 'smelting' process where the mineral ores are heated for long time and metals were extracted, but it is rather pollution generating process as sulfur dioxide is the byproduct which in turn causes acidic rain (Patterson & Passino, 1987). However, the modern smelting technology diminishes the generation of sulfur dioxide byproduct as it is converted in to an essential byproduct sulfuric acid during the copper extraction. Up till now, metals including nickel (Watling, 2008), copper (Schnell, 1997), gold (Olson, 1994), zinc (Konishi et al., 1992), cobalt (Olson et al., 1990) etc. enormously extracted by bioleaching approach. The microorganisms used for extraction of metals are generally acidophilic in nature that require acidic environment for optimum growth and metabolism. During bioleaching, the ores are crushed and sprayed with the dilute sulphuric acid to maintain the optimum acidic environment that enhances the colonization of bacteria that oxidizes insoluble metal sulfides to soluble metal sulfates (Domic, 2007). In addition to *Thiobacillusthiooxidans* and *Thiobacillusferrooxidans* (Bosecker, 1997), other microbes involved in recovery of metals by bioleaching are *Bacillus licheniformis*(Sukla et al., 1993), *B. luteus*(Patra et al., 2011), *B. megaterium*(Castro et al., 2000), *Sulfolobusacidocaldarius*(Seh lin&Lindström, 1992), etc. The process of bioleaching of iron is represented in following reaction.

$$2FeS_2 + 7O_2 + 2H_2O \xrightarrow{\textit{Thiobacillus sp.}} 2FeSO_4 + 2H_2SO_4$$

Microbes extract the metal from their ores as shown in Figure 1.

Figure 1. Schematic representation of microbial extraction of metals

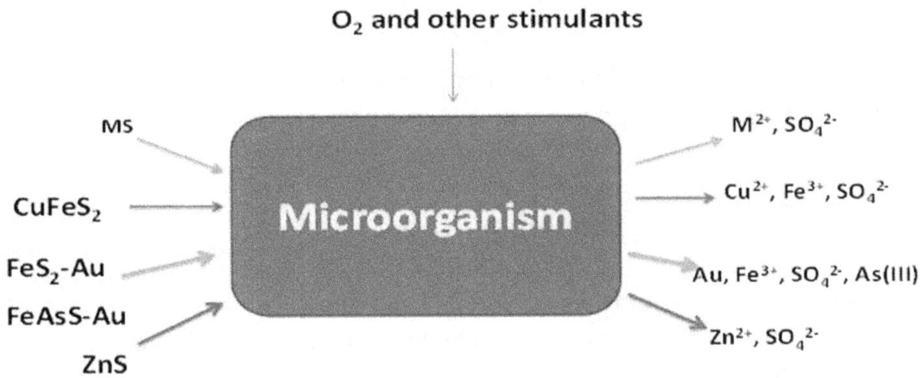

BIOMINING APPROACHES

Biomining or bioleaching can be operated with the help of following possible approaches:

In Situ Leaching (ISL)

In this case, the extraction process is carried out at the site of mine. The bioleaching process can be initiated by pumping leaching solution containing *Thiobacillusferrooxidans*into the mine where microbes get injected into the mineral ore. By this approach, uranium was extracted by bacterial flood leaching of ores in Denison mine, Canada (Tsezoset al., 1989). However, in this case the permeability of ore is essential; if the ore is not sufficiently porous then it must be cracked by explosives. The leaching reagents are pumped to dissolve the desired metal. The porosity of ore is directly proportional to the leaching as it improves the flowing ability of injected reagents (Dixon & Hendrix, 1993). Here leaches can be collected from lower down the mine, pumped to the surface where the metal is pulled through. One of the major disadvantages of this approach is percolation of liquid to the ore body. Also, ground water may get contaminate if the solution flow is not under control.

However, the major advantage of this operation is that the ore does not need to be removed from the ground. This technique was started in 1960s, for the recovery of uranium by spraying acid mine drainage and by in situ irrigation of fractured underground ore deposits (Mehta et al., 2011). Further the recovery of other metal including copper was achieved by this approach etc.

Heap Leaching

In this study, ores are added as it is in to the large heap tank and water containing microbial inoculum is continuously sprinkled on it. Followed by extraction of metals by collecting water from bottom and microbes in oxidation pond (Petersen & Dixon, 2002).

Stirred Tank Reactor Leaching

Here bioleaching is achieved under controlled optimum process variables in stirred tank reactors. The microorganism used for bioleaching can be grown under optimum conditions. This maintains the uniformity in their growth.

FACTORS INFLUENCING BIOLEACHING

There are some physic-chemical parameters which affects the bioleaching process as described:

pH

The growth and metabolism of microorganisms depends on the pH of the environment. Generally bioleaching is carried out by acidophilic bacteria that require pH 1-2.5 for their optimum growth and metabolism. So, to render bioleaching the pH should be maintained as acidic by adding H2SO4 prior to inoculum. This will stimulate the enzymes which indeed bioleaching.

Temperature

Temperature is also affecting parameter of leaching process, as it affects the growth and metabolism of microorganisms if not optimum. The temperature at which maximum bioleaching occurs is called optimum temperature for that microorganism. Increase in temperature above the optimum point denatures the proteins and enzymes which then inhibit the growth of microorganism and bioleaching.

Oxygen

Oxygen is required for the growth and viability of microorganism as most of the bioleaching microorganisms are aerobic in nature and the extraction of metals is an oxidizing reaction.

Microbial Diversity

The efficiency of bioleaching is more when mixed microbial culture is used than the pure culture. Due to the ability of microorganisms to secrete different enzymes and metabolites, mixed culture results into significantly improved in bioleaching. Recently, the mesophilic and thermophilic microbes are commonly applied for the biomining operations.

Biomining operations are more promptly efficient as compared to the traditional methods in a way that, it does not need high amounts of energy as used during the roasting and smelting strategies (Rawlings et al., 2003). The traditional methods roasting and smelting is not applicable now days as these processes generates highly environmentally disastrous sulfur dioxide gases. Due to this emission of environmental polluting gases these methods are obsolete in many countries (Hong et al., 1996).

ROLE OF MICROORGANISMS IN METAL EXTRACTION

Usually the microorganisms involved in biomining are chemolithoautotrophic and extremophilic in nature, since they are living in highly acidic environment (pH 1-3.0) containing heavy concentrations of metals (Valenzuela et al., 2006). Bacteria involved in leaching of metals are firstly observed in Spain. These bacteria extract metal from its ores by secreting some chemicals that chelate metals from its ores. In nature usually gold is found to be in crude mixture having arsenic and pyrite minerals purified by using *Acidithiobacillusferrooxidans.* The free gold metal is released as represented in following reaction:

$$2FeAsS\ [Au] + 7O_2 + 2H_2O + H_2SO_4 \rightarrow Fe\ (SO4)_3 + 2H_3AsO_4 + [Au]$$

The purification of gold from crude extracts is operated at large scale industrial plants by *A. ferrooxidans*.

Moreover, there are some hyperthermophilic iron-reducing archea *Pyrococcusfuriosus* and *Pyrobaculumislandicum* used to obtain insoluble gold from gold chloride can be extensively utilized at large scale industrial operations. Biomining renders the increased improvement in the extraction of metals with reduced cost. Using microorganisms the metals extraction can be achieved from their low grade ores.

Molecular Profiling of Microorganisms

The other archea including *Pyrococcusfuriosus, Pyrobaculumislandicum, Acidianusmanzaensis, Sulfolobus spp.* etc. are prominently involved in chalcopyrite leaching. The molecular profiling of these microbes is represented in Figure 2 (Valenzuela et al., 2006).

Figure 2. Schematic representation ofmolecularprofiling ofbiomining microorganisms

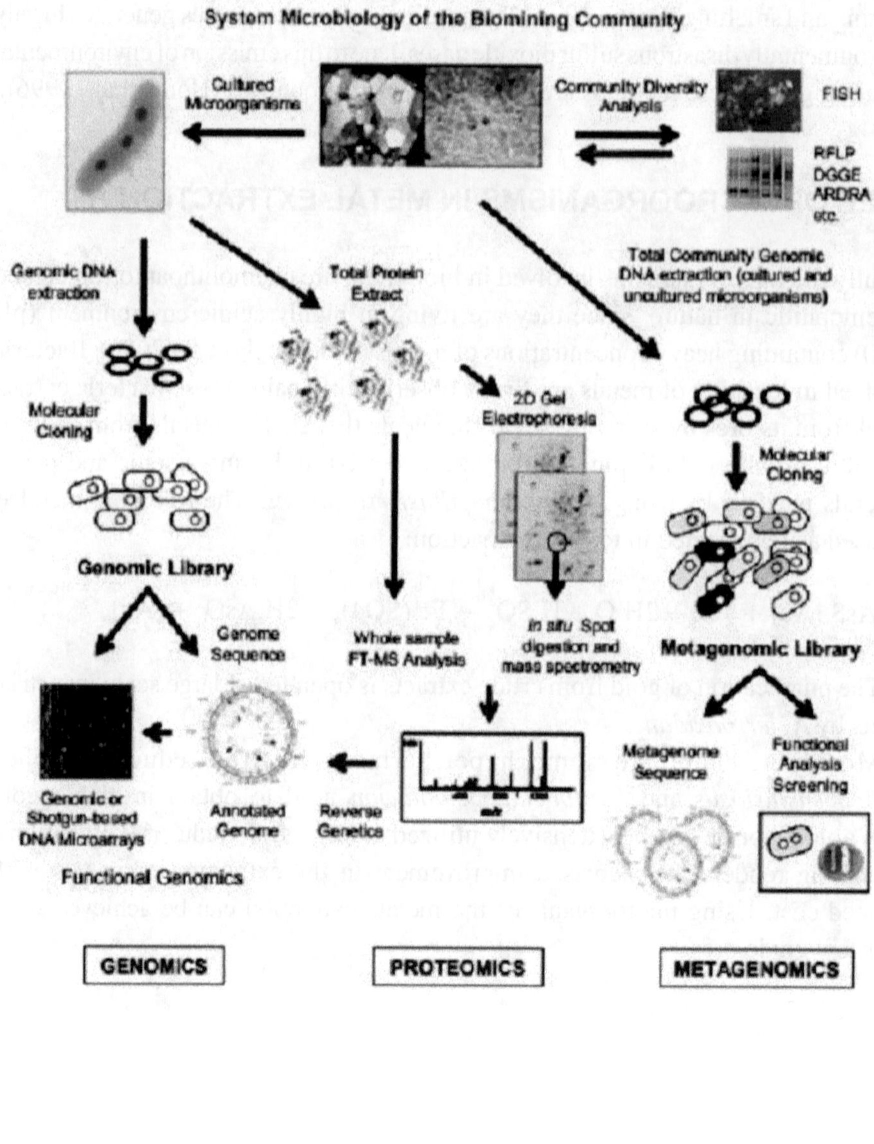

APPLICATIONS IN BIOTECHNOLOGY

Using these microorganisms, several biotechnological applications can be fulfilled. Due to the extremophilic nature of the biomining microorganisms, these can be employed for to overcome several following environmental pollution aspects.

Desulfurification

In order to recycle the waste rubber or tyres, sulfur bonds present in the polymer chain need to be removed which leads to unvulcanization. The unvulcanization can be sought either by chemical treatment or by biological treatments. The chemical treatment uses the reagent triphenylphospine that converts polysulfide into monosulfides. In addition to this, piperidine and propanethiol has efficiency to remove the polysulfide bonds. However, these chemical reagents causes severe environmental pollution, inefficient reactions, reagents cost, uneconomic. In contrast to this thermophilic microorganism *Thiobacillus spp.* particularly is most effective for unvulcanization reaction (Aytar et al., 2013). Due to difficult in handling the microbes and slow action chemical treatment is more popular.

Coal Decontamination

As coal acts as fossil fuel which is present abundantly in environment. The combustion process of coal leads to generate highly environment pollutant gas. Using biomining microorganisms, the sulfur which is associated with coal can be removed (Taylor et al., 2012). With biodesulfurization, the removal of sulfur of around 90 percent has been achieved ensuring the efficiency of pollution reduction. These microorganisms are found in natural environment of coal where they play desulfurization naturally. So can be advantages to extract these microbes and use at commercial scale. The most common and efficient microbe extracted from coal mines is *Acidithiobacillusferrooxidans*.

Bioremediation

Bioremediation is defined as the use of biological entities to detoxify or degrades environmental contaminants or pollutants. The process of bioremediation could be accomplished by microbe or microbial consortium, plants, or enzymes. *In situ* bioremediationcan be economical as well as safety for the degradation as compared to the other physicochemical methods. The microbes extracted from the mines are having enormous physic-chemical diversity. So these are efficient candidates for cleaning environmental pollutants. As the habitat of microorganisms living naturally in mines are extremophilic i.e. tolerant to highly acidic environment (acidophilic), can be survived at 100°C temperature (hyperthermophilic), and have ability to withstand at high concentrations of metals (metallotolerant). The major use of these microorganisms for bioremediation processes are described as:

Acidophiles

These are microorganisms which are usually grown under highly acidic environment of pH 2 or less than pH 2. The biomining microorganisms are naturally found to be present in such environment are so called as acidophiles or acidophilic (Johnson, 1998). These microbes play essential role in maintaining biogeochemical cycle by oxidizing sulfur minerals. Due to this natural habitat, these are commercially used for bioremediation of pollutants present in the environment. In contrast to acidophiles, other microorganisms cannot withstand at such environment. The mechanism behind the survival of acidophiles is the passive and active regulation in acidic environment. Some acidophiles withstand due to the passive regulation by forming biofilm around their membrane. This could protect from harsh high hydrogen ion concentration by slowing down the diffusion of molecules entering in to the interior of the cell. In addition to this, other renders survival by incorporating some fatty acid molecules into the cellular membrane which is also responsible for protection in acidic environment. However some are able to secrete buffer molecules to facilitate passive regulation by increasing the pH. In contrast to these mechanisms, some other microbes have built up capacity to regulate pH actively by pumping out hydrogen ions to the exterior. By which the inside pH can be maintained at 6.5 to 7.0. The acidophiles play major role in acid mine drainage.

Acid mine drainage (AMD) is the outflow of acidic solutions generated due to natural oxidation by chemolithotrophicacidophiles from mining areas. AMD is mostly responsible for water pollution causing severe impact on the aquatic as well as other life forms. AMD mainly contains sulfur bearing minerals, H_2SO_4 which are detrimental to other life forms. Acidophiles convert sulfide compounds present in metal ores to sulphuric acid. The well-studied acidophile responsible for AMD is *Ferroplasma spp.* which can withstand at pH 0 in AMD found in Iron Mountain in California (Nordstrom et al., 2000).

Hyperthermophiles

These are the microorganisms which require of around 100°C for their optimum growth and are commonly found in mines. These microbes possess stable proteins and cellular membrane, hence stable at extreme temperatures. Biotechnological applications of these thermophile lies in the isolation of enzymes that can sustain the high temperature, e.g. Taq DNA Polymerase isolated from a thermophile *Thermusaquaticus* used in PCR (Polymerase Chain Reaction) has optimum temperature in the range 75-80°C. Due to the thermostable nature of these isolated proteins, these are emphasizing the attention towards the medical therapeutic preparations. These Hyperthermophiles are efficient towards the bioremediation process at high temperature e.g. biomass conversion.

Metallotolerant

As the microorganisms present in the mines are living under the high concentrations of various metals like uranium, copper, magnesium, aluminum, iron etc. So they maintain biogeochemical cycles in ores. These microbes especially archea group facilitates metabolism of metals by nitrogen fixation, energy generation, oxidoreductions etc. These archea possess genes for metabolism, resistance and detoxification of metals (Lloyd &Lovley, 2001). The property of detoxifying the metals renders their wide applications in bioremediation of metals. The metals at high concentrations are responsible for environmental damages causing water, soil pollution. As high metal concentrations inhibits the growth of other organisms including plant, bacteria, mammals etc. At high concentration the metals are reliably toxic for the humans too causing various types of metal poisoning. Removal of metals from contaminated domestic as well as industrial effluent can be efficiently carried out by consortium of metal resistant microorganisms (Rawlings, 2005). Consortium of microbes includes *Enterobacter spp., Stenotrophomonas spp., Viridibacillusarenosi, Sporosarcina soli, Enterobacter cloacae etc.*

Disadvantages of Biomining

Although extraction of metals through microorganisms is environment friendly, the process has some demerits as mentioned below:

- The metals extraction from their ores is a slow driving process.
- Sometimes the microbes cannot acclimatize the concentrated metal area.
- Optimum physical environment is necessary to colonize the microbial community.

CONCLUSION

Biomining microorganisms with diverse structural and functional characteristics are prone to perform the various commercial processes efficiently like metal extractions, removal of any mineral (desulfurization) as compared to the chemical extractions. Due to the extremophilic nature, these are used to oxidize sulfur minerals (Acidophiles), to isolate the thermostable proteins for diagnostic as well as therapeutic purposes (Hyperthermophiles), to remove metals at the metal contaminated sites by biostimulation process etc. The biomining microorganisms are used for accelerating the removal of pollutants in the environment.

REFERENCES

Andrews, M. (2010). Transit on iron radar. *Australia's Paydirt, 1*(178), 60.

Angleviel, F. (2003). Restoring the Economic Balance: The Nickel Stakes in New Caledonia. *The New Pacific Review, 155*.

Arévalo, A., Skewes, A., Stern, C., Floody, R., Goldfarb, R., Nielsen, R., Zúñiga, R. (2002). The giant El Teniente, breccia deposit: hypogene copper distribution and emplacement. *Integrated methods for discovery: Global exploration in the Twenty-first Century, 9*, 299-332.

Aroca, P. (2001). Impacts and development in local economies based on mining: The case of the Chilean II region. *Resources Policy*, *27*(2), 119–134. doi:10.1016/S0301-4207(01)00013-7

Association, A. (1984). *Aluminum: Properties and physical metallurgy*. ASM International.

Aytar, P., Kay, C. M., Mutlu, M. B. I., & Çabuk, A. (2013). Coal Desulfurization with Acidithiobacillusferrivorans, from Balya Acidic Mine Drainage. *Energy & Fuels*, *27*(6), 3090–3098. doi:10.1021/ef400360t

Ballard, J. R., Palin, J. M., Williams, I. S., Campbell, I. H., & Faunes, A. (2001). Two ages of porphyry intrusion resolved for the super-giant Chuquicamata copper deposit of northern Chile by ELA-ICP-MS and SHRIMP. *Geology*, *29*(5), 383–386. doi:10.1130/0091-7613(2001)029<0383:TAOPIR>2.0.CO;2

Bond, A. R., & Levine, R. M. (2001). Noril'sk Nickel and Russian platinum-group metals production. *Post-Soviet Geography and Economics*, *42*(2), 77–104.

Bosecker, K. (1997). Bioleaching: Metal solubilization by microorganisms. *FEMS Microbiology Reviews*, *20*(3-4), 591–604. doi:10.1111/j.1574-6976.1997.tb00340.x

Bouché, C., Brandt, C., Broussaud, A., & Drunick, W. (2005). Advanced control of gold ore grinding plants in South Africa. *Minerals Engineering*, *18*(8), 866–876. doi:10.1016/j.mineng.2005.02.003

Castro, I. M., Fietto, J. L. R., Vieira, R. X., Trópia, M. J. M., Campos, L. M. M., Paniago, E. B., & Brandão, R. L. (2000). Bioleaching of zinc and nickel from silicates using Aspergillus niger cultures. *Hydrometallurgy*, *57*(1), 39–49. doi:10.1016/S0304-386X(00)00088-8

Chen, G.Z., Fray, D.J., & Farthing, T.W. (2000). Direct electrochemical reduction of titanium dioxide to titanium in molten calcium chloride. *Nature*, *407*(6802), 361.

Craig, J. R., & Rimstidt, J. D. (1998). Gold production history of the United States. *Ore Geology Reviews*, *13*(6), 407–464. doi:10.1016/S0169-1368(98)00009-2

Dixon, D., & Hendrix, J. (1993). A general model for leaching of one or more solid reactants from porous ore particles. *Metallurgical and Materials Transactions. B, Process Metallurgy and Materials Processing Science*, *24*(1), 157–169.

Domic, E. M. (2007). A review of the development and current status of copper bioleaching operations in Chile: 25 years of successful commercial implementation. In *Biomining* (pp. 81–95). Springer. doi:10.1007/978-3-540-34911-2_4

Dutta, M., & Mukherjee, S. (2010). An outlook into energy consumption in large scale industries in India: The cases of steel, aluminium and cement. *Energy Policy*, *38*(11), 7286–7298. doi:10.1016/j.enpol.2010.07.056

Evans, J., & Kvande, H. (2008). Sustainability, climate change, and greenhouse gas emissions reduction: Responsibility, key challenges, and opportunities for the aluminum industry. *JOM Journal of the Minerals. Metals and Materials Society*, *60*(8), 25–31. doi:10.1007/s11837-008-0103-2

Falk, J., Green, J., & Mudd, G. (2006). Australia, uranium and nuclear power. *The International Journal of Environmental Studies*, *63*(6), 845–857. doi:10.1080/00207230601047131

Fyodorov, G. (2005). *Uranium deposits of the Inkay-Mynkuduk ore field, Kazakhstan. In Developments in uranium resources, production, demand and the environment TECDOC-1425* (pp. 95–112). Vienna: IAEA.

Golding, B., & Golding, S. D. (2017). *Copper and Coal Resources. In Metals, Energy and Sustainability* (pp. 21–35). Springer. doi:10.1007/978-3-319-51175-7_2

Gupta, V. K., Gupta, M., & Sharma, S. (2001). Process development for the removal of lead and chromium from aqueous solutions using red mud—an aluminium industry waste. *Water Research*, *35*(5), 1125–1134. doi:10.1016/S0043-1354(00)00389-4 PMID:11268832

HILSON, G. (2001). Putting theory into practice: How has the gold mining industry interpreted the concept of sustainable development? *Mineral Resources Engineering*, *10*(04), 397–413.

Hong, S., Candelone, J.-P., Patterson, C. C., & Boutron, C. F. (1996). History of ancient copper smelting pollution during Roman and medieval times recorded in Greenland ice. *Science*, *272*(5259), 246–249. doi:10.1126/science.272.5259.246

Jindal, A.K. (2011). *Comparison of financial investment & return on the capital in the mining industries of the developing countries–case studies*. Academic Press.

Johnson, D. B. (1998). Biodiversity and ecology of acidophilic microorganisms. *FEMS Microbiology Ecology*, *27*(4), 307–317. doi:10.1111/j.1574-6941.1998.tb00547.x

Johnson, D. B. (2014). Biomining—biotechnologies for extracting and recovering metals from ores and waste materials. *Current Opinion in Biotechnology*, *30*, 24–31. doi:10.1016/j.copbio.2014.04.008 PMID:24794631

Johnson, D. B., Rolfe, S., Hallberg, K. B., & Iversen, E. (2001). Isolation and phylogenetic characterization of acidophilic microorganisms indigenous to acidic drainage waters at an abandoned Norwegian copper mine. *Environmental Microbiology*, *3*(10), 630–637. doi:10.1046/j.1462-2920.2001.00234.x PMID:11722543

Jordanova, N., Petrovský, E., Kapicka, A., Jordanova, D., & Petrov, P. (2017). Application of magnetic methods for assessment of soil restoration in the vicinity of metallurgical copper-processing plant in Bulgaria. *Environmental Monitoring and Assessment*, *189*(4), 158. doi:10.1007/s10661-017-5834-5 PMID:28285437

Kerrich, R., Goldfarb, R., Groves, D., Garwin, S., & Jia, Y. (2000). The characteristics, origins, and geodynamic settings of supergiant gold metallogenic provinces. *Science in China. Series D, Earth Sciences*, *43*(1), 1–68. doi:10.1007/BF02911933

Konishi, Y., Kubo, H., & Asai, S. (1992). Bioleaching of zinc sulfide concentrate by Thiobacillusferrooxidans. *Biotechnology and Bioengineering*, *39*(1), 66–74. doi:10.1002/bit.260390111 PMID:18600888

Liu, W., Yang, J., & Xiao, B. (2009). Review on treatment and utilization of bauxite residues in China. *International Journal of Mineral Processing*, *93*(3), 220–231. doi:10.1016/j.minpro.2009.08.005

Lloyd, J. R., & Lovley, D. R. (2001). Microbial detoxification of metals and radionuclides. *Current Opinion in Biotechnology, 12*(3), 248–253. doi:10.1016/S0958-1669(00)00207-X PMID:11404102

Lovley, D. R. (1995). Bioremediation of organic and metal contaminants with dissimilatory metal reduction. *Journal of Industrial Microbiology & Biotechnology, 14*(2), 85–93. doi:10.1007/BF01569889 PMID:7766214

Massot, P., & Chen, Z.-M. (2013). China and the global uranium market: Prospects for peaceful coexistence. *The Scientific World Journal.* PMID:23606818

Mathur, A., Viswamohan, K., Mohanty, K., Murthy, V., & Seshadrinath, S. (2000). Uranium extraction using biogenic ferric sulfate a case study on quartz chlorite ore from Jaduguda, Singhbhum Thrust Belt (STB), Bihar, India. *Minerals Engineering, 13*(5), 575–579. doi:10.1016/S0892-6875(00)00037-6

McCready, A., Parnell, J., & Castro, L. (2003). Crystalline placer gold from the Rio Neuquen, Argentina: Implications for the gold budget in placer gold formation. *Economic Geology and the Bulletin of the Society of Economic Geologists, 98*(3), 623–633.

Mehta, K., Kumar, V., Pandey, B., & Tamrakar, P. (2011). Bioleaching-an alternate uranium ore processing technology for India. *Energy Procedia, 7*, 158–162. doi:10.1016/j.egypro.2011.06.021

Moneim, M. A. (2009). Staff suggestion scheme (3Ss) within the UAE context: Implementation and critical success factors. *Education, Business and Society, 2*(2), 153–167. doi:10.1108/17537980910960717

Muscatello, J., Belknap, A., & Janz, D. (2008). Accumulation of selenium in aquatic systems downstream of a uranium mining operation in northern Saskatchewan, Canada. *Environmental Pollution, 156*(2), 387–393. doi:10.1016/j.envpol.2008.01.039 PMID:18346828

Nordstrom, D. K., Alpers, C. N., Ptacek, C. J., & Blowes, D. W. (2000). Negative pH and extremely acidic mine waters from Iron Mountain, California. *Environmental Science & Technology, 34*(2), 254–258. doi:10.1021/es990646v

Olson, G. J. (1994). Microbial oxidation of gold ores and gold bioleaching. *FEMS Microbiology Letters, 119*(1-2), 1–6. doi:10.1111/j.1574-6968.1994.tb06858.x

Olson, G. J., Sakai, C. K., Parks, E., & Brinckman, F. (1990). Bioleaching of cobalt from smelter wastes byThiobacillusferrooxidans. *Journal of Industrial Microbiology & Biotechnology, 6*(1), 49–52. doi:10.1007/BF01576176

Patra, A., Pradhan, D., Kim, D., Ahn, J., & Yoon, H. (2011). Review on bioleaching of uranium from low-grade ore. *Journal of the Korean Institute of Resources Recycling*, *20*(2), 30–44. doi:10.7844/kirr.2011.20.2.030

Patterson, J. W., & Passino, R. (1987). *Metals separation and recovery*. Metal Speciation Separation and Recovery.

Petersen, J., & Dixon, D. (2002). Thermophilic heap leaching of a chalcopyrite concentrate. *Minerals Engineering*, *15*(11), 777–785. doi:10.1016/S0892-6875(02)00092-4

Philip, T. P. (2007). Process control in metallurgical plants–from an xstrata perspective. *IFAC Proceedings Volumes*, *40*(11), 377-389.

Radwanek-Bak, B. (2014). Opportunities and barriers of critical minerals management in Poland. *International Multidisciplinary Scientific GeoConference: SGEM: Surveying Geology & mining Ecology Management*, *3*, 411.

Raghavan, R., Mohanan, P., & Swarnkar, S. (2000). Hydrometallurgical processing of lead-bearing materials for the recovery of lead and silver as lead concentrate and lead metal. *Hydrometallurgy*, *58*(2), 103–116. doi:10.1016/S0304-386X(00)00108-0

Rawlings, D. E. (2005). Characteristics and adaptability of iron-and sulfur-oxidizing microorganisms used for the recovery of metals from minerals and their concentrates. *Microbial Cell Factories*, *4*(1), 13. doi:10.1186/1475-2859-4-13 PMID:15877814

Rawlings, D. E., Dew, D., & du Plessis, C. (2003). Biomineralization of metal-containing ores and concentrates. *Trends in Biotechnology*, *21*(1), 38–44. doi:10.1016/S0167-7799(02)00004-5 PMID:12480349

Riekkola-Vanhanen, M. (2010). TalvivaaraSotkamo mine—bioleaching of a polymetallic nickel ore in subarctic climate. *New Biotechnology*, *10*(1), 7–14. PMID:20709631

Schippers, A., Hedrich, S., Vasters, J., Drobe, M., Sand, W., & Willscher, S. (2013). Biomining: metal recovery from ores with microorganisms. In *Geobiotechnology I* (pp. 1–47). Springer. doi:10.1007/10_2013_216

Schnell, H. A. (1997). *Bioleaching of copper.in: Biomining*. Springer. doi:10.1007/978-3-662-06111-4_2

Sehlin, H. M., & Lindström, E. B. (1992). Oxidation and reduction of arsenic by Sulfolobusacidocaldarius strain BC. *FEMS Microbiology Letters*, *93*(1), 87–92. doi:10.1111/j.1574-6968.1992.tb05045.x PMID:1381326

Seidel, D. (1976). Stored technology for possible use in uranium ore processing. *Uranium ore processing (Proceedings of the Mortgage Advisory Group Washington, DC, 1975),* 79-89.

Sukla, L., Panchanadikar, V., & Kar, R. (1993). Microbial leaching of lateritic nickel ore. *World Journal of Microbiology & Biotechnology*, *9*(2), 255–257. doi:10.1007/BF00327850 PMID:24419960

Sukla, L. B., Kar, R. N., & Panchanadikar, V. (1992). Leaching of copper converter slag withAspergillusniger culture filtrate. *Biometals*, *5*(3), 169–172. doi:10.1007/BF01061324

Taylor, M.P., Van Zyl, L., Tuffin, M., Cowan, D. (2012). Extremophiles and biotechnology: how far have we come. *Extremophiles: microbiolbiotechnol*, *1*, 1-24.

Tsezos, M., McCready, R., & Bell, J. (1989). The continuous recovery of uranium from biologically leached solutions using immobilized biomass. *Biotechnology and Bioengineering*, *34*(1), 10–17. doi:10.1002/bit.260340103 PMID:18588046

Tuncuk, A., Stazi, V., Akcil, A., Yazici, E. Y., & Deveci, H. (2012). Aqueous metal recovery techniques from e-scrap: Hydrometallurgy in recycling. *Minerals Engineering*, *25*(1), 28–37. doi:10.1016/j.mineng.2011.09.019

Underhill, D. (1992). In situ leach uranium mining in the United States of America: past, present and future. *Uranium in situ leaching*.

Valenzuela, L., Chi, A., Beard, S., Orell, A., Guiliani, N., Shabanowitz, J., ... Jerez, C. A. (2006). Genomics, metagenomics and proteomics in biomining microorganisms. *Biotechnology Advances*, *24*(2), 197–211. doi:10.1016/j.biotechadv.2005.09.004 PMID:16288845

Vilks, P., Cramer, J., Bachinski, D., Doern, D., & Miller, H. (1993). Studies of colloids and suspended particles, Cigar Lake uranium deposit, Saskatchewan, Canada. *Applied Geochemistry*, *8*(6), 605–616. doi:10.1016/0883-2927(93)90016-A

Wang, J., Shen, S., Kang, J., Li, H., & Guo, Z. (2010). Effect of ore solid concentration on the bioleaching of phosphorus from high-phosphorus iron ores using indigenous sulfur-oxidizing bacteria from municipal wastewater. *Process Biochemistry*, *45*(10), 1624–1631. doi:10.1016/j.procbio.2010.06.013

Watling, H. (2008). The bioleaching of nickel-copper sulfides. *Hydrometallurgy*, *91*(1), 70–88. doi:10.1016/j.hydromet.2007.11.012

APPENDIX

Table 1. Some Worldwide leaching companies

S.No.	Company Name	Metal	Country	Reference
1.	BHP Billiton	Silver, lead	Australia	(Rawlings et al., 2003)
2.	Codelco	Copper	Chile	(Ballard et al., 2001)
3.	El Teniente	Copper	Chile	(Arévalo et al., 2002)
4.	Cameco corporation	Uranium	Canada	(Muscatello et al., 2008)
5.	Crow Butte	Uranium	United States	(Underhill, 1992)
6.	Inkai	Uranium	Kazakhstan	(Fyodorov, 2005)
7.	Cigar Lake Mine	uranium	Canada	(Vilks et al., 1993)
8.	ARMZ Uranium Holding	Uranium	Russia	(Massot& Chen, 2013)
9.	Heathgate Resources	Uranium	Australia	(Falk et al., 2006)
10.	RUSAL	aluminium	Russia	(Evans &Kvande, 2008)
11.	CHALCO	aluminium	China	(Liu et al., 2009)
12.	Alcoa	aluminium	America	(Association, 1984)
13.	Freeport- McMoRan	Copper, gold	United States	(Kerrich et al., 2000)
14.	Dubai Aluminium Co. Ltd.	aluminium	Dubai	(Moneim M, 2009)
15.	Glencoreplc	Copper	Switzerland	(Golding & Golding, 2017)
16.	KGHM PolskaMiedz S. A.	Copper, gold, silver nickel, molybdenum rhenium, selenium, platinum etc.	Poland	(Radwanek-Bak, 2014)
17.	Cortez Gold Mine	Gold	United States	(Craig &Rimstidt, 1998)
18.	Cerro Gold Mine	Gold	Argentina	(McCready et al., 2003)
19.	AngloGold Ashanti	Gold	South Africa	(Bouché et al., 2005)
20.	Newcrest Mining Ltd.	Gold	Australia	(HILSON, 2001)
21.	Sibplaz	Cobalt, selenium, rhodium, palladium, silver, titanium etc.	Russia	(Jindal, 2011)
22.	Xstrata plc	nickel, copper, zinc, vanadium etc.	Switzerland	(Philip, 2007)
23.	Nornickel	Palladium, nickel	Russia	(Bond & Levine, 2001)
24.	Jupiter Mines Ltd.	manganese	Australia	(Andrews, 2010)
25.	Bharat Aluminium Company	Aluminium	India	(Dutta& Mukherjee, 2010)
26.	Hindustan Zinc Ltd.	Zinc, cadmium, silver, lead	India	(Raghavan et al., 2000)
27.	Hindalco Industries Ltd.	aluminium	India	(Gupta et al., 2001)
28.	Eramet	Manganese, nickel	France	(Angleviel, 2003)
29.	Aurubis	Copper	Germany	(Jordanova et al., 2017)
30.	Hindustan Copper Ltd.	Copper	India	(Sukla et al., 1992)

Chapter 2
Prospective Sustainability of Utilization of Effective Techniques for Remediation of Heavy Metals From Textile Effluents

Bipin Jagdishprasad Agrawal
The M. S. University of Baroda, India

ABSTRACT

Textile industries are one of the prevalent water consumers and contaminators ensuing high generation of waste water. Wastewater from dyeing and printing units is often rich in organic compounds, colours, and heavy metals containing residues of various dyes and chemicals. Among these waste water pollutants, heavy metals are of serious environmental concern in recent years. Metals are extensively used for manufacture of synthetic dyestuffs and for colouration of natural dyes on textiles to achieve different shades and hues. Heavy metals such as lead (Pb), chromium (Cr), cadmium (Cd), iron (Fe), zinc (Zn), copper (Cu), etc. are widely used for production of colour pigments of textile dyes. These heavy metals, highly toxic, get conveyed to the environment, and can bio-accumulate in the human body, aquatic life, natural water-bodies, and also possibly become trapped in the soil. The chapter deals with the utilization of heavy metals in textile wet processing, their important characteristics, various toxic aspects, and different methods for their elimination from the textile effluent liquors.

DOI: 10.4018/978-1-5225-4162-2.ch002

Copyright © 2018, IGI Global. Copying or distributing in print or electronic forms without written permission of IGI Global is prohibited.

INTRODUCTION

A vital role is played by the textile industry not only in the day-to-day activities of human beings but also in the world economy. The textile industries have been categorized as one of the major consumer of water (up to 150 L of water to dye 1 kg of cotton) and is also the nastiest delinquents of pollution contributors since more than 2000 types of chemicals and over 7000 types of synthetic dyestuffs are being consumed by these industries (Nemerow, 1978; Ghoreishi and Haghighi, 2003; Hai et al., 2006). Among various textile production processes, the chemical wet processing of textiles consumes relatively large amount of water. The textile chemical wet processing industries, particularly printing and dyeing industries, have expanded rapidly in recent years. Their effluents are discharged either directly (mostly untreated/partially treated) or along with domestic wastewater. Apart from enormous heat evolved from the discharged liquors and high pH value, the textile wastewater is often saturated with a mixture of colourants (dyes and pigments) containing carcinogenic amines, de-foamers, chlorine bleaching agents, halogen carriers, detergents, optical brighteners, free formaldehyde, biocides, fire retardants, softeners, equalizers and various organic compounds used as cleaning solvents, plasticizers, etc. (Correia et al., 1994). It also contains high concentrations of toxic heavy metals, total dissolved solids, and possess high chemical and biological oxygen demand. Sediments, suspended and dissolved solids are important depositories for noxious dyes and heavy metals, instigating rapid depletion of dissolved oxygen and leading to oxygen sag in the receiving water (Ali et al., 2008).

Environmental pollution is one of the major and most important problems of the modern world. Among the various textile wastewater pollutants, heavy metals are of serious environmental concern in the recent years. This is because of their toxicity, bio-accumulation and bio-concentration in the living organisms. Also, their persistence in the environment and non-biodegradable nature has intensified its concern. The treated and untreated wastewater effluent from textiles processing units contain toxic metals as well as metal chelates; and among the chemical pollutants, heavy metals, being non-biodegradable, can be concentrated along the food chain, producing their toxic effects at points after far removal from the source of pollution. "Heavy metal" is demarcated as any metallic element that has a comparatively high density and is toxic or poisonous at even low concentrations. They are naturally occurring elements of the Earth's crust, present in varying concentrations in all ecosystems, and they cannot be degraded or destroyed. There is a huge number of heavy metals, found in elemental form as well as variety of other chemical compounds; each form or compound has different characteristics. Stimulatingly, minor amounts of these elements are common in our environment and nutrition and are fundamentally desirable for good health. It has been testified that metals such as cobalt (Co), copper (Cu), chromium (Cr), iron (Fe), magnesium (Mg), manganese (Mn), molybdenum (Mo), nickel (Ni), selenium (Se) and zinc (Zn) are vital nutrients that are mandatory for numerous biochemical and physiological purposes. Inadequate supply of these micro-nutrients may subsequently result in a variety of deficiency diseases or syndromes (World Health Organization, 1996). However, human activities have significantly changed the biochemical cycles and equilibrium of some heavy metals. Consequently, when our bodies have to deal with large amounts of these heavy metals, then the complications may arise and many diseases are instigated (Halimoon and Yin, 2010). Environmental exposure to high concentrations of heavy metals may be associated with the development of various forms of cancers and kidney damage in human beings. Further, it is toxic to microorganism and fishes and hence appropriate treatment is essential before it is being released into the environment.

The metals are generally present in the form of soluble organic or inorganic salts, and are different from non-metallic pollutants in that they cannot be disintegrated by biological or chemical treatments. No matter what supervenes to the organic molecule containing a metal, i.e. a dye or some other organic metal salt or a complex, the metal will persist in its metallic form or as an inorganic salt. Metals can be chemically precipitated or coagulated as salts or reduced to the metallic form. In any case, they must be separated from the liquid phase by filtration, settling, centrifuging or electrodeposition. Unless a metal is economically recuperated, it may eventually end up as a sludge that must be disposed of safely. The chapter deals with the utilization of heavy metals in textile wet processing, their important characteristics, various toxic aspects and different physico-chemical and bioremediation techniques for their elimination from the textile effluent liquors.

UTILIZATION OF HEAVY METALS IN THE PRODUCTION OF DYE INTERMEDIATES AND SYNTHETIC DYES

Colouration of textiles is one of our oldest industries; earlier natural dyes, extracted from leaves, flowers, berries, roots of various vegetation, were used for dyeing of natural fibres, but with the advent of synthetic dyes in 1856 by William Henry Perkin (Robinson, 1957), the use of natural dyes on a commercial scale has almost disappeared due to ease of production and economical aspects associated with the synthetic dyestuffs. Today, the production of synthetic chemical dyestuffs has become gigantic business, but injudiciously the manufacture and usage of these synthetic dyes is one of the world's most polluting industries. Conventional synthetic dyes present health menaces to those working with them and also to those who wear them, apart from damaging the environment in a number of ways.

The synthetic dyestuffs, used for colouration (dyeing and printing) of textiles, have been characterized under the groups as anionic, cationic and nonionic dyes. Anionic dyes mostly comprise of the direct, acid and reactive dyes. Basic dyes are the only class of cationic dyes used in the textile industry. Non-ionic dyes refer to disperse dyes, which do not ionize in an aqueous medium. The manufacture of synthetic dyes is achieved from various dye intermediates, which themselves are prepared from basic organic natural raw materials by simple organic reactions. From these intermediates, the manufacture of the synthetic dyestuffs is performed in a reactor, after which the content is filtered, dried, and blended with other additives to produce the final product. Various steps involved in the synthesis include chemical reactions such as sulphonation, halogenation, nitration, reduction, amination, diazotization followed by coupling, etc.; after which various separation processes, namely distillation, precipitation, crystallization, etc. are performed. On completion of the manufacture of actual dyestuff, in order to maintain the consistency of the product quality, several finishing set-ups, viz. drying, grinding and standardization are executed. The preparation of various dye-intermediates and synthetic dyestuff involve the utilization of metals, directly in the metallic form as a catalyst or in the form of salts and other metallic compounds (Allen, 1971; Zollinger, 1971, Fierz-David and Blangey, 1979; Gordon and Gregory, 1983).

Apart from the effluent liquor produced in gigantic magnitude after completion of the manufacture, major solid wastes of concern include filtration sludge, process and effluent treatment sludge, and container residues. Examples of wastes considered toxic include wastewater treatment sludge, spent acids, and process residues from the manufacture of chrome yellow and orange pigments, molybdenum orange pigments, zinc yellow pigments, chrome and chrome oxide green pigments, iron blue pigments, and azo dyes (Joshi et al., 2004).

The environmental impact of metals in wastewater effluents is also an important issue faced by the dye manufacturing and application industries today. This is an important point because a significant number of dyestuffs are metallized dyes. The synthetic dyes generally containing the metallic compounds are azoic colours, direct dyes, mordant or chrome dyes, metal-complex dyes, etc. Commonly employed metals are aluminium, iron, chromium, cobalt, nickel and copper, all of which are designated as priority pollutants by the US Environmental Protection Agency (EPA) (Freeman et al., 1996).

Utilization of Heavy Metals for Dye-Intermediates' Synthesis

All synthetic dyes are generally prepared from various dye-intermediates, which in turn are obtained from basic organic compounds, such as benzene, toluene, naphthalene, anthracene, etc. The preparation methods of these dye-intermediates involve various reactions, such as sulphonation, nitration, reduction, oxidation, hydrogenation, halogenation, dioazotization, alkali fusion, Friedel-Crafts (alkylation and acylation), etc. Some basic reactions utilizing heavy metals during the synthesis of dye intermediates are illustrated in Table 1.

Utilization of Heavy Metals in the Manufacture of Synthetic Dyestuffs

The synthetic dyestuffs also require use of various metals/metallic compounds during their manufacture. The metals utilized for the dye manufacture have been listed in the Table 2 along with the name of the dye and its chemical configuration.

Table 1. Use of heavy metals for synthesis of dye intermediates

Name of the Reaction	Chemical Representation	Heavy Metal(s) Used	Reference
Béchamp reduction	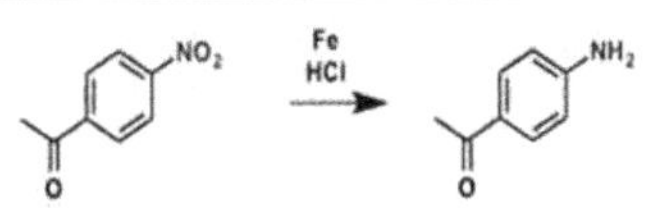	Iron, zinc, tin	Béchamp, 1854; Othmer, 2004
Preparation of Phenyl hydroxylamine	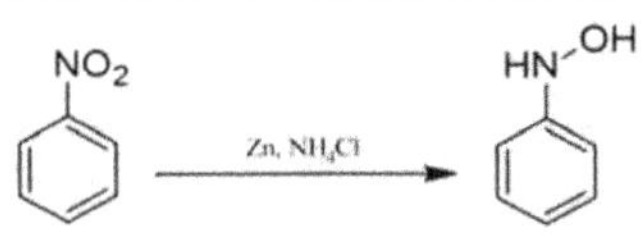	Zinc, Raney nickel	Feuer et al., 1965
Hydrogenation	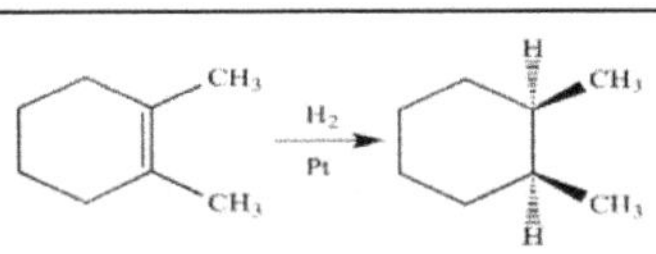	Platinum, palladium, rhodium, and ruthenium	Hudlický, 1996; de Vries and Elsevier, 2007
	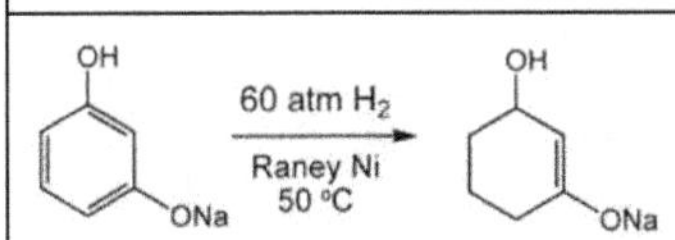	Raney nickel	
Friedel-crafts (Acylation) reaction	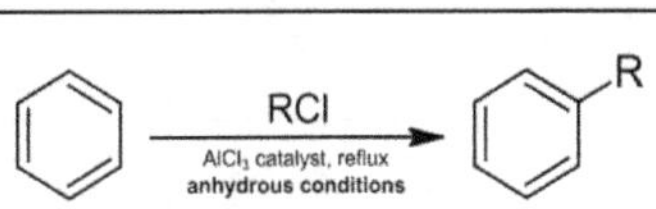	Aluminum, boron, silicon, tin, titanium, zirconium, iron, copper, zinc	Rueping and Nachtsheim, 2010; Michael and Jerry, 2007
Friedel-crafts, Cyclization and sulphonation	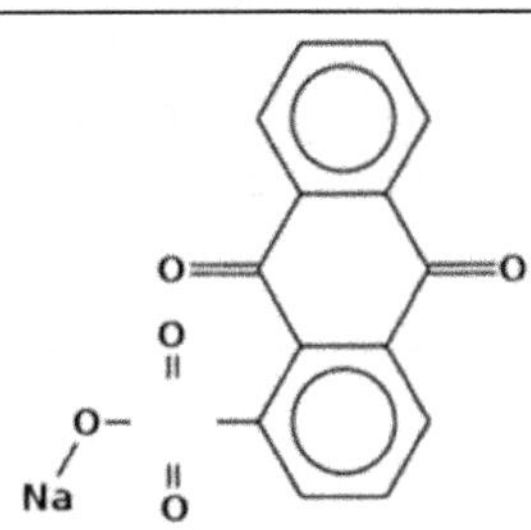	Aluminium chloride and mercuric oxide	

Table 2. Metals and metallic compounds used in the manufacture of synthetic dyestuffs

Commercial/ Trade-Name of the Dye	C. I. Generic Name (CC Number)	Chemical Structure	Metal or Metallic Compound Used for Manufacture	Reference
		Vat Dyes		
Indanthrene Golden Yellow GK	C.I. Vat Yellow 4 (C. I. 59100)	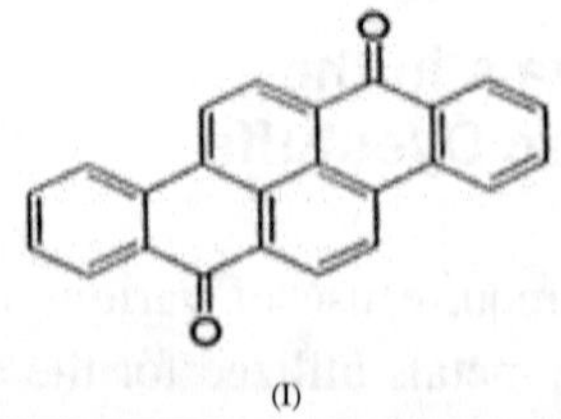 (I)	Aluminium chloride	
Novatic Golden Orange 3G	C I Vat Orange 15 (C I 69025)	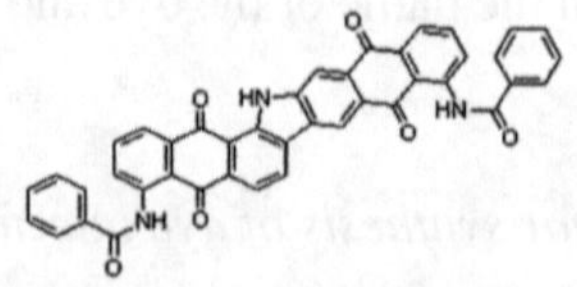 (II)	Mercury	Cain, and Thorpe, 1946; Venkatraman, 1952-1972
Violanthrone	CI Vat Blue 20 (C I 59800)	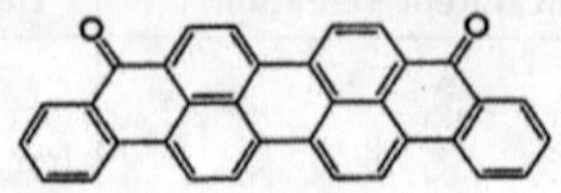 (III)	Iron, manganese dioxide	
Indanthrene Corinth RK	C.I. Vat Violet 16 (C I 65020)	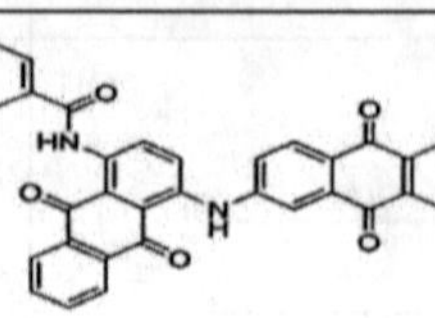 (IV)	Copper powder and copper acetate	
Indanthrene Khaki GG	C I Vat Green 8 (C I 71050)	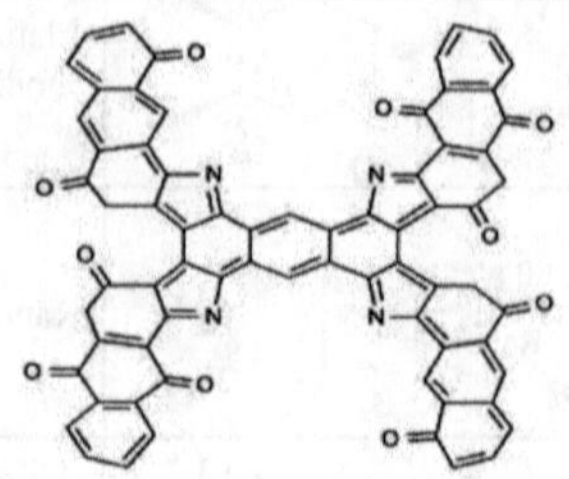 (V)	Copper powder and aluminium chloride	
Caledone Jade Green XB	C I Vat green 1 (128-58-5)	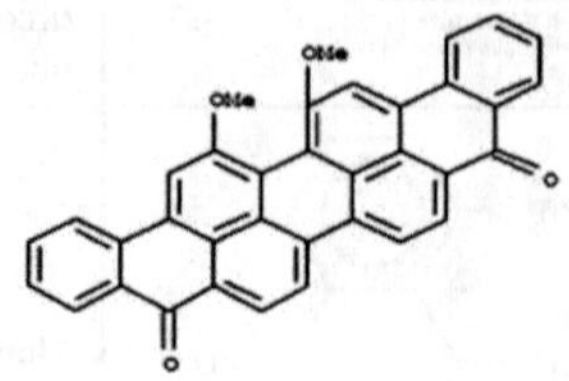 (VI)	Manganese oxide	

continued on following page

Table 2. Continued

Commercial/ Trade-Name of the Dye	C. I. Generic Name (CC Number)	Chemical Structure	Metal or Metallic Compound Used for Manufacture	Reference
Phthalocyanine Pigments				
Monastral Fast Blue G	CI Pigment Blue 15	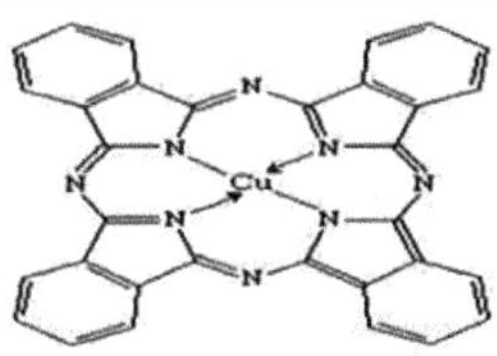 (VII)	Copper	
Monastral Fast Green G	C.I. Pigment Green 7	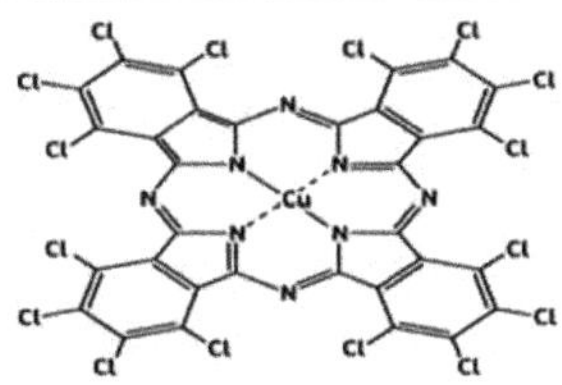 (VIII)	Copper	Löbbert, 2002; Chaidogiannos, 2009
Monastral Fast Green 6Y	C. I. Pigment Green 36	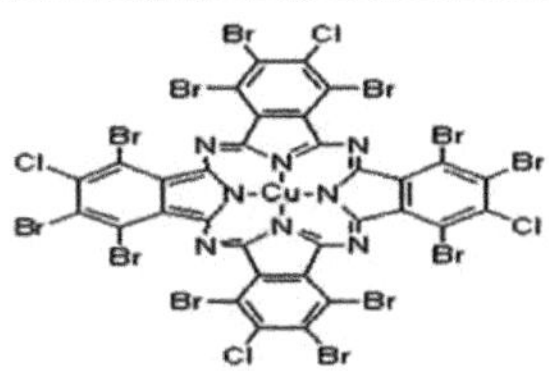 (IX)	Copper	
Chlorantine Fast Turquoise Blue GLL	C.I. Direct Blue 86	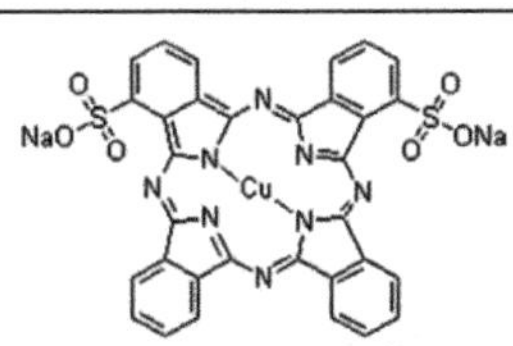 (x)	Copper	
Cibacron Turquoise Blue G-F	C I Reactive Blue 15 (CAS No. 12225-39-7)	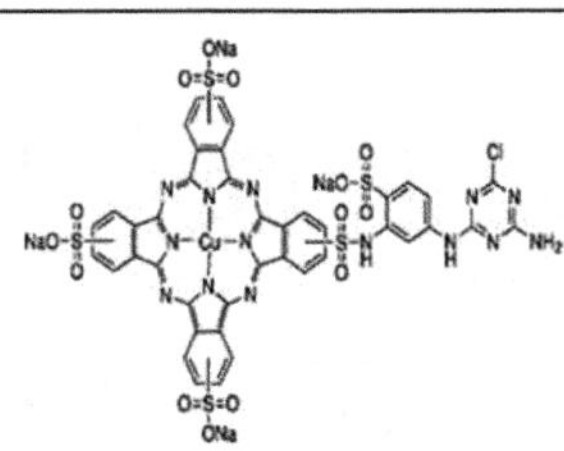 (XI)	Copper	Venkatraman, 1952-1972
Procion Turquoise MX-G	C I Reactive Blue 140 (CAS No. 71872-74-7)	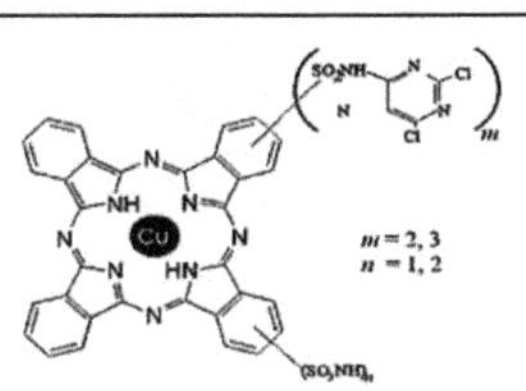 (XII)	Copper	

continued on following page

Table 2. Continued

Commercial/ Trade-Name of the Dye	C. I. Generic Name (CC Number)	Chemical Structure	Metal or Metallic Compound Used for Manufacture	Reference
		Mordant Dyes		
Eriochrome Black T	C.I. Mordant Black 11 (C.I. 14645)	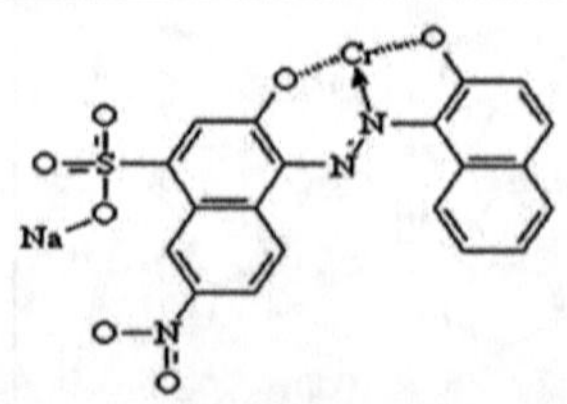 (XIII)	Chromium	
		Metal Complex Acid Dyes		
1:1 metal complex (pre-metallized) acid dyes	--	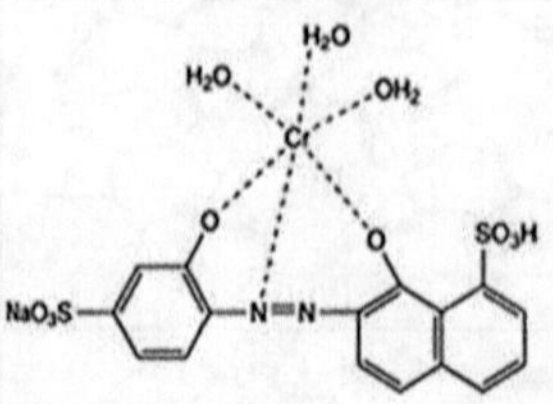 (XIV)	Chromium	
1:2 metal complex (pre-metallized) acid dyes	--	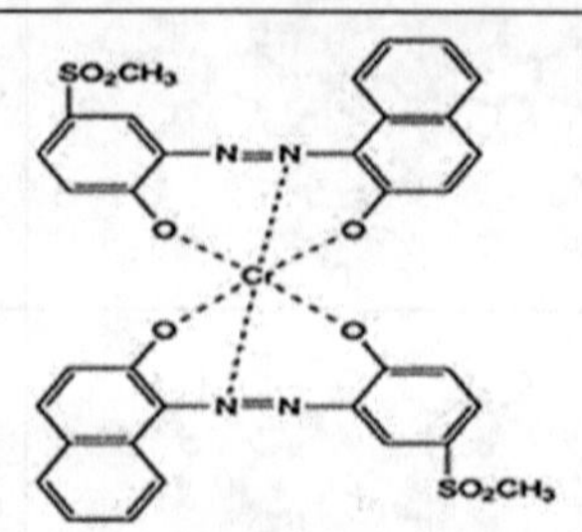 (XV)	Chromium	
		Basic (Cationic) Dyes		
Malachite Green	C I Basic Green 4 (C I 42000)	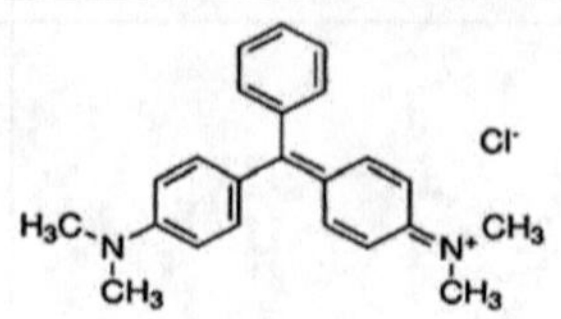 (XVI)	Lead oxide, zinc chloride	Vishwanathan, 2011
Methylene Blue	C I Basic Blue 9 (C I 52015)	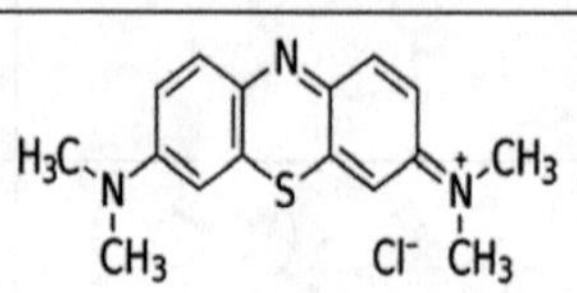 (XVII)	Iron, sodium dichromate, zinc chloride	

A large number of vat dyes, phthalocyanine pigments, reactive dyes, metal-complex dyes, basic dyes etc. include different metals and/or their salts during their various stages of manufacture.

The synthesis of phthalocyanine pigments is of great importance to a textile dyer. Phthalocyanines are fundamentally made up of four molecules of isoindole which are linked to each other by nitrogen atoms. The phthalocyanines co-ordinate with metal atoms, particularly copper metal. Phthalocyanines became commercially available in the 1930s, with the parent compound and its copper complex promoted as Monastral Fast Blue B and Monastral Fast Blue G respectively. Today, approximately 25% of all synthetic organic pigments are phthalocyanine derivatives (Lobbert, 2002; Chaidogiannos et al., 2009).

Copper phthalocyanines (CuPc), the most imperative among numerous recognized metal complexes, can also be molded directly on cotton. Amongst synthetic textile substrates, some complexes are employed with nylon but are not suitable for polyester and acrylics textiles. These pigments exhibit resilient, bright blue to green shades with notable chemical stability. Copper phthalocyanine sublimes remain unchanged at 580° C and dissolves in concentrated sulphuric acid without undergoing any change in chemical structure. They also display excellent lightfastness; however, quite disparity appears in their features as compared with those of natural pigments (i.e., hemoglobin and chlorophyll) that are easily destroyed by intense light or heat and mild chemical reagents. The extraordinary stability, strength, and brightness of the shades makes phthalocyanines commercially cost-effective.

Figure 1. Molecular structures of (a) Isoindone and (b) Monastral Fast Blue B (Phthalocyanine)

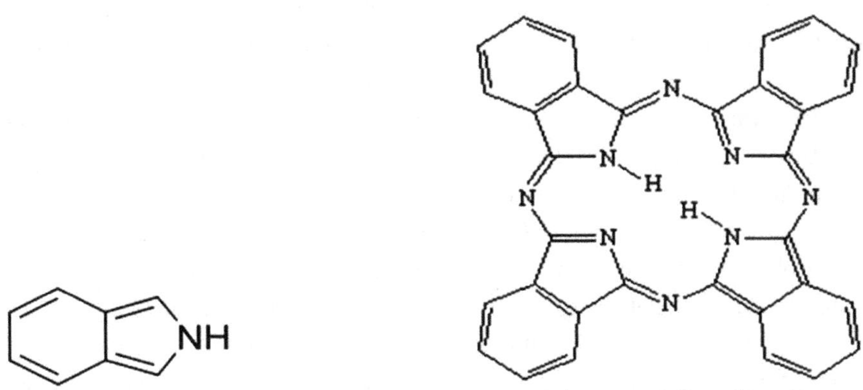

(a) (b)

APPLICATION OF HEAVY METALS IN COLOURATION OF TEXTILE SUBSTRATES

Colouration of Natural Dyes on Textiles

Till the advent of synthetic dyes in 1856 by William Henry Perkin, natural dyes were the only means for the colouration (dyeing and printing) of textile materials in various forms, viz. fibre, yarn and/or fabric. The word 'natural dye' encompasses various colouring components, viz. pigments/dyes derived from the natural sources like plants, animal and minerals. Natural dyes are mostly employed for dyeing of natural fiber textiles to enhance their eco-friendly characteristics. Apart from indigo, other natural dyes are usually not used for printing directly. For producing printed fabrics, the printing is usually done with mordant and the whole material is dyed whereby only the area printed with mordants picks up the color. Natural dyes can be directly applied to the textile materials; however, dyeing of fabrics with natural dyes often leads to problems such as narrow range of shades, and low colour fastness of dyed textiles. Attempts to overcome these problems have been mainly focused on the use of mordants in order to achieve good fastness properties and to obtain variety of shades. In addition to creating affinity between dye and fibre, the use of mordants also change the hue of certain dyes. Metals/metallic salts, natural tannin based products, oil mordants, etc. are the well-known mordants for the application of natural dyes (Shukla, 2006). Transition metal ions generally possess strong co-ordinating power and/or are capable of forming weak to medium forces of attraction. They act as bridging material to generate substantivity of natural dyes/colourants when a textile material being impregnated with such metallic salt (i.e. mordanted) is subjected to dyeing with different natural dyes. The metallic mordants forms an insoluble precipitate or lake after coalescing with dye in the fibre; thus, both the dye and the mordant get fixed on the substrate and substantially improves the wash and light fastness properties to a reasonable level (Samanta and Konar, 2011).

Natural dyes, like synthetic dyes, can also be used to dye textiles at all stages such as fiber, yarn, or fabric. Dyeing is normally carried out by the traditional dyers/artisans by hand in large vessels, preferably made of iron, stainless steel, copper, and aluminum. Generally, copper vessels are utilized for producing bright shades, aluminium vessel may be utilized only for dyeing one particular type of dye due to staining problems of the vessel, whereas in the recent times, stainless steel vessels are most preferred for the natural dyeing process. Different parts of the plants are used for different shades, depending on the chemical constitution of the chromophoric group present in that particular part. The practice for the dyeing of textiles with natural dyes differs from the synthetic dye application process because only some natural dyes can be applied directly to textiles. In most of the cases the dye is not substantive to the fibre on which it is being dyed; therefore, a supplementary step of mordanting is involved, making it a two-step process (Gulrajani, 1999; Agrawal and Patel, 2000; Gulrajani, 2001; Alemayehu and Teklemariam, 2014; Saxena and Raja, 2014).

Mordanting of Textile Substrates

Many textile substrates, particularly cellulosics, lack substantivity for majority of the natural colourants; hence they require an additional mordanting step before actual dyeing operation. Mordanting is the treatment of textile substrate with various mordants, which possess affinity for both textile fibres and the dyestuffs. The mordant act as an associating link between the fibre and the dyestuff, thereby assisting the binding of the natural mordantable dyestuffs onto the textile fibres. They substantially increase the dye uptake, various fastness characteristics of the dyed fabrics and also help in producing varieties of shades of different hues and tones (Samanta and Agarwal 2009). The mordant may belong to the category of heavy metals/metallic salts and other complex forming intermediaries, oil mordants or tannins. In the present chapter, only the characteristics of various important heavy metals and their corresponding salts have been taken into consideration.

Metallic Mordants

Metal mordants are frequently used for the colouration (dyeing and printing) of textiles with natural dyes. Metal salts of aluminum, chromium, tin, copper, and iron have been in use as mordants by traditional dyers since time immemorial. Different mordants produce diversified range of brilliant shades with the same natural dyestuff. For example, Cochineal, when mordanted with alum gives a crimson shade; with iron, purple; with tin, scarlet; and with chrome or copper, purple shades. Logwood also gives mauve and blue shades on treatment with alum and chrome mordants respectively; while alizarin forms a red lake or complex with aluminum and a violet lake with iron. Similarly, onion natural dye which is yellow in color will change to orange color with stannous chloride mordant and grey color with ferrous sulphate. Fustic, weld, and most of the yellow natural dyes, give a greenish yellow with alum, but a bright golden shade with chrome; and fawns of various shades with other mordants. Sometimes, two or more mordants may be combined together to get still richer shades of secondary and tertiary levels. However, there are some natural dyes available, such as indigo, which can be applied to the textile substrates without mordanting.

Different colours from the same dyestuff can be obtained by use of different metallic mordants because colour obtained with many natural dyes is due to the formation of coloured insoluble dye complexes with metal salts or mordants (Mohanty et al., 1984)]. Dye complexes with different metals have different colours and may also differ in fastness properties. For example, alizarin forms a red lake or complex with aluminum and a violet lake with iron. Similarly, onion natural dye which is yellow in colour will change to orange color with stannous chloride mordant and grey color with ferrous sulphate.

The list of various metallic mordants used for the application of natural dyes on textile materials is as follows:

Conventional Metallic Mordants

- Alum
- Aluminium sulphate
- Aluminum chloride
- Aluminium nitrate
- Ferrous sulphate
- Ferric chloride
- Stannous chloride
- Stannous sulphate
- Stannic chloride
- Copper sulphate
- Copper chloride
- Potassium dichromate
- Cobalt sulphate
- Magnesium sulphate
- Zinc sulphate
- Zinc Chloride
- Manganese sulphate
- Nickel sulphate

Newly Discovered Metallic Mordants

- Zinc tetrafluoroborate
- Zirconium oxy chloride
- Rhenium trichloride
- Neodymium trichloride
- Lanthanum Oxide

The presence of small amounts of heavy metals in our environment and diet is very common and their presence is actually considered essential for good health. However, for some heavy metals, toxic levels can be just above the background concentrations naturally found in nature. Therefore, it is important for us to inform ourselves about the heavy metals and to take protective measures against excessive exposure. In the present scenario of eco-regulations posed by many countries, chromium has been categorized as a red-listed mordant; therefore, it should be used carefully to preserve the eco-friendly nature of dyed textile material as well as the discharged effluent. Copper is also in the restricted category but its permissible levels are higher; hence, it may be used in small amounts so as not to cross the permissible limit on the dyed textiles. Lead accounts for most of the cases of pediatric heavy metal poisoning, according to the Agency for Toxic Substances and Disease Registry (ATSDR). Tin is not restricted under eco-standard labels but its existence in effluent is not considered appropriate from the environmental viewpoint. The mordants which are considered environmentally safe are alum and iron because they also exist naturally in the environment in great amounts.

Mordanting Techniques

Basically, there are three different methods for application of the mordants depending on the time of their usage. The complex between dye, fibre and the mordant may be formed by first applying the mordant or by simultaneous application of the mordant and the dye or by after treatment of the dyed material with mordant. These methods are categorized as (i) pre-mordanting, (ii) post-mordanting and (iii) meta-mordanting or simultaneous mordanting (Agrawal and Patel, 2000; Samanta and Agarwal, 2009).

Chemical Care of Mordants

The mordants or modifiers must be kept tightly-lidded container/bottle, out of direct sun, away from heat until use, and out of reach of children and pets. These chemicals often arrive in plastic bags for easy shipping and should be placed in a lidded glass or good plastic jar as soon as possible. It is essential to transfer all label information, including warnings, to the new container and to dispose the plastic bag immediately.

Identification of Mordant During Application

This is extremely essential when differently-mordanted fibres are introduced in the same dye pot and to know which mordant/dye has been used on the fibre. Such mordanted substrates, after application of the mordant, may be subsequent dyed at a later stage. One method is to cut tags from white express mail envelopes and to mark them with a permanent black felt-nib pens or a marker. Usage of other marker colours is not worthwhile since they are not water-proof. The tags are tied or attached with a safety pin on the fibrous material; it is essential to remove the safety pin after dyeing otherwise staining of the fibre may occur due to reaction of the metal with the chemicals.

Knotting System

- **No Mordant:** No knot
- **Alum Mordant:** 1 knot
- **Copper Mordant:** 2 knots
- **Iron Mordant:** 3 knots
- **Tin Mordant:** 4 knots

In another more frequently used method, a system of knots can also be used, particularly when dyeing of yarn in hank form is to be performed. The knotting system is represented schematically in Figure 2. The following knots are used for understanding the application of mordant:

Toxicity of Heavy Metals

Heavy metals are metallic elements which have a high atomic weight and a density much greater (at least 5 times) than water. There are more than 20 heavy metals, but four are of particular apprehension to human health: lead (Pb), cadmium (Cd), mercury (Hg), and arsenic (As). According to the U.S. Agency for Toxic Substances and Disease Registry, these four heavy metals are four of the top six threats present in toxic waste sites. They are highly toxic and can cause damaging effects even at very low concentrations (Inoue, 2013). They tend to accumulate in the food chain and in the body and can be stored in soft (e.g., kidney) and hard tissues (e.g., bone).

Figure 2. Knot system for understanding mordant application

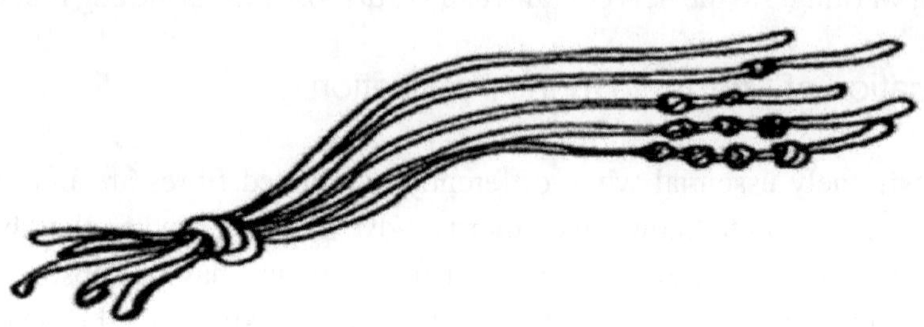

Being metals, they often exist in a positively-charged form and can bind on to negatively-charged organic molecules to form complexes.

Human body has the requirement for approximately 70 friendly trace elemental heavy metals, but there are another 12 poisonous heavy metals, such as Lead, Mercury, Aluminum, Arsenic, Cadmium, Nickel, etc., that act as poisonous intrusion to the enzyme systems and metabolism of the body. The overload of the heavy metal will be a hindrance to the natural healing functions of the body irrespective of the good health supplements or precautions taken (Morais et al., 2012). Some metals are naturally found in the body and are essential to human health. Iron, for example, prevents anemia, and zinc is a cofactor in over 100 enzymatic reactions. Magnesium and copper are other familiar metals, which are essential for the occurrence of proper metabolic rate. Heavy or toxic metals normally occur at low concentrations and are termed as trace metals that are at least five times denser than water. As such, they are stable elements (meaning they cannot be metabolized by the body) and bio-accumulative (passed up the food chain to humans). These include: mercury, nickel, lead, arsenic, cadmium, aluminum, platinum, and copper (metallic form versus ionic form). Heavy metals are taken into the human body via inhalation, ingestion, and skin absorption. If heavy metals enter and accumulate in body tissue faster than the body's detoxification pathways can dispose of them, a gradual buildup of these toxins will occur (Nagajyoti et al., 2010; Jan et al., 2015).

High-concentration exposure is not necessary to produce a state of toxicity in the body tissues and, over time, can reach toxic concentration echelons. The surplus of heavy metal in the walls of coronary arteries seems to decrease levels of nitric oxide, a compound known as "Endothelial Relaxing Factor,"— without this substance, the normal blood flow is impeded thereby increasing the risk of vascular blockages. Heavy metal overload in the adrenal glands reduces the production of hormones, which causes early aging, stress, decreased sex drive and aggravation of menopausal symptoms (Apostoli and Catalani, 2011). It can also lead to unresponsiveness of diabetics to their medications, various neurological diseases such as depression and loss of thinking power, and aggravate conditions such as osteoporosis and hypothyroidism. Therefore, it is quite essential to remove the heavy metals from the human body.

Analysis of Heavy Metals

Heavy metals play a key role in the health of the aquatic environment as well as human health but become toxic if their concentration surpasses recommended limit, and have prospective to bio magnification. There are numerous qualitative and quantitative analytical methods reported for the assessment of heavy metals in textile industry, but it is a challenging field due to the wide range of analytical tasks.

The determination of heavy metals is a very complex process and does not only comprise final analyses. Usually, it is not possible to analyze the whole specimen depending on the kind and amount of the sample. Therefore, only a small fragment (or portion) of the system under study is selected for the analytical procedure either in a laboratory or directly on the site. The common steps involved in the analytical process are: sample collection, preservation and preparation, analysis of the sample by a selective analytical procedure, and finally evaluation of the data. Various analytical methods (Zeiner et al., 2007) used for the qualitative and quantitative determination of heavy metals in the samples of wastewater, textile materials and reagent solutions are Ultraviolet Visible Spectrometric analysis (Pobozy et al., 2003; Patnaik, 2004), Thin layer chromatography (Zeiner et al. 2006), Atomic absorption spectrometry (Manzoor et al., 2006), Optical emission spectrometry (Basaran et al., 2006; Zeiner et al., 2006) Inductively coupled plasma- mass spectrometry (Szynkowska et al., 2006; Chen et al., 2007), etc. Apart from these analytical methods, various other techniques have also been used widely to evaluate the degree of metal contamination or pollution in terrestrial, aquatic and marine environment sediments (Long and MacDonald, 1998; Long et al. 2000). These techniques include – Geo-accumulation index (Muller, 1969), Contamination Factor and Degree of Contamination (Hakanson, 1980), Enrichment Factor (Buat-Menerd and Chesselt, 1979), mean ERM quotient (MERM-Q) (Long et al., 1998), Potential Ecological Risk Index (Hakanson, 1980), and Pollution Load Index (Tomlinson et al., 1980).

TECHNIQUES FOR REMOVING HEAVY METALS FROM TEXTILE EFFLUENTS

Heavy metals are the main group of inorganic contaminants and a considerable large area of land and water is contaminated with them (Halim et al., 2002). Metal concentration in soil and effluent liquors typically ranges from less than one to as high as 100,000 mg/kg. Irrespective of the origin of the metals in the soil, excessive levels of many metals can result in soil quality degradation, crop yield reduction, and poor quality of agricultural products, posing significant hazards to human, animal, and ecosystem health (Long et al., 2002).

Conventional Physical and Chemical Processes for Removal of Heavy Metals From Textile Effluents

The conventional processes for removing heavy metals from textile wastewater involve many processes such as chemical precipitation, flotation, adsorption, ion exchange, and electrochemical deposition.

Chemical Precipitation

This is the most widely used for elimination of heavy metal from inorganic effluent. The conceptual mechanism of heavy metal removal by chemical precipitation is presented in the following equation (Wang et al., 2004):

$$M_2^+ + 2(OH)^- \leftrightarrow M(OH)_2\downarrow$$

where M_2^+ and OH^- represent the dissolved metal ions and the precipitant respectively, while $M(OH)_2$ is the insoluble metal hydroxide. The main consideration that significantly improves heavy metal removal by chemical precipitation is proper adjustment of pH to the basic conditions (pH 9–11). Lime and limestone are the most commonly employed precipitating agents due to their availability and low-cost in most countries (Mirbagherp and Hosseini, 2004; Aziz et al., 2008). Other advantages of using lime precipitation include the simplicity of the process, convenient and safe operations, and inexpensive equipment requirement. However, chemical precipitation requires a large amount of chemicals to reduce metals to an acceptable level for discharge. Other drawbacks are its excessive sludge production that requires further treatment, slow metal precipitation, poor settling, the aggregation of metal precipitates, and the long-term environmental impacts of sludge disposal (Aziz et al., 2008).

Adsorption

Sorption is transfer of ions from water to the soil i.e. from solution phase to the solid phase. Sorption actually describes a group of processes, which includes adsorption and precipitation reactions. Recently, adsorption has become one of the alternative treatment techniques for wastewater laden with heavy metals. Various low-cost adsorbents, derived from agricultural waste, industrial by-product, natural material, or modified biopolymers, have been recently developed and applied for the removal of heavy metals from metal-contaminated wastewater (Kurniawan and Babel, 2003). Activated carbon has shown to be an efficient adsorbent for the removal of a wide variety of organic and inorganic contaminants present in the aquatic environment. Because of its high surface area that range from 500 to 1500 m2/g, it is widely used in the treatment of wastewaters. The effectiveness of activated carbon in cleaning up polluted water is due to its well-developed porosity structure as well as the presence of a wide spectrum of surface functional groups. This makes it capable of distributing pollutants on its large internal surface, making them accessible to reactants (Liu et al., 2010; Bhatnagar et al., 2013; Karnib et al., 2014).

Natural zeolites have also gained a significant interest, mainly due to their valuable properties as ion exchange capability. Among the most frequently studied natural zeolites, clinoptilolite has shown to have high selectivity for certain heavy metal ions such as Pb(II), Cd(II), Zn(II), and Cu(II). The ability of different types of synthetic zeolite for heavy metals removal was recently investigated. The role of pH is very important for the selective adsorption of different heavy metal ions (Basaldella, 2007; Barakat, 2008; Ríos et al. 2008). Industrial by-products such as fly ash, waste iron, iron slags, hydrous titanium oxide, can be chemically modified to enhance its removal performance for metal removal from wastewater. Recently, a great deal of interest in the research for the removal of heavy metals from industrial effluent has been focused on the use of agricultural by-products as adsorbents. The use of agricultural by-products in bioremediation of heavy metal ions, is known as bio-sorption. This utilizes inactive (non-living) microbial biomass to bind and concentrate heavy metals from waste streams by purely physico-chemical pathways (mainly chelation and adsorption) of uptake (Igwe, 2005). The bio-remediation technique based on adsorption principle for removal of heavy metals from textile effluents using micro-organisms has been discussed in the next sub-section. New resources such as hazelnut shell, rice husk, pecan shells, jackfruit, maize cob or husk can be used as an adsorbent for heavy metal uptake after chemical modification or conversion by heating into activated carbon.

Adsorption on modified biopolymers and hydrogels has also gained importance in the recent times. Biopolymers are industrially attractive because they are, capable of lowering transition metal ion concentrations to sub-part per billion concentrations, widely available, and environmentally safe. Another attractive feature of biopolymers is that they possess a number of different functional groups, such as hydroxyls and amines, which increase the efficiency of metal ion uptake and the maximum chemical loading possibility. New polysaccharide-based-materials are described as modified biopolymer adsorbents (derived from chitin, chitosan, and starch) for the removal of heavy metals from the wastewater.

Hydrogels, which are crosslinked hydrophilic polymers, are capable of expanding their volumes due to their high swelling in water. Accordingly they are widely used in the purification of wastewater. Various hydrogels, viz. poly(ethyleneglycol dimethacrylate-co-acrylamide), poly(vinylpyrrolidone-co-methylacrylate), poly(3-acrylamidopropyl) trimethyl ammonium chloride, etc. have been synthesized and their adsorption behavior for heavy metals was investigated. The removal is basically governed by the water diffusion into the hydrogel, carrying the heavy metals inside especially in the absence of strongly binding sites. Maximum binding capacity increases with pH increase to > 6.

Ion Exchange

A metal deposit that is strippable and recoverable. A noticeable disadvantage was that corrosion could become a significant limiting factor, where electrodes would frequently have to be replaced (Kurniawan et al., 2006).

Membrane Filtration

Membrane filtration has received considerable attention for the treatment of inorganic effluent, since it is capable of removing not only suspended solid and organic compounds, but also inorganic contaminants such as heavy metals (Agrawal, 2014). Depending on the size of the particle that can be retained, various types of membrane filtration such as ultrafiltration, nano-filtration and reverse osmosis can be employed for heavy metal removal from wastewater. Polymer-supported ultrafiltration (PSU) technique has been shown recently to be a promising alternative for the removal of heavy metal ions from industrial effluent (Rether and Schuster, 2003). This method employs proprietary water-soluble polymeric ligands to bind metal ions of interest, and the ultrafiltration technique to concentrate the formed macromolecular complexes and produce an effluent, essentially free of the targeted metal ions. Advantages of the PSU technology over ion exchange and solvent extraction are the low-energy requirements involved in ultrafiltration, the very fast reaction kinetics, all aqueous based processing and the high selectivity of separation if selective bonding agents are applied.

Electrodialysis

Electrodialysis (ED) is a specialized membrane separation technique in which ionized species in the solution are passed through an ion exchange membrane by applying an electric potential. The membranes are thin sheets of plastic materials with either anionic or cationic characteristics. When a solution containing ionic species passes through the cell compartments, the anions migrate toward the anode and the cations toward the cathode, crossing the anion exchange and cation-exchange membranes (Chen, 2004), ED process was modeled based on basic electrochemistry rules and copper ion separation experimental data (Mohammadi et al., 2005). In spite of its limitation, ED offers advantages for the treatment of wastewater laden with heavy metals such as the ability to produce a highly concentrated stream for recovery and the rejection of undesirable impurities from water. Moreover, valuable metals such as Cr and Cu can be recovered. Since ED is a membrane process, it requires clean feed, careful operation, periodic maintenance to prevent any stack damages.

Photocatalysis

In recent years, photocatalytic process in aqueous suspension of semiconductor has received considerable attention in view of solar energy conversion. This photocatalytic process was achieved for rapid and efficient destruction of environmental pollutants. Upon illumination of semiconductor–electrolyte interface with light energy greater than the semiconductor band gap, electron–hole pairs (e^-/h^+) are formed in the conduction and the valence band of the semiconductor, respectively (Herrmann, 1999). These charge carriers, which migrate to the semiconductor surface, are capable of reducing or oxidizing species in solution having suitable redox potential. Various semiconductors have been used: TiO_2, ZnO, CeO_2, CdS, ZnS, etc. As generally observed, the best photocatalytic performances with maximum quantum yields are always obtained with titanium dioxide.

Advance Bioremediation Techniques for Removal of Heavy Metals From Textile Effluents

The conventional methods of removal of heavy metals from textile wastewaters are effective only if the effluent volume was small. However, physico-chemical methods are ineffective or expensive when the concentration of heavy metals is very low. Therefore, innovative treatment technologies need to be investigated for textile effluent treatment for removal of particularly heavy metals. On this basis, biological remediation of untreated textile effluent is an emerging technology and studied elaborately by environmentalists. Biological treatment of textile effluents may be either aerobic, anaerobic or a combination of both depending on the type of microbe being employed (Keharia and Madamwar, 2004; Ayangbenro and Babalola, 2017). Microbial decolorization and degradation is an ecofriendly and inexpensive alternative practice to chemical decomposition processes. Many microorganisms belonging to different taxonomic groups of bacteria, fungi, actinomycetes and algae have been reported for their ability to decolorize the dyes (Chen et al., 2003).

Bioremediation is an innovative and promising technology accessible for elimination of heavy metals and reclamation of the heavy metals in polluted water and lands. Since microorganisms have established various strategies for their endurance in heavy metal-polluted locales, these organisms are identified to develop and adopt different detoxifying mechanisms such as biosorption, bioaccumulation, biotransformation and biomineralization, which can be exploited for bioremediation either *ex situ* or *in situ* (Kapoor and Viraraghvan, 1995; Gadd, 2000; Lim et al., 2003; C. Lin and H. Lin, 2005; Malik, 2004).

Phytoremediation Technique for Remediation of Heavy Metal Removal

Phytoremediation essentially states the utilization of plants and associated microorganisms to partially or completely remediate selected contaminants (such as radionuclides, organic pollutants, heavy metals, etc.) from soil, sludge, sediments, wastewater and ground water (Karami et al., 2010; Ali et al., 2013). Using plants and their coexisting microbial system to remove heavy metals is a new technology. It is a cost-effective, efficient and eco-friendly in situ remediation technology driven by solar energy. The technique of phytoremediation comprises of various processes such as phytoextraction, phytofiltration, phytostabilization, phytovolatilization and phytodegradation. The initial step of phytoremediation is phytoextraction, the uptake of contaminants from soil or water by plant roots and their translocation to and accumulation in biomass, i.e., shoots. The next important process of phytoremediation is phytofiltration, which includes rhizofiltration (use of plant roots), blastofiltration (use of seedlings) or caulofiltration (use of expurgated plant shoots). In addition to the above process, phytostabilization or phytoimmobilization occurs, which reduces the mobility and bioavailability of metals in the environment and thus prevents their migration into groundwater or the food chain. Several heavy metals absorbed by plants get converted into volatile forms and subsequently released into the atmosphere by the process called phytovolatilization. This process has been used for removal of some volatile heavy metals like Hg and Se from polluted soils.

Bioremediation of Heavy Metals Using Genetic Engineering Technique

Genetic engineering has now become integrated with bioremediation since many microbes can be artificially designed which can consume the toxic waste and pollutants that are usually not taken in by normal microbes. This is done by first genetically altering the sequence of the desired microbe and enhancing its ability to digest the toxic particles of the pollutant or by genetically engineering a new microbe which has extraordinary ability to take in, consume and digest the pollutant. Thus, micro-organisms are designed specifically for bioremediation (Agrawal, 2017). Bioremediation using genetic engineering operation is a relatively cheap, environmental-friendly and socially acceptable technology which can eliminate waste permanently (Eapen et al., 2007; Doty, 2008; Fulekar, 2009; Wei et al., 2014).

Genetic engineering permits introduction of anticipated qualities into cells to metal clean up techniques, and this approach has already been used to construct cells for the bioremediation of mercury. The recombinant *staphylococcus xylosus* and *staphylococcus carnosus*, had gained Ni^2 and Cd^2 binding capacity and advocated that they could be used in bioremediation of heavy metals. There are numerous progresses in genetic engineering micro-organisms (GEMs), which overcome the limits of using wild type microbes. *Escherichia coli* strain accumulated Ni^2 by introducing the nixA gene from *helicobacter pylori* into *E. coli* JM 109 that expressed a glutathione S-transferase pea metalothioneine fusion protein. The recombinant *E. coli* strain accumulated four times more nickel than the wild type. Metal resistant *R. eutropha* isolate, engineered to produce metalothionein accumulated more Cd^2 than its wild type and offered tobacco plants some protection form Cd^2 when inoculated into contaminated soil. Nickel (Ni) is possibly the most recalcitrant pollutant and can be accumulated by the GE *Escherichia coli SE500* strain from an aqueous solution. Arsenic (As) is a very toxic metal which can be found in nature. GM bacteria expressing the ArsR gene can promote the bioaccumulation of As when present in contaminated water (Zhao et al., 2005; Kiyono and Pan-Hou, 2006; Patel et al., 2010).

Nano-Based Technologies for Remediation of Heavy Metals

The remediation, particularly bioremediation, of heavy metals and trace elements achieved by nanotechnology is an innovative emerging technology (Sarikaya et al, 2003). Compared to other conventional methods, various nanoparticles or nanomaterials have been found to be very effective for the removal of a wide range of toxic metals from the environment since they enhance the microbial activity to remove heavy metals due to the very small size and high reactivity. Nano-based technologies reduce the costs of cleaning up contaminated sites at a large scale and the process time as well (Rizwan et al., 2014; Yadav et al, 2017). Nanoparticles can be synthesized using various plants and microorganisms, such as bacteria, algae and fungi; and are commendably used to clean-up the sites contaminated with toxic metals. Such technique in which the nanoparticles are used to enhance the microbial activity for removal of toxic pollutants is called "nanobioremediation". "Bionanotechnology" or "nanotechnology through biotechnology" is the bio-fabrication of nano-objects or bifunctional macromolecules used as tools to build or manipulate nano-objects. This state-of-the-art technique would be an encouraging means to address the intensifying problem of heavy metal as well as organic pollutants in the environment.

CONCLUSION

The contamination of water by toxic heavy metals in the form of wastes effluents from industries is a worldwide environmental problem. Textile industries are the major sources of pollution and contributors of metal contaminants to the environment. Industrial effluents discharged from textiles contain a higher amount of metals. Heavy metals hold a specific position in textile industry, right from the synthesis of various synthetic dyes and dye intermediates to their utilization as mordants for improving the affinity of the dye towards the fibre. It is anticipated to use these metal/metallic compounds in the permissible limits endorsed by the Pollution Control Board so as to preserve the environment. The toxic behaviour of some of these heavy metals make their contamination in the effluent liquors harmful to the environment. These effluents are released on the land as well as discharged into the surface water which ultimately end up in ground water by leaching and lead to contamination of it due to accumulation of toxic metallic components. The presence of such toxic components in the industrial effluents induce a wide range of environmental problems, associated with more complex and critical health hazards. Industrial effluent is also a main source of direct and often continuous input of pollutants into aquatic ecosystems with long-term insinuations on ecosystem functioning. The toxicity level of these heavy metals can be assessed by various analytical approaches. Their removal from the effluent liquor is necessary and various primary, secondary and tertiary treatment methods are utilized for their elimination. Numerous physico-chemical as well as bioremediation techniques have been designed for the removal of the heavy metals from the effluent system as well as contaminated land and water bodies.

REFERENCES

Agrawal, B. J. (2014). Designing efficient zero liquid discharge system for a textile effluent treatment plant. In *Recent Trends in Water Science and Technology* (pp. 32–38). Anand, India: Charotar Publishing House Pvt. Ltd.

Agrawal, B. J. (2017). *Textile Biological Decolourization and Effluent Remediation Techniques.* In A. K. Rathoure (Ed.), *Bioremediation: Current Research and Applications* (pp. 335–358). IK International Publishing House.

Agrawal, B.J., & Patel, B.H. (2000). The reincarnation of Natural Colourants – A Review. *Chemical Weekly*, 139-146.

Alemayehu, T., & Teklemariam, Z. (2014). Application of natural dyes on textile: a review. *International Journal of Research – Granthaalayah*, *2*(2), 61-68.

Ali, H., Khan, E., & Sajad, M. A. (2013). Phytoremediation of Heavy Metals – Concepts and Applications. *Chemosphere*, *91*(7), 869–881. doi:10.1016/j.chemosphere.2013.01.075 PMID:23466085

Ali, N., Hameed, A., & Ahmed, S. (2008). Physicochemical characterization and bioremediation perspective of textile effluent, dyes and metals by indigenous bacteria. *Journal of Hazardous Materials*, *164*(1), 322–328. doi:10.1016/j.jhazmat.2008.08.006 PMID:18818017

Allen, R. L. M. (1971). *Colour Chemistry*. London: Thomas Nelson & Sons. doi:10.1007/978-1-4615-6663-2

Apostoli, P., & Catalani, S. (2011). Metal ions affecting reproduction and development. *Metal Ions in Life Sciences*, *8*, 263–303. PMID:21473384

Ayangbenro, A. S., & Babalola, O. O. (2017). A New Strategy for Heavy Metal Polluted Environments: A Review of Microbial Biosorbents. *International Journal of Environmental Research and Public Health*, *14*(1), 94–110. doi:10.3390/ijerph14010094 PMID:28106848

Aziz, H. A., Adlan, M. N., & Ariffin, K. S. (2008). Heavy metals (Cd, Pb, Zn, Ni, Cu and Cr (III)) removal from water in Malaysia: Post treatment by high quality limestone. *Bioresource Technology*, *99*(6), 1578–1583. doi:10.1016/j.biortech.2007.04.007 PMID:17540556

Barakat, M. A. (2008). Adsorption of heavy metals from aqueous solutions on synthetic zeolite. *Research Journal of Environmental Sciences*, *2*(1), 13–22. doi:10.3923/rjes.2008.13.22

Basaldella, E., Vázquez, P. G., Iucolano, F., & Caputo, D. (2007). Caputo Chromium removal from water using LTA zeolites: Effect of pH. *Journal of Colloid and Interface Science*, *313*(2), 574–578. doi:10.1016/j.jcis.2007.04.066 PMID:17553515

Basaran, B., Iscan, M., Bitlisli, B. O., & Aslan, A. (2006). Heavy metal contents of various finished leathers. *Journal of the Society of Leather Technologists and Chemists*, *90*(6), 229–234.

Béchamp, A. (1854). De l'action des protosels de fer sur la nitronaphtaline et la nitrobenzine. nouvelle méthode de formation des bases organiques artificielles de Zinin. *Annales de chimie et de physique*, *42*, 186–196.

Bhatnagar, A., Hogland, W., Marques, M., & Sillanpaa, M. (2013). An overview of the modification methods of activated carbon for its water treatment applications. *Chemical Engineering Journal*, *219*, 499–511. doi:10.1016/j.cej.2012.12.038

Buat-Menerd, P., & Chesselt, R. (1979). Variable influence of the atmospheric flux on the trace metal chemistry of oceanic suspended matter. *Earth and Planetary Science Letters*, *42*, 398–411.

Cain, J. C., & Thorpe, J. F. (1946). *The Synthetic Dyestuffs and Intermediate Products from which they are derived*. London: Charles Griffin & Co.

Chaidogiannos, G., Petraki, F., Glezos, N., Kennou, S., & Nešpůrek, S. (2009). Low voltage operating OFETs based on solution-processed metal phthalocyanines. *Applied Physics. A, Materials Science & Processing*, *96*(3), 763–767. doi:10.1007/s00339-009-5268-1

Chavan, R. B. (2001). Indian Textile Industry – Environmental Issues. *Indian Journal of Fibre and Textile Research*, *26*, 11–21.

Chen, G. H. (2004). Electrochemicals technologies in wastewater treatment. *Separation and Purification Technology*, *38*(1), 11–41. doi:10.1016/j.seppur.2003.10.006

Chen, K. C., Wu, J. Y., Liou, D. J., & Hwang, S. C. J. (2003). Decolorization of the textile dyes by newly isolated bacterial strains. *Journal of Biotechnology*, *101*(1), 57–68. doi:10.1016/S0168-1656(02)00303-6 PMID:12523970

Chen, Z., Megharaj, M., & Naidu, R. (2007). Speciation of chromium in waste water using ion chromatography inductively coupled plasma mass spectrometry. *Talanta*, *72*(2), 394–400. doi:10.1016/j.talanta.2006.10.041 PMID:19071630

Correia, V. M., Stephenson, T., & Judd, S. J. (1994). Characterization of textile wastewaters: A review. *Environmental Technology*, *15*(10), 917–919. doi:10.1080/09593339409385500

de Vries, J. G., & Elsevier, C. J. (2007). *The Handbook of Homogeneous Hydrogenation* (Vols. 1–3). Weinheim: Wiley-VCH.

Doty, S. L. (2008). Enhancing phytoremediation through the use of transgenics and endophytes. *The New Phytologist*, *179*(2), 318–333. doi:10.1111/j.1469-8137.2008.02446.x PMID:19086174

Eapen, S., Singh, S., & D'Souza, S. (2007). Advances in development of transgenic plants for remediation of xenobiotic pollutants. *Biotechnology Advances*, *25*(5), 42–451. doi:10.1016/j.biotechadv.2007.05.001 PMID:17553651

Feuer, H., Bartlett, R. S., Vincent, B. F. Jr, & Anderson, R. S. (1965). Diborane Reduction of Nitro Salts. A New Synthesis of N-Monosubstituted Hydroxylamines. *The Journal of Organic Chemistry*, *30*(9), 2880–2882. doi:10.1021/jo01020a002

Fierz-David, H. E., & Blangey, L. (1979). *Fundamental Processes in Dye Chemistry* (5th ed.). New York: Interscience Publications.

Freeman, H. S., Hinks, D., & Esancy, J. (1996). Genotoxicity of Azo Dyes: Bases and Implications. In A. T. Peters & H. S. Freeman (Eds.), *Physico Chemical Principles of Colour Chemistry* (pp. 254–292). Chapman & Hall. doi:10.1007/978-94-009-0091-2_7

Fulekar, M., Singh, A., & Bhaduri, A. M. (2009). Genetic engineering strategies for enhancing phytoremediation of heavy metals. *African Journal of Biotechnology*, *8*, 529–535.

Gadd, G. M. (2000). Bioremedial potential of microbial mechanisms of metal mobilization and immobilization. *Current Opinion in Biotechnology*, *11*(3), 271–279. doi:10.1016/S0958-1669(00)00095-1 PMID:10851150

Ghoreishi, S. M., & Haghighi, R. (2003). Chemical catalytic reaction and biological oxidation for treatment of non-biodegradable textile effluent. *Chemical Engineering Journal*, *95*(1-3), 163–169. doi:10.1016/S1385-8947(03)00100-1

Gordon, P. E., & Gregory, P. (1983). *Organic Chemistry in Colour*. Berlin: Springer-Verlag.

Gulrajani, M. L. (1999). Present status of natural dyes. In Book of papers of the convention on natural dyes. Department of Textile Technology, IIT Delhi.

Gulrajani, M. L. (2001). Present status of natural dyes. *Indian Journal of Fibre and Textile Research*, *26*, 191–201.

Hai, F. I., Yamamoto, K., & Fukushi, K. (2006). Development of a submerged membrane fungi reactor for textile wastewater. *Desalination*, *192*(1-3), 315–322. doi:10.1016/j.desal.2005.06.050

Hakanson, L. (1980). An ecological risk index aquatic pollution control: A sedimentological approach. *Water Research*, *14*(8), 975–1001. doi:10.1016/0043-1354(80)90143-8

Halim, M., Conte, P., & Piccolo, A. (2002). Potential availability of heavy metals to phytoextraction from contaminated soils induced by exogenous humic substances. *Chemosphere*, *52*, 26–75. PMID:12729711

Halimoon, N., & Yin, R. G. S. (2010). Removal of Heavy Metals from Textile Wastewater using Zeolite. *Environment Asia*, *3*(SI), 124–130. Available online at www.tshe.org/EA

Herrmann, J. M. (1999). Heterogeneous photocatalysis: Fundamentals and applications to the removal of various types of aqueous pollutants. *Catalysis Today*, *53*(1), 115–129. doi:10.1016/S0920-5861(99)00107-8

Hudlický, M. (1996). *Reductions in Organic Chemistry*. Washington, DC: American Chemical Society.

Igwe, J. C., Ogunewe, D. N., & Abia, A. A. (2005). Competitive adsorption of Zn (II), Cd (II) and Pb (II) ions from aqueous and non-aqueous solution by maize cob and husk. *African Journal of Biotechnology*, *4*(10), 1113–1116.

Inoue, K.I. (2013). Heavy Metal Toxicity. *Journal of Clinical Toxicology*. doi: 10.4172/2161-0495.S3-007

Jan, A.T., Azam, M., Siddiqui, K., Ali, A., Choi, I., & Rizwanul-Haq, Q.M. (2015). Heavy Metals and Human Health: Mechanistic Insight into Toxicity and Counter Defense System of Antioxidants. *International Journal of Molecular Sciences*, *16*, 29592–29630. doi:10.3390/ijms161226183

Joshi, M., Bansal, R., & Purwar, R. (2004). Colour removal from Textile Dyes. *Indian Journal of Fibre and Textile Research*, *29*(6), 239–259.

Kapoor, A., & Viraraghvan, T. (1995). Fungal biosorption – An alternative treatment option for heavy metal bearing wastewater: A review. *Bioresource Technology*, *53*, 195–206.

Karami, A., & Shamsuddin, Z. H. (2010). Phytoremediation of heavy metals with several efficiency enhancer methods. *African Journal of Biotechnology*, *9*, 3689–3698.

Karnib, M., Kabbani, A., Holail, H., & Olama, Z. (2004). Heavy metals Removal using activated carbon, silica and silica activated carbon composite. *Energy Procedia*, *50*, 113–120. doi:10.1016/j.egypro.2014.06.014

Keharia, H., & Madamwar, D. (2004). Article. In A. Pandey (Ed.), Textile dye effluent in concise encyclopedia of Bioresearch Technology (pp. 167–175). New York: Food Product Press, The Haworth Preference Press Inc.

Kiyono, M., & Pan-Hou, H. (2006). Genetic engineering of bacteria for environmental remediation of mercury. *Journal of Health Science*, *52*(3), 199–204. doi:10.1248/jhs.52.199

Kurniawan, T. A., & Babel, S. (2003). A research study on Cr (VI) removal from contaminated wastewater using low-cost adsorbents and commercial activated carbon. *Second Int. Conf. on Energy Technology towards a Clean Environment (RCETE)*, 2, 1110-1117.

Kurniawan, T. A., Chan, G. Y. S., Lo, W. H., & Babel, S. (2006). Physico-chemical treatment techniques for wastewater laden with heavy metals. *Chemical Engineering Journal*, *118*(1-2), 83–98. doi:10.1016/j.cej.2006.01.015

Lim, P. E., Mak, K. Y., Mohamed, N., & Noor, A. M. (2003). Removal and speciation of heavy metals along the treatment path of wastewater in subsurface-flow constructed wetlands. *Water Science and Technology*, *48*, 307–313. PMID:14621178

Lin, C. C., & Lin, H. L. (2005). Remediation of soil contaminated with the heavy metal (Cd^{2+}). *Journal of Hazardous Materials*, *122*(1-2), 7–15. doi:10.1016/j.jhazmat.2005.02.017 PMID:15943924

Liu, Q., Zheng, T., Wang, P., Jiang, J., & Li, N. (2010). Adsorption isotherm, kinetic and mechanism studies of some substituted phenols on activated carbon fibers. *Chemical Engineering Journal*, *157*(2-3), 348–356. doi:10.1016/j.cej.2009.11.013

Löbbert, G. (2002). Phthalocyanines. In Ullmann's Encyclopedia of Industrial Chemistry. Wiley-VCH Verlag GmbH & Co. KGaA.

Long, E. R., Field, L. J., & McDonald, D. D. (1998). Predicting toxicity in marine sediments with numerical sediment quality guidelines. *Environmental Toxicology and Chemistry*, *17*(4), 714–727. doi:10.1002/etc.5620170428

Long, E. R., & MacDonald, D. D. (1998). Recommended uses of empirically derived, sediment quality guidelines for marine and estuarine ecosystems. *Human and Ecological Risk Assessment*, *4*(5), 1019–1039. doi:10.1080/10807039891284956

Long, E. R., MacDonald, D. D., Severn, C. G., & Hong, C. B. (2000). Classifying probabilities of acute toxicity in marine sediments with empirically derived sediment quality guidelines. *Environmental Toxicology and Chemistry*, *19*(10), 2598–2601. doi:10.1002/etc.5620191028

Long, X. X., Yang, X. E., & Ni, W. Z. (2002). Current status and prospective on phytoremediation of heavy metal polluted soils. *Journal of Applied Ecology*, *13*, 757–762. PMID:12216411

Malik, A. (2004). Metal bioremediation through growing cells. *Environment International*, *30*(2), 261–278. doi:10.1016/j.envint.2003.08.001 PMID:14749114

Manzoor, S., Shah, M.H., Shaheen, N., Khalique, A., & Jaffar, M. (2006). Multivariate Analysis of Trace Metals in Textile Effluents in Relation to Soil and Groundwater. *Journal of Hazardous Materials*, *137*, 31-37. 10.1016/j.jhazmat.2006.01.077

Michael, B. S., & Jerry, M. (2007). Advanced Organic Chemistry: Reactions, Mechanisms, and Structure (6th ed.). New York: John Wiley & Sons.

Mirbagherp, S. A., & Hosseini, S. (2004). Pilot plant investigation on petrochemical wastewater treatment for the removal of copper and chromium with the objective of reuse. *Desalination*, *171*(1), 85–93. doi:10.1016/j.desal.2004.03.022

Mohammadi, T., Mohebb, A., Sadrzadeh, M., & Razmi, A. (2005). Modeling of metal ion removal from wastewater by electrodialysis. *Separation and Purification Technology*, *41*(1), 73–82. doi:10.1016/j.seppur.2004.04.007

Mohanty, B. C., Chandranouli, K. V., & Nayak, N. D. (1984). Natural dyeing processes of India. Calico Museum of Textiles.

Morais, S., Costa, F. G., & Pereira, M. L. (2012). Heavy Metals and Human Health. In *Environmental Health - Emerging Issues and Practice* (pp. 227-246). Academic Press. Available from: http://www.intechopen.com/books/environmental-health-emerging-issuesand-practice/heavy-metals-and-human-health

Muller, G. (1969). Index of Geo-accumulation in sediments of the Rhine River. *Geological Journal*, *2*, 109–118.

Nagajyoti, P. C., Lee, K. D., & Sreekanth, T. V. M. (2010). Heavy metals, occurrence and toxicity for plants: A review. *Environmental Chemistry Letters*, *8*(3), 199–216. doi:10.1007/s10311-010-0297-8

Nemerow, N. L. (1978). *Industrial Water Pollution: Origins, Characteristics and Treatment*. Reading, MA: Addison-Wesley.

Othmer, K. (2004). Amine by Reduction. *Encylopedia of Chemical Technology*, *2*, 476–490.

Patel, J., Zhang, Q., & Michael, R. (2010). Genetic engineering of *Caulobacter crescentus* for removal of cadmium from water. *Applied Biochemistry and Biotechnology*, *160*(1), 232–243. doi:10.1007/s12010-009-8540-0 PMID:19214794

Patnaik, P. (2004). *Dean's Analytical Chemistry Handbook*. New York: McGraw-Hill.

Pobozy, E., Knell, M., Kilian, K., Kataky, R., & Trojanowicz, M. (2003). Capillary electrophoresis speciation of chromium in leather tanning liquor. *Electrophoresis*, *24*(1213), 2259–2263. doi:10.1002/elps.200305461 PMID:12858399

R'ıos, C. A., Williams, C. D., & Roberts, C. L. (2008). Removal of heavy metals from acid mine drainage (AMD) using coal fly ash, natural clinker and synthetic zeolites. *Journal of Hazardous Materials*, *156*(1–3), 23–35. doi:10.1016/j.jhazmat.2007.11.123 PMID:18221835

Rether, A., & Schuster, M. (2003). Selective separation and recovery of heavy metal ions using water-soluble N-benzoylthiourea modified PAMAM polymers. *Reactive & Functional Polymers*, *57*(1), 13–21. doi:10.1016/j.reactfunctpolym.2003.06.002

Rizwan, Md., Singh, M., Mitra, C.K., & Morve, R.K. (2014). Ecofriendly Application of Nanomaterials: Nanobioremediation. *Journal of Nanoparticles*. 10.1155/2014/431787

Robinson, R. (1957). The Life and Work of Sir William Henry Perkin. In *Proceedings of the Perkin Centennial 1856-1956*. AATCC.

Rueping, M., & Nachtsheim, B. J. (2010). A review of new developments in the Friedel–Crafts alkylation – From green chemistry to asymmetric catalysis. *Beilstein Journal of Organic Chemistry*, *6*(6). doi:10.3762/bjoc.6.6 PMID:20485588

Samanta, A. K., & Agarwal, P. (2009). Application of natural dyes on textiles. *Indian Journal of Fibre and Textile Research*, *34*(4), 384–399.

Samanta, A. K., & Konar, A. (2011). *Dyeing of Textiles with Natural Dyes*. Available from https://www.researchgate.net/publication/221919685_Dyeing_of_Textiles_with_Natural_Dyes

Sarikaya, M., Tamerler, C., Jen, A. K., Schulten, K., & Baneyx, F. (2003). Molecular biomimetics: Nanotechnology through biology. *Nature Materials*, *2*(9), 577–585. doi:10.1038/nmat964 PMID:12951599

Saxena, S., & Raja, A. S. M. (2014). Natural Dyes: Sources, Chemistry, Application and Sustainability Issues. In Roadmap to Sustainable Textiles and Clothing, Textile Science and Clothing Technology. Springer Science + Business Media.

Shukla, S. R. (2006). Heavy metals. *Colourage*, *53*(8), 74–77.

Szynkowska, M. I., Czerski, K., Paryjczak, T., Rybicki, E., & Wlochowicz, A. (2006). Testing Textiles Using the LA-ICP-MS-TOF Method. *Fibres & Textiles in Eastern Europe*, *14*, 87–90.

Tomlinson, D. L., Wilson, J. G., Harris, C. R., & Jeffrey, D. W. (1980). Problem in the assessment of heavy metals levels in estuaries and the formation of a pollution index. *Helgoländer Meeresuntersuchungen*, *33*(1-4), 566–575. doi:10.1007/BF02414780

Venkatraman, K. (1952-1972). *The Chemistry of Synthetic Dyes* (Vols. 1-8). Academic Press Inc.

Vishwanathan, M. (2011). *Principles of Organic Chemistry*. Kollam: Jai Sai Publications.

Wang, L. K., Vaccari, D. A., Li, Y., & Shammas, N. K. (2004). Chemical precipitation. In Physicochemical Treatment Processes (vol. 3, pp. 141-198). Humana Press.

Wei, W., Liu, X., Sun, P., Wang, X., Zhu, H., Hong, M., ... Zhao, J. (2014). Simple whole-cell biodetection and bioremediation of heavy metals based on an engineered lead-specific operon. *Environmental Science & Technology*, *48*(6), 3363–3371. doi:10.1021/es4046567 PMID:24564581

World Health Organization. (1996). *Trace Elements in Human Nutrition and Health.* WHO.

Yadav, K. K., Singh, J. K., Gupta, N., & Kumar, V. (2017). A Review of Nanobioremediation Technologies for Environmental Cleanup: A Novel Biological Approach. *Journal of Materials and Environmental Sciences*, *8*(2), 740–757.

Zeiner, M., Reziæ, I., & Steffan, I. (2006). TLC determination of allergenic and carcinogenic dyes. *Toxicology Letters*, *164*, S184. doi:10.1016/j.toxlet.2006.07.040

Zeiner, M., Reziæ, I., & Steffan, I. (2007). Analytical Methods for the Determination of Heavy Metals in the Textile Industry. *Kem. Ind.*, *56*(11), 587–595.

Zhao, X. W., Zhou, M. H., Li, Q. B., Lu, Y. H., He, N., Sun, D. H., & Deng, X. (2005). Simultaneous mercury bioaccumulation and cell propagation by genetically engineered Escherichia coli. *Process Biochemistry*, *40*(5), 1611–1616. doi:10.1016/j.procbio.2004.06.014

Zollinger, H. (1991). *Color Chemistry* (2nd ed.). Weinheim: VCH.

Chapter 3
Low Cost Absorbents, Techniques, and Heavy Metal Removal Efficiency

Harendra Kumar Sharma
Jiwaji University, India

Irfan Rashid Sofi
Jiwaji University, India

Khursheed Ahmad Wani
ITM University, India

ABSTRACT

Heavy metal contamination in water is a serious concern to the environment and human health. High concentrations of heavy metals in the environment can be toxic to a variety of living species. Natural bio-absorbents are abundant and inexpensive and considered a waste if not managed properly. The role of bio-absorbents has been widely studied and has been utilized for the removal of heavy metals. The objective of the chapter is to search the database for different absorbents and their efficiency for the removal of heavy metals. Key words related to the study have been used to select different papers published by the researchers all over the world. A rigorous three-tier process has been utilized by the authors to select the papers from the database for the current study. This chapter has identified a few research gaps in the field of heavy metal removal by using different low cast absorbents that need to be taken into account in future research.

DOI: 10.4018/978-1-5225-4162-2.ch003

Copyright © 2018, IGI Global. Copying or distributing in print or electronic forms without written permission of IGI Global is prohibited.

INTRODUCTION

Due to rapid industrialization and population growth the load of different pollutants is increasing and causing damage to almost every sphere of life (Banerjee, Chen, 2007; Zhou et al., 2009). The toxic pollutants, especially heavy metals are cause of serious concern for the environment and human health (Kanan & Sundarbam, 2001). The different heavy metals like chromium, nickel, zinc Cadmium and zinc are used for various industrial purposes and their importance can't be ignored. However, their excess concentration is equally responsible for great damage to different ecosystem and human beings. The use of heavy metals in wood preservation, tanning, of leather, fabrication of steel, and finishing of metals have changed the scenario of industrialization over a period of time (Jaing, 2013; Zongo et al., 2009). The minimum permissible limit Cr (VI) of drinking water is 50ppm and its adverse effects on the human health is related to deadly diseases like cancer and hence their removal becomes significant (Jaing et al., 2013). Adsorption is highly effective when compared to other conventional methods for the removal of heavy metals from waste water. Adsorption method is used for removing a large number of pollutants from waste water along with low concentration pollutants. The different adsorbents that were used for removal of heavy metals include few biomaterials (Li et al., 2003; Ekmekyaper et al,. 2006) zeolites (Biskup & Subotic, 2010), manganese oxides (Sublet et al., 2003), kaolinite (Arias et al., 2002), resins (Diniz et al., 2002), peanuts hulls (Brown et al., 2000), algal biomass (Gupta et al., 2010) and recycled alum sludge (Chu, 1999) etc due to low cost efficiency of these materials considered a hindrance for their use in commercial sector. Hence, a search to develop new adsorbents, that are cost effective and can be tested for commercial scale in recent years. Although coagulation, filtration, ion exchange reverse osmosis, adsorption are used to reduce the concentration of heavy metals from waste water but adsorption on activated carbon is one of the most practical and economical ways for water treatment (Girgis, 1997).

Biosorption is an ideal alternative for decontamination of metal containing effluents (Modak &Natarajan, 1995) and biosorption is a fast method of passive metal uptake sequestration by non-growing biomass (Beveridge & Doyle, 1989). Further, the efficiency of removal is convincing and binding capacity of certain biomass is comparable with the commercial synthetic cation exchange resins (Wase & Foster, 1997). The biosorption efficiency depends upon many factors, including the capacity, affinity and specificity of the biosorbents and their physical and chemical conditions in effluents (Sarbjeet & Goyal 2007).

The different physical forces with exception of chemisorptions are involved in the process of adsorption that may be due to the presence of various minerals and carbon moieties. The process of adsorption is controlled by van der waals forces, hydrophobicity, hydrogen bonds, polarity and steric interaction, dipole induced interaction, pep interaction etc. The degree of rigid packing that can occur in pores of adsorbent will determine the adsorption capacity (Rubio & Kitchener, 1976). The surface of adsorbent molecule and its pore size must be comparable for effective adsorption. It is observed that adsorption capacity increases with increase in the concentration and low molecular weight pollutants are adsorbed initially are replaced by high molecular species in due course of time during the process of adsorption. It is due to the fact that activated carbon easily adsorbs propane as compare to methane (Tinge et al., 1987).

The demand on water usage is high as the domestic and industrial activities are increasing rapidly especially in the manufacturing sectors. However, the waste litters are not managed properly and at the same time the amount of waste being dumped is increasing greatly. This problem has affected the water quality causing the water quality to deteriorate (UNESCAP, 2000). Discharge of effluent from the industrial and agriculture activities in high volume loaded with heavy metals, including organic compounds and some effluents were drained without treatment that affects the water quality (UNESCAP, 1999). Surface water and groundwater are also affected by the nonpoint source pollutants contributed by agricultural activities, where the usage of chemical fertilizers leaches into groundwater and runoff. The high demand on the agricultural products led to the high utilization of fertilizers in order to yield a high number of productions (Dzikiewicz, 2000). The removal of heavy metals in a cost effective and scientific way is the need of hour. Using the low cost techniques and ecofriendly strategies will not only increase water budget but will save the environment from contamination. On the other hand, this will prove very helpful in designing other techniques at very cheap cost. The aim of the present study is to review the scientific literature available in the removal of heavy metals by the process of bioadsorption.

CONVENTIONAL METHODS FOR HEAVY METAL REMOVAL FROM INDUSTRIAL EFFLUENTS

To mitigate the heavy metal pollution, many processes like precipitation, coagulation, ion exchange, cementation, electro-dialysis, electro-winning, electro-coagulation and reverse osmosis have been developed.

Precipitation

One of the oldest methods for the removing of toxic heavy metals up to parts per million (ppm) levels from water is precipitation. Some metals are insoluble in water and which get precipitated by the addition of anions. However, the process is cost effective; its efficiency is affected by low pH and the presence of other salts (ions). Further, the addition of other chemicals results in the generation of high water content sludge and the disposal of which is costly (Gray, 1999). Precipitation with other chemicals like lime, bisulphide or ion exchange lacks the specificity and is not effective in removal of metal ions at low concentration. (Fu et al., 2012).

Ion Exchange

The process of ion exchange has been used successfully in the industry for the removal of heavy metals from effluents. Over the years, this process is relatively expensive as compared to the other methods but is more efficient and clean up heavy metals to a greater extent. The exchange of cations and anions are involved during the process. The contamination of water by organics and other solids in waste water for highly concentration solution are the major disadvantage for this method. The ion exchange is non-selective in nature and is dependent upon pH of the solution. (Wang et al., 2014; Gupta et al., 2005).

Electro Winning

Leaching and acid mine drainage in mining and metallurgic industry is achieved by electro winning. The removal and recovery of metals in the metal transformation, electronics and electrical industries is also being done by electro winning. The electro-degradation technique is used to recover metals like Ag, Au, Cd, Co, Cr, Ni, Pb, Sn, and Zn from the effluents (Barakat, 2011).

Electro-Coagulation

The process of electro-coagulation uses electric current to remove metals from solution. This process is also effective in removing the solids, dissolved metals, tannins, and dyes. The different contaminants that are part of waste water are maintained in solution by electrical charges. The precipitation occurs when these ions and other charged particles are neutralized with ions of opposite electrical charges provided by electro coagulation system (Mendez et al., 2015; Zu, Ma, and Yang, 2007).

Cementation

The metals with higher oxidation state are recovered by a process known as cementation e.g. oxidation of metallic iron, Fe(III) to ferrous iron(II) to replace a metal having a lower oxidation potential. The noble metals like Ag, Au, Pb as well as As, Cd, Ga, Pb, Sb and Sn can be recovered by the process of cementation (Abdel-Aziz, 2011).

Reverse Osmosis

The use of semi- permeable membrane for the recovery of metal ions from dilute wastewater is known as Reverse Osmosis. However, the use of modified electro-dialysis selective membranes that are fitted between the electrodes in electrolytic cells and under continuous electric current helps in the recovery of metals (Benoit, Barriuso, & Calvet 1998; Zhaung et al. 2009).

BIOMATERIALS IN USE FOR REMOVAL OF HEAVY METALS

Various workers have worked on different biomaterials to tap heavy metals from wastewater. The efficiency of heavy metal removal depends on the selection of bio absorbent and the type of heavy metal under investigation. Almost very heavy metal has been removed with greater efficiency and most of these fitted with pseudo first, and pseudo second models and Langmuir and Fruendlich models. A detailed account of bio materials used to remove heavy metals along with efficiency is detailed below:

Da´vila-Jime´nez et al., (2003) conducted batch adsorption studies with cactus to sorb Cu^{2-}, Cd^{2-}, Cr^{3-}, CrO_4^{2-}, Fe^{3-}, Ni^{2-}, Pb^{2-} and Zn^{2-} with different concentrations. Under similar experimental conditions the copper ions were retained more than other cations with a maximum efficiency of 84% for 1000 ml/l Cu solutions. The surface adsorption complexation mechanism was followed by the metal cations due to interaction of respective oxalate complexes. Al-Khaldi et al., (2015) reported the adsorption characteristics of cadmium(II) on modified structures of carbon nanotubes (CNTs), carbon nanofibres (CNFs), activated carbon (AC), and fly ash (FA). The investigation was carried with several experimental parameters like contact time of 24 h, pH of 7, adsorbent dosage of 50mg and rotational speed of 150rpm were the optimum conditions for the adsorption of cadmium(II). First-order, second-order and pseudo second order models fitted to the experimental data, and pseudo-second-order kinetics described the adsorption of cadmium(II) on these adsorbents. The Langmuir models indicated that the maximum adsorption capacities of the modified adsorbents were observed to be 2.02 mg/g, 1.98 mg/g, 1.22 mg/g and 1.58 mg/g, for CNTs, AC, CNFs and FA, respectively. Hence, it is concluded that the carbon nanotubes are promising adsorbents with respect to water purification applications. The kinetics and equilibrium of phenol bio sorption on Ca-alginate beads solution containing mineral and vitamin materials with complex composition by batch technique were studied at pH of 5.5. The study concluded that the blank Ca- alginate, free and immobilized fungal biomass is described by Langmuir, anti- Langmuir and Fruendlich isotherm models using nonlinear least-squares estimation. Moreover, Bohli et al., (2016) investigated the performance of a micro porous activated carbon prepared chemically from olive stones for removing Cu(II), Cd(II) and Pb(II) from single and binary aqueous solutions via the batch technique by using N_2 adsorption–desorption isotherms, SEM, XRD, FTIR, and Boehm titration. The effect of initial pH and contact time were studied. Adsorption kinetic rates were found to be fast and kinetic experimental data fitted very well with the pseudo-second-order equation. The adsorption isotherms fit the Redlich–Peterson model very well and maximum adsorption amounts of single metal ions solutions follow the trend Pb(II) > Cd(II) > Cu(II). The adsorption behavior of binary solution systems shows a relatively high affinity to Cu(II) at the activated carbon surface from the mixture of Cd(II) or Pb(II). An antagonistic competitive adsorption phenomenon was observed. Desorption experiments indicated that about 59.5% of Cu(II) and 23% of Cd(II) were desorbed using a diluted sulfuric acid solution.

The combination of fruit wastes, coconut shell, scrap tires, bark and other tannin-rich materials, sawdust and other wood type materials, rice husk, petroleum wastes, fertilizer wastes, fly ash, sugar industry wastes, blast furnace slag, chitosan and seafood processing wastes, seaweed and algae, peat moss, clays, red mud, zeolites, sediment and soil, ore minerals have been found to remove various organic pollutants ranging from 80 to 99.9% (Ali, Asim, & Khan, 2012). The hybrid Eucalyptus loaded with activated carbon were used in removal of cadmium from waste water by (Venkatshen, Senthilnathan, & Rajam 2014). Similarly Wu et al., (2013) removed Cr(VI) and Ni(II) by coconut activated carbon with a efficiency of 94.5% for Cr(VI) and 58.92% for Ni(II). Moreno-Piraján, Garcia-Cuello, & Giraldo, (2011) synthesized carbon from coconut shells were used for the removal of metals ions from aqueous solution. The copper ion exhibit the greatest adsorption on activated carbon obtained from coconut shells because of their size and pH conditions. Kyzas et al., (2015), used hemp based material either in raw form or by chemical modified bioadsorbents for the removal of Ni(II) from aqueous solution. The adsorption experiment fitted to linear pseudo first, second order and Elovich model. The characterization study by FTIR XRD, SEM, EDAX, crystality and cellulose content revealed the surface morphology of prepared hemp material along with some possible adsorption interaction between them and ions. The material was regenerated by desorption studies that revealed the reuse potential of the low cost hemp based material

Ahluwalia, & Goyal, (2007) studied the existing literature and argued that biomass of *Aspergillus niger, Penicillium chrysogenum, Rhizopus nigricans, Ascophyllum nodosum, Sargassum natans, Chlorella fusca, Oscillatoria anguistissima, Bacillus firmus and Streptomyces* sp. have highest metal adsorption capacities ranging from 5 to 641 mg/g mainly for Pb, Zn, Cd, Cr, Cu and Ni. However, the modification of chitosan with 4-((1,3-dioxoisoindolin-2-ylimino) methyl) benzaldehyde was analyzed by diverse techniques for adsorption capacity of CS-B for metal ions Co^{2+}, Hg^{2+}, Cu^{2+}, Zn^{2+}, and $Pb^{2+}$from aqueous systems at various pH values. The highest metal adsorption capacity was shown for Co^{2+} and Hg^{2+}, while Cu^{2+}, Zn^{2+}, Pb^{2+} showed moderate adsorption capacity (Kandile, Mohammad, & Mohamed, 2015).

Cutillas-Barreiro et al., (2014) used pine bark to remove five metals by stirring flow and batch experiments. Adsorption was between 98% and 99% for $Pb^{2+,}$ 83 - 84% for $Cu^{2+,}$ 78- 84% for $Cd^{2+,}$ 77 -83% for $Zn^{2+,}$ and 70-75% for Ni^{2+} and it was faster for low concentrations, with Pb suffering the highest retention, followed by Cu, Cd, Ni and Zn. Monier et al., (2010) cross-linked magnetic chitosan–diacetylmonoxime Schiff's base resin (CSMO) was also prepared for different adsorption of metal ions viz Cu^{2+}, Co^{2+}, and Ni^{2+}. The maximum adsorption capacities were 95 ± 4, 60 ± 1.5, and 47 ± 1.5 mg/g for Cu^{2+}, Co^{2+} and Ni^{2+} ions, respectively and cross-linked magnetic CSMO displayed higher adsorption capacity for Cu^{2+} in all pH ranges studied. However, the adsorption capacity of the metal ions decreased with increasing temperature. By using EDTA the metal ion-loaded cross-linked magnetic CSMO were regenerated with an efficiency of greater than 84%.

Ali, Saeed, & Mabood, (2016) used banana peels grafted with acrylonitrile for the removal of hexavalent chromium and Fruendlich and Langmuir isotherm model and pseudo- second order kinetic model was fully satisfied and on the other hand (Dehghani, Sanaei & Ali, 2016) used treated waste news paper (TWNP) to remove chromium(VI) from aqueous solution using batch experiments and the percentage removal of Cr(VI) was 59.88 mg/g (64% at pH 3). Javaid et al., (2011) explored the adsorption potential of *Pleurotusostreatus* (a macro-fungus) to remove copper, nickel, zinc and chromium from water all together. The maximum biosorption capacity of fungus was 8.06, 20.40, 3.22 and 10.75 mg/g/1 for Cu(II), Ni(II), Zn(II) and Cr(VI) in that order. For real effluents of electroplating, biosorption capacities were 2.73, 8.45, 0.88 and 4.45 mg/g for Cu(II), Ni(II), Zn(II) and Cr(VI) ions, respectively. Further, Milojkovi 'ca et al., (2016) studied the effects of adsorbent dose, duration of sorption and solution concentration on the sorption of heavy metals on *Myriophyllum spicatum* by Scanning electron microscope SEM) and thermo gravimetric and differential thermal analysis (TG-DTA). The experimental results revealed that compost has a high biosorption capacity for the removal of selected metals from wastewater. On the other hand, lingo cellulose precursors obtained from crops and fruits that include the biomass of plum kernel and Jacaranda for waste water treatment and this study indicated that both carbonization and activation processes improved the uptake of lead, acid blue 25 and methylene blue upto 10 times with respect to adsorption results obtained for raw precursors. In comparison the adsorption capacities of plum kernel adsorbents are higher (up to 100%) than those obtained for adsorbents prepared from jacaranda (Cordero et al., 2013).

The investigation of duckweed Lemna gibba and coontail *Ceratophyllum demersum* for the removal of Cd and Ni showed that increasing heavy metal concentrations in nutrient solution caused a decrease to 91% for Cd and 50% for Ni, respectively

(Parnian et al., 2016). The preliminary experiments conducted by AL. Qahtani, (2016) using Kiwi cortex, material with a 1 mm particle size showed a higher retention capability (up to 10–16% of $Cd^{2+,}$ Zn^{2+} and Cr^{3+} than material with 2 mm particles. A strong dependence of the adsorption capacity on the initial metal concentration was observed and the capacity increased as the initial concentrations decreased. The maximum removal of Cd^{2+}, Zn^{2+}, and Cr^{2+} was observed at an adsorbent dosage of 2.0 g and an initial concentration of 10 mg/l. The removal of Cu (II) ions by black tea powder using bark solution of 0.1 m NAOH was investigated by Weng et al., (2014). The study confirmed that OH-BTLP exhibit highest removal efficiency with maximum adsorption capacity of 43.18 mg/g at pH 4.4. The synergistic effects of *T. reesei cellulases* on tea waste revealed that cellulosic material provides more accessibility area for absorption of heavy metal. This study also provides an alternative approach to remove toxic Cr(VI) from aqueous solutions and extend the utilization of agricultural tea waste (Ng et al., 2012). The toxic metals in the water can't be removed completely by conventional treatment; hence, attractive purification methods that are inexpensive and easily available have been explored by researchers (Lim, & Aris, 2014).

Onundi et al., (2010) used granular activated carbon obtained from kernel shells used as adsorbent to remove different heavy metals from waste water. The equilibrium was reached within 30 minutes for lead, 75 minutes for copper and nickel with lead recording 100%, copper 97% and nickel 55% removal. In addition to that, Doum palm shell and activated carbon fibers were used to remove the Cd(II) and Pb(II) from waste water by Gayal, Otenel, & Abdullah, (2015). However, batch adsorption indicated that the process of adsorption is dependent on adsorption dosage, agitation time, pH and adsorbate concentration with high adsorption efficiency by keeping all other parameters constant. Yin, Aroua, & Duad et al., (2008) impregnated palm shell activated carbon with poly ethyleneimine (PEI) and the effect of impregnation on batch adsorption of Ni^{2+}, Cd^{2+}or Pb^{2+} this process enhanced the single metal adsorption capacities of Ni^{2+} or Cd^{2+}except for Pb^{2+}, where its adsorption capacities were reduced by 16.67% and 19.55% for initial solution pH of 3 and 5 respectively.

Abdulrazzak, Hussaini, & Sani, (2016) removed heavy metals cadmium copper nickel and lead from waste \water effluent with the help of carbon produced from African palm fruit. The observed percentage removal was at the temperature of 80°C (93.23±0.35, 96.71±0.097, 92.01±0.018 and 95.42± 0.067% for Cd, Cu, Ni, and Pb respectively and at an optimum contact time of 60 minutes. The maximum adsorption capacities were achieved at pH 6 with contact time of 180 min at dosage of 0.3 g/l when activated mixed waste was used as adsorbent (AL-othaman et al., 2016). On the other hand, pH 8 showed maximum sorption for same metal Cu (II) and Pb (II) when cassava peels were used. However sorption efficiency increased from 12 to 73% for Pb (II) and 26 to 79% for Cu (II) at the dosage of 2 to 12 g (Owamah et al., 2014).

Sencan, Karaboyaci, & Kilic, (2015) determined the efficiency of hazelnut shell and activated carbon for the removal of Pb^{2+}, the pseudo first and second order kinetics studies were performed when the hazelnut was pyrolyzed at 250 and 700 °C. The surface of hazel wood shell at 700°C after being activated with Zncl was found to be 736.49 m^2/g and the physical adsorption is dominant for the activated carbon pyrolysis at lower degrees and for raw hazelnut shell. Young et al., (2013) studied the effects of solution pH, contact time, initial concentration and temperature on the immobilization of persimmon tannin. 1.0 gm activated carbon modified by 17.32 mg persimmon tannin showed best results as it was found to be more effective than the plain activated carbon in the removal of Pb(II).By using nitric acid to activate olive stones carbon for its ability to remove metal ions Bohli, & Ouederni, (2016) found an important maximum adsorption from binary system as compared to single system.

Batch experiments were conducted by Boudrahem, Aissani, & Soualah, (2013) to study the effects of the main parameters such as contact time, initial concentration of Pb(II), adsorbent dose, solution pH, agitation speed, ionic strength and temperature on the sorption of Pb(II) by activated carbon. The maximum removal of lead(II) was found to be 94% at pH5.8, initial Pb(II) concentration of 10mg dm^{-3}, adsorbent dose of 1 g/l, agitation speed of 200 rpm, ionic strength of 0.005 M, and temperature of 25 °C. Ali, & Alrafi, (2016) showed the optimum conditions for 99% removal of chromium (VI) ions from aqueous solutions with activated leaves of *F. nitida* at 25 min, and 50.0 mg/l of initial concentration of chromium (VI). Hence, it is suggested that the chemically activated leaves of *F. nitida* can be used as low cost, eco-friendly and effective sorbent for the removal of chromium (VI) from aqueous solutions. Okoye, Ejikeme, & Onukwuli, (2010) found that the amount of lead (II) ion adsorbed on *Telfairia occidentalis* at equilibrium from a 200 mg/l solute concentration was 14.286 mg/g by confirming very well to the Pseudo second order equation where equilibrium adsorption capacities increased with increasing initial lead (II) concentration. The study carried out by Chowdhury et al., (2012) revealed that kenaf fibers that were semi carbonized and impregnated with potassium hydroxide and finally activated with CO_2 gas can be used for heavy metal removal like copper (II). This biomaterial poses functional groups and modifying these biomaterials by chemical will enhance the removal capacity for heavy metals (Dhir, 2014).

Badmus, Audu, & Anyata, (2007) removed copper from industrial waste water by using activated periwinkle shell carbon (PSC). The result revealed that activated periwinkle shell carbon competes favorably with commercial activated carbon and maximum adsorption capacity for (PSC): (CAC) with 88.12% removal at an optimum pH of 8. The (PSC) and (CAC) had 84.19% removal and 85.15% removal rate, respectively. Liu et al., (2014) impregnated humic acid onto powdered activated carbon to improve its Cu(II) adsorption capability. The maximum adsorption capacity of humic acid impregnated activated carbon was up to 5.98 mg/g/1, which is five times the capacity of virgin activated carbon. Ucar et al., (2015) used rapeseed oil cake to prepare activated carbon by chemical activation with sodium carbonate Na_2Co_3 at 600 and 800 ^{0}C. The adsorption data for both ions were well correlated with Langmuir isotherm. 129.87 and 133.33 mg/g of lead (II) and nickel was removed by the monolayer of rapeseed oil cake. The chemical activation process was used to activate water melon and walnut shell for the removal of Pb(II) and Zn(II) ions from aqueous solution with the yield of 85 and 80% respectively (Moreno-Barbosa et al., 2013). Apricot stones were modified with 2-hydroxyl -2- mercaptopyrimidine (2-thiouracil) for the removal of different heavy metals with appropriate results (Vassileva et al., 2008).

Lemon peels were studied by Singh, & Shukla, (2016) for bioadsorption of Co(II) from aqueous cobalt nitrate solution at different concentrations. This study revealed that lemon peels have a good potential for treating industrial waste water at low cobalt concentration. The removal percentage of Ni(II) decreased from 85.0 to 70.0% for Citrus limettioides peel and remained constant (90%) for CLPC when the initial concentration was increased from 10 to 50 mg/l (Sudha et al., 2015). Chen et al., (2014) prepared high quality carbon from *Eupatorium adenophorum* and Bucket straw for the removal of Cr (VI) and has found that these adsorbents after simple treatment had a great efficiency for the removal of heavy metals like chromium (VI). On the other hand, pomegranate seeds were used for the removal of Chromium with high efficiency (Ghanien et al., 2016). Mishra, Balomajumder, & Agarwal et al., (2010) delineated the biosorption potential of eucalyptus bark saw dust, mango bark saw dust and pineapple fruit peel with respect to Zn(II) ion removal from liquid phase through batch experiments. On an average, about 84.4%, 51.4%, and 22.9% Zn(II) was removed by eucalyptus bark, saw dust and mango tree respectively.

Li, Law, & Chan, (2015) investigated the adsorption and removal behaviour of copper(II) and cadmium(II) ions using rice husk and rice straw in aqueous solutions with 90% removal of Cu(II) and Cd(II) by rice husk at pH 9 and 90% by rice straw at pH 6. 99.88% removal of Cr(VI) at pH 1.16 and at adsorbent dose of 6 g/l was achieved by using *Phaseolus Vulgaris* husk (Srivastav, Agarwal, & Mondal, 2016). The removal efficiency of Cr(II) and Cu(II) enhanced by increasing waste water initial concentration and then decreased due to saturation of rice husk particles (Sobhanardakani, Parvizi, Mosaed, & Olyaie, 2013). The removal of Cu(II) from aqueous solution by *Acacia nilotica* seed pods that were chemically activated revealed that the metal uptake decreased by the presence of other metals (Thenmozhi, &Santhi, 2015). El-Said, Badawy, & Attyat, (2011) used rice husk for the removal of Zn (II) and Se (IV) ions from waste water at different concentration. The results were quite appreciable and desorption process utilized recovered the heavy metals by using different aqueous media. Asif, & Chan, (2015) removed 90.7% of arsenic with risk husk while as Fan et al., (2016) removed Cr (VI) with the help of rice husk from waste water. Water melon shell, lemon and banana peels were used by Mohammed et al (2016) for the removal copper ion. The elimination of the watermelon shell was 90% compared with that of lemon peel (78%) while for banana it was 65% with highest elimination of 9.54% for water melon shell 8.24% for lemon shell and 7.65mg/g for banana peel. The column test confirmed that reloading of the column could enhance removing efficiencies by 15- 30%. Abedi, Salamani, & Mozaffari, (2016) studied experimental conditions for preparation of pomegranate peel carbon and Fe(III) modified carbon. The adsorption capacity for Cd(II) species on Fe(III) modified pomegranate peel carbon was 22.72 mg/g and the adsorption kinetic presented the Pseudo second order kinetic model. The result indicated that the sorption of iron ions onto the pomegranate peel carbon was an exothermic process that could be well fitted with the Langmiur adsorption model and pseudo first order kinetic model. The bark of *Acacia albida* was used to remove Cr(VI) from aqueous solution. pH was found to be most important parameter among the parameters studied for maximum removal of chromium (98.47)% indicating that bark of Acacia can be utilized as low cost alternative to commercial adsorbent for the treatment of effluents contaminated with Cr(VI) (Gebrehawaria et al., 2013).

CONCLUSION

The presence of different heavy metals in water is considered very dangerous not for only aquatic life but also for humans as well. In order to control toxicity of heavy metals and to recover heavy metals from water, efforts must be made to completely eliminate the risk of heavy metals. This can be achieved by creating public awareness about toxicity of heavy metals and government policies need to be enforced strictly to regulate the discharge of effluents contaminated with heavy metals. Only few studies regenerated the bioadsorbents and this need to be researched further. However, the identification of various techniques that are cheap and sustainable may help in decreasing the burden of metal pollution in water. Various absorbents have been already researched by various workers that can remove heavy metals from water with great efficiency and are best suited to different isotherm models.

REFERENCES

Abdel-Aziz, M. H. (2011). Production of copper powder from wastewater containing CuSO 4 and alcoholic additives in a modified stirred tank reactor by cementation. *Hydrometallurgy*, *109*(1), 161–167. doi:10.1016/j.hydromet.2011.06.007

Abdulrazak, S., Hussaini, K., & Sani, H. M. (2017). Evaluation of removal efficiency of heavy metals by low-cost activated carbon prepared from African palm fruit. *Applied Water Science*, *7*(6), 3151–3155. doi:10.1007/s13201-016-0460-x

Abedi, M., Salmani, M. H., & Mozaffari, S. A. (2016). Adsorption of Cd ions from aqueous solutions by iron modified pomegranate peel carbons: Kinetic and thermodynamic studies. *International Journal of Environmental Science and Technology*, *13*(8), 2045–2056. doi:10.1007/s13762-016-1002-7

Adekola, F. A., Hodonou, D. S. S., & Adegoke, H. I. (2016). Thermodynamic and kinetic studies of biosorption of iron and manganese from aqueous medium using rice husk ash. *Applied Water Science*, *6*(4), 319–330. doi:10.1007/s13201-014-0227-1

Ahluwalia, S. S., & Goyal, D. (2007). Microbial and plant derived biomass for removal of heavy metals from wastewater. *Bioresource Technology*, *98*(12), 2243–2257. doi:10.1016/j.biortech.2005.12.006 PMID:16427277

Al-Khaldi, F. A., Abusharkh, B., Khaled, M., Atieh, M. A., Nasser, M. S., Saleh, T. A., ... Gupta, V. K. (2015). Adsorptive removal of cadmium (II) ions from liquid phase using acid modified carbon-based adsorbents. *Journal of Molecular Liquids*, *204*, 255–263. doi:10.1016/j.molliq.2015.01.033

Al-Qahtani, K. M. (2016). Water purification using different waste fruit cortexes for the removal of heavy metals. *Journal of Taibah University for Science*, *10*(5), 700–708. doi:10.1016/j.jtusci.2015.09.001

Ali, A., & Saeed, K. (2015). Decontamination of Cr (VI) and Mn (II) from aqueous media by untreated and chemically treated banana peel: A comparative study. *Desalination and Water Treatment*, *53*(13), 3586–3591. doi:10.1080/19443994.2013.876669

Ali, A., Saeed, K., & Mabood, F. (2016). Removal of chromium (VI) from aqueous medium using chemically modified banana peels as efficient low-cost adsorbent. *Alexandria Engineering Journal*, *55*(3), 2933–2942. doi:10.1016/j.aej.2016.05.011

Ali, I., Asim, M., & Khan, T. A. (2012). Low cost adsorbents for the removal of organic pollutants from wastewater. *Journal of Environmental Management*, *113*, 170–183. doi:10.1016/j.jenvman.2012.08.028 PMID:23023039

Ali, I. H., & Alrafai, H. A. (2016). Kinetic, isotherm and thermodynamic studies on biosorption of chromium (VI) by using activated carbon from leaves of Ficus nitida. *Chemistry Central Journal*, *10*(1), 36. doi:10.1186/s13065-016-0180-1 PMID:27252777

ALOthman, Z. A., Habila, M. A., Al-Shalan, N. H., Alfadul, S. M., Ali, R., Rashed, I. G. A., & Alfarhan, B. (2016). Adsorptive removal of Cu (II) and Pb (II) onto mixed-waste activated carbon: Kinetic, thermodynamic, and competitive studies and application to real wastewater samples. *Arabian Journal of Geosciences*, *9*(4), 315. doi:10.1007/s12517-016-2350-9

Arias, M., Barral, M. T., & Mejuto, J. C. (2002). Enhancement of copper and cadmium adsorption on kaolin by the presence of humic acids. *Chemosphere*, *48*(10), 1081–1088. doi:10.1016/S0045-6535(02)00169-8 PMID:12227513

Arshadi, M., Amiri, M. J., & Mousavi, S. (2014). Kinetic, equilibrium and thermodynamic investigations of Ni (II), Cd (II), Cu (II) and Co (II) adsorption on barley straw ash. *Water Resources and Industry*, *6*, 1–17. doi:10.1016/j.wri.2014.06.001

Asberry, H. B., Kuo, C. Y., Gung, C. H., Conte, E. D., & Suen, S. Y. (2014). Characterization of water bamboo husk biosorbents and their application in heavy metal ion trapping. *Microchemical Journal*, *113*, 59–63. doi:10.1016/j.microc.2013.11.011

Asubiojo, O. I., & Ajelabi, O. B. (2009). Removal of heavy metals from industrial wastewaters using natural adsorbents. *Toxicological and Environmental Chemistry*, *91*(5), 883–890. doi:10.1080/02772240802614721

Badmus, M. O. A., Audu, T. O. K., & Anyata, B. (2007). Removal of copper from industrial wastewaters by activated carbon prepared from periwinkle shells. *Korean Journal of Chemical Engineering*, *24*(2), 246–252. doi:10.1007/s11814-007-5049-5

Banerjee, S. S., & Chen, D. H. (2007). Fast removal of copper ions by gum arabic modified magnetic nano-adsorbent. *Journal of Hazardous Materials*, *147*(3), 792–799. doi:10.1016/j.jhazmat.2007.01.079 PMID:17321674

Barakat, M. A. (2011). New trends in removing heavy metals from industrial wastewater. *Arabian Journal of Chemistry*, *4*(4), 361–377. doi:10.1016/j.arabjc.2010.07.019

Benoit, P., Barriuso, E., & Calvet, R. (1998). Biosorption characterization of herbicides, 2, 4-D and atrazine, and two chlorophenols on fungal mycelium. *Chemosphere*, *37*(7), 1271–1282. doi:10.1016/S0045-6535(98)00125-8

Beveridge, T. J., & Doyle, R. J. (1989). *Metal ions and bacteria*. Wiley.

Biškup, B., & Subotić, B. (2005). Removal of heavy metal ions from solutions using zeolites. III. Influence of sodium ion concentration in the liquid phase on the kinetics of exchange processes between cadmium ions from solution and sodium ions from zeolite A. *Separation Science and Technology*, *39*(4), 925–940. doi:10.1081/SS-120028454

Bohli, T., & Ouederni, A. (2016). Improvement of oxygen-containing functional groups on olive stones activated carbon by ozone and nitric acid for heavy metals removal from aqueous phase. *Environmental Science and Pollution Research International*, *23*(16), 15852–15861. doi:10.1007/s11356-015-4330-0 PMID:25794582

Bohli, T., Ouederni, A., Fiol, N., & Villaescusa, I. (2015). Evaluation of an activated carbon from olive stones used as an adsorbent for heavy metal removal from aqueous phases. *Comptes Rendus. Chimie*, *18*(1), 88–99. doi:10.1016/j.crci.2014.05.009

Boudrahem, F., Aissani-Benissad, F., & Soualah, A. (2013). Kinetic and equilibrium study of the sorption of lead (II) ions from aqueous phase by activated carbon. *Arabian Journal for Science and Engineering*, *38*(8), 1939–1949. doi:10.1007/s13369-013-0614-1

Bouhamed, F., Elouear, Z., Bouzid, J., & Ouddane, B. (2016). Multi-component adsorption of copper, nickel and zinc from aqueous solutions onto activated carbon prepared from date stones. *Environmental Science and Pollution Research International*, *23*(16), 15801–15806. doi:10.1007/s11356-015-4400-3 PMID:25843824

Brown, P., Jefcoat, I. A., Parrish, D., Gill, S., & Graham, E. (2000). Evaluation of the adsorptive capacity of peanut hull pellets for heavy metals in solution. *Advances in Environmental Research*, *4*(1), 19–29. doi:10.1016/S1093-0191(00)00004-6

Chen, J., Yang, P., Song, D., Yang, S., Zhou, L., Han, L., & Lai, B. (2014). Biosorption of Cr (VI) by carbonized Eupatorium adenophorum and Buckwheat straw: Thermodynamics and mechanism. *Frontiers of Environmental Science & Engineering*, *8*(6), 960–966. doi:10.1007/s11783-013-0612-2

Chowdhury, Z. Z., Zain, S. M., Khan, R. A., & Islam, M. S. (2012). Preparation and characterizations of activated carbon from kenaf fiber for equilibrium adsorption studies of copper from wastewater. *Korean Journal of Chemical Engineering*, *29*(9), 1187–1195. doi:10.1007/s11814-011-0297-9

Chu, W. (1999). Lead metal removal by recycled alum sludge. *Water Research*, *33*(13), 3019–3025. doi:10.1016/S0043-1354(99)00010-X

Cordero, H. C., Trevino, H., Juárez-Aguilar, H. L. G., Castillo, D. I. M., Mendoza, D. I., Montoya, V. H., ... Morán, M. A. M. (2013). Synthesis and adsorption properties of activated carbons from biomass of Prunusdomestica and Jacaranda mimosifolia for the removal of heavy metals and dyes from water. *Industrial Crops and Products*, *42*, 315–323. doi:10.1016/j.indcrop.2012.05.029

Cutillas-Barreiro, L., Ansias-Manso, L., Fernández-Calviño, D., Arias-Estévez, M., Nóvoa-Muñoz, J. C., Fernández-Sanjurjo, M. J., & Núñez-Delgado, A. (2014). Pine bark as bio-adsorbent for Cd, Cu, Ni, Pb and Zn: Batch-type and stirred flow chamber experiments. *Journal of Environmental Management*, *144*, 258–264. doi:10.1016/j.jenvman.2014.06.008 PMID:24973614

Da'vila-Jime'nez, M. M., Marı'a, P., Elizalde-Gonza'lez, W., Geyer, J., & Mattusch, R. (2003). Adsorption of metal cations from aqueous solution onto a natural and a model biocomposite. *Physicochem. Eng. Aspects*, *219*(1-3), 243–252. doi:10.1016/S0927-7757(03)00052-9

Dehghani, M. H., Sanaei, D., Ali, I., & Bhatnagar, A. (2016). Removal of chromium (VI) from aqueous solution using treated waste newspaper as a low-cost adsorbent: Kinetic modeling and isotherm studies. *Journal of Molecular Liquids*, *215*, 671–679. doi:10.1016/j.molliq.2015.12.057

Dhir, B. (2014). Potential of biological materials for removing heavy metals from wastewater. *Environmental Science and Pollution Research International*, *21*(3), 1614–1627. doi:10.1007/s11356-013-2230-8 PMID:24185905

Diniz, C. V., Doyle, F. M., & Ciminelli, V. S. (2002). Effect of pH on the adsorption of selected heavy metal ions from concentrated chloride solutions by the chelating resin Dowex M-4195. *Separation Science and Technology*, *37*(14), 3169–3185. doi:10.1081/SS-120006155

Dzikiewicz, M. (2000). Activities in nonpoint pollution control in rural areas of Poland. *Ecological Engineering*, *14*(4), 429–434. doi:10.1016/S0925-8574(99)00066-X

Ekmekyapar, F., Aslan, A., Bayhan, Y. K., & Cakici, A. (2006). Biosorption of copper (II) by nonliving lichen biomass of Cladonia rangiformis hoffm. *Journal of Hazardous Materials*, *137*(1), 293–298. doi:10.1016/j.jhazmat.2006.02.003 PMID:16530938

El-Said, A. G., Badawy, N. A., Abdel-Aal, A. Y., & Garamon, S. E. (2011). Optimization parameters for adsorption and desorption of Zn (II) and Se (IV) using rice husk ash: Kinetics and equilibrium. *Ionics*, *17*(3), 263–270. doi:10.1007/s11581-010-0505-3

Fan, Y., Yang, R., Lei, Z., Liu, N., Lv, J., Zhai, S., ... Wang, L. (2016). Removal of Cr (VI) from aqueous solution by rice husk derived magnetic sorbents. *Korean Journal of Chemical Engineering*, *33*(4), 1416–1424. doi:10.1007/s11814-015-0248-y

Farhan, A. M., Al-Dujaili, A. H., & Awwad, A. M. (2013). Equilibrium and kinetic studies of cadmium (II) and lead (II) ions biosorption onto Ficus carcia leaves. *International Journal of Industrial Chemistry*, *4*(1), 24. doi:10.1186/2228-5547-4-24

Fu, F., Xie, L., Tang, B., Wang, Q., & Jiang, S. (2012). Application of a novel strategy—Advanced Fenton-chemical precipitation to the treatment of strong stability chelated heavy metal containing wastewater. *Chemical Engineering Journal*, *189*, 283–287. doi:10.1016/j.cej.2012.02.073

Gaya, U. I., Otene, E., & Abdullah, A. H. (2015). Adsorption of aqueous Cd (II) and Pb (II) on activated carbon nanopores prepared by chemical activation of doum palm shell. *SpringerPlus*, *4*(1), 458. doi:10.1186/s40064-015-1256-4 PMID:26339560

Gebrehawaria, G., Hussen, A., & Rao, V. M. (2015). Removal of hexavalent chromium from aqueous solutions using barks of Acacia albida and leaves of Euclea schimperi. *International Journal of Environmental Science and Technology*, *12*(5), 1569–1580. doi:10.1007/s13762-014-0530-2

Ghaneian, M. T., Bhatnagar, A., Ehrampoush, M. H., Amrollahi, M., Jamshidi, B., Dehvari, M., & Taghavi, M. (2017). Biosorption of hexavalent chromium from aqueous solution onto pomegranate seeds: Kinetic modeling studies. *International Journal of Environmental Science and Technology*, *14*(2), 331–340. doi:10.1007/s13762-016-1216-8

Girgis, B. S., & Hendawy, A. N. A. (1997). Capacity of activated carbon from date pits in the removal of organic pollutants and heavy metals. *1st Inter. conference on chem. edu.*

Gray, N. F. (1999). *Water Technology*. New York: John Wiley & Sons.

Gupta, H., & Gogate, P. R. (2016). Intensified removal of copper from waste water using activated watermelon based biosorbent in the presence of ultrasound. *Ultrasonics Sonochemistry*, *30*, 113–122. doi:10.1016/j.ultsonch.2015.11.016 PMID:26627740

Gupta, V. K., Ali, I., Saini, V. K., Van Gerven, T., Van der Bruggen, B., & Vandecasteele, C. (2005). Removal of dyes from wastewater using bottom ash. *Industrial & Engineering Chemistry Research*, *44*(10), 3655–3664. doi:10.1021/ie0500220

Gupta, V. K., Rastogi, A., & Nayak, A. (2010). Biosorption of nickel onto treated alga (Oedogonium hatei): Application of isotherm and kinetic models. *Journal of Colloid and Interface Science*, *342*(2), 533–539. doi:10.1016/j.jcis.2009.10.074 PMID:20004906

Jalali, M., & Aboulghazi, F. (2013). Sunflower stalk, an agricultural waste, as an adsorbent for the removal of lead and cadmium from aqueous solutions. *Journal of Material Cycles and Waste Management*, *15*(4), 548–555. doi:10.1007/s10163-012-0096-3

Javaid, A., Bajwa, R., Shafique, U., & Anwar, J. (2011).Removal of heavy metals by adsorption on Pleurotus ostreatus. *Biomass and Bioenergy, 35*(5), 1675-1682.

Jiang, W., Pelaez, M., Dionysiou, D. D., Entezari, M. H., Tsoutsou, D., & O'Shea, K. (2013). Chromium (VI) removal by maghemite nanoparticles. *Chemical Engineering Journal*, *222*, 527–533. doi:10.1016/j.cej.2013.02.049

Kandile, N. G., Mohamed, H. M., & Mohamed, M. I. (2015). New heterocycle modified chitosan adsorbent for metal ions (II) removal from aqueous systems. *International Journal of Biological Macromolecules*, *72*, 110–116. doi:10.1016/j.ijbiomac.2014.07.042 PMID:25128098

Kannan, N., & Sundaram, M. M. (2001). Kinetics and mechanism of removal of methylene blue by adsorption on various carbons—a comparative study. *Dyes and Pigments*, *51*(1), 25–40. doi:10.1016/S0143-7208(01)00056-0

Kobya, M., Demirbas, E., Senturk, E., & Ince, M. (2005). Adsorption of heavy metal ions from aqueous solutions by activated carbon prepared from apricot stone. *Bioresource Technology*, *96*(13), 1518–1521. doi:10.1016/j.biortech.2004.12.005 PMID:15939281

Kumar, B., Smita, K., Sánchez, E., Stael, C., & Cumbal, L. (2016). Andean Sacha inchi (Plukenetia volubilis L.) shell biomass as new biosorbents for Pb^{2+} and Cu^{2+} ions. *Ecological Engineering*, *93*, 152–158. doi:10.1016/j.ecoleng.2016.05.034

Kyzas, G. Z., Terzopoulou, Z., Nikolaidis, V., Alexopoulou, E., & Bikiaris, D. N. (2015). Low-cost hemp biomaterials for nickel ions removal from aqueous solutions. *Journal of Molecular Liquids*, *209*, 209–218. doi:10.1016/j.molliq.2015.05.060

Li, W. C., Law, F. Y., & Chan, Y. H. M. (2017). Biosorption studies on copper (II) and cadmium (II) using pretreated rice straw and rice husk. *Environmental Science and Pollution Research International*, *24*(10), 8903–8915. doi:10.1007/s11356-015-5081-7 PMID:26206132

Li, Y. H., Wang, S., Luan, Z., Ding, J., Xu, C., & Wu, D. (2003). Adsorption of cadmium (II) from aqueous solution by surface oxidized carbon nanotubes. *Carbon, 41*(5), 1057–1062. doi:10.1016/S0008-6223(02)00440-2

Lim, A. P., & Aris, A. Z. (2014). A review on economically adsorbents on heavy metals removal in water and wastewater. *Reviews in Environmental Science and Biotechnology, 13*(2), 163–181. doi:10.1007/s11157-013-9330-2

Liu, H., Feng, S., Zhang, N., Du, X., & Liu, Y. (2014). Removal of Cu (II) ions from aqueous solution by activated carbon impregnated with humic acid. *Frontiers of Environmental Science & Engineering, 8*(3), 329–336. doi:10.1007/s11783-013-0553-9

Mendez, J. O., Melian, J. H., Arana, J., Rodriguez, J. D., Diaz, O. G., & Pena, J. P. (2015). Detoxification of waters contaminated with phenol, formaldehyde and phenol–formaldehyde mixtures using a combination of biological treatments and advanced oxidation techniques. *Applied Catalysis B: Environmental, 163*, 63–73. doi:10.1016/j.apcatb.2014.07.032

Milojković, J., Pezo, L., Stojanović, M., Mihajlović, M., Lopičić, Z., Petrović, J., & Kragović, M. (2016). Selected heavy metal biosorption by compost of Myriophyllum spicatum—a chemometric approach. *Ecological Engineering, 93*, 112–119. doi:10.1016/j.ecoleng.2016.05.012

Mise, S., & Patil, T. N. (2015). Adsorption Studies of Chromium (VI) on Activated Carbon Derived from Mangifera indica. *Journal of The Institution of Engineers (India): Series A, 96*(3), 237-247.

Mishra, V., Balomajumder, C., & Agarwal, V. K. (2010). Biosorption of Zn (II) onto the surface of non-living biomasses: A comparative study of adsorbent particle size and removal capacity of three different biomasses. *Water, Air, and Soil Pollution, 211*(1-4), 489–500. doi:10.1007/s11270-009-0317-0

Modak, J. M., & Natarajan, K. A. (1995). Biosorption of metals using nonliving biomass-A review. *Minerals & Metallurgical Processing, 12*(4), 189–196.

Moghadam, M. R., Nasirizadeh, N., Dashti, Z., & Babanezhad, E. (2013). Removal of Fe (II) from aqueous solution using pomegranate peel carbon: Equilibrium and kinetic studies. *International Journal of Industrial Chemistry, 4*(1), 19. doi:10.1186/2228-5547-4-19

Mohammed, T. J., & Ibrahim, R. I. (2016). Remediation of Cu (II) from Well Water of Iraq by Using Cortex of Fruits and Agricultural Waste. *Arabian Journal for Science and Engineering, 41*(2), 345–355. doi:10.1007/s13369-015-1848-x

Monier, M., Ayad, D. M., Wei, Y., & Sarhan, A. A. (2010). Preparation and characterization of magnetic chelating resin based on chitosan for adsorption of Cu (II), Co (II), and Ni (II) ions. *Reactive & Functional Polymers*, *70*(4), 257–266. doi:10.1016/j.reactfunctpolym.2010.01.002

Moreno-Barbosa, J. J., López-Velandia, C., del Pilar Maldonado, A., Giraldo, L., & Moreno-Piraján, J. C. (2013). Removal of lead (II) and zinc (II) ions from aqueous solutions by adsorption onto activated carbon synthesized from watermelon shell and walnut shell. *Adsorption*, *19*(2-4), 675–685. doi:10.1007/s10450-013-9491-x

Moreno-Piraján, J. C., Garcia-Cuello, V. S., & Giraldo, L. (2011). The removal and kinetic study of Mn, Fe, Ni and Cu ions from wastewater onto activated carbon from coconut shells. *Adsorption*, *17*(3), 505–514. doi:10.1007/s10450-010-9311-5

Ng, I. S., Wu, X., Yang, X., Xie, Y., Lu, Y., & Chen, C. (2013). Synergistic effect of Trichoderma reesei cellulases on agricultural tea waste for adsorption of heavy metal Cr (VI). *Bioresource Technology*, *145*, 297–301. doi:10.1016/j.biortech.2013.01.105 PMID:23419991

Okoye, A. I., Ejikeme, P. M., & Onukwuli, O. D. (2010). Lead removal from wastewater using fluted pumpkin seed shell activated carbon: Adsorption modeling and kinetics. *International Journal of Environmental Science and Technology*, *7*(4), 793–800. doi:10.1007/BF03326188

Onundi, Y. B., Mamun, A. A., Al Khatib, M. F., & Ahmed, Y. M. (2010). Adsorption of copper, nickel and lead ions from synthetic semiconductor industrial wastewater by palm shell activated carbon. *International Journal of Environmental Science and Technology*, *7*(4), 751–758. doi:10.1007/BF03326184

Owamah, H. I. (2014). Biosorptive removal of Pb (II) and Cu (II) from wastewater using activated carbon from cassava peels. *Journal of Material Cycles and Waste Management*, *16*(2), 347–358. doi:10.1007/s10163-013-0192-z

Pan, Y. M., Jiang, R. X., Yang, J. L., Zheng, H., & Yin, E. Q. (2016). Removal of Pb (II) ions from aqueous solutions by litchi pericarp and its leachate. *Journal of Central South University*, *23*(7), 1626–1632. doi:10.1007/s11771-016-3217-7

Parnian, A., Chorom, M., Jaafarzadeh, N., & Dinarvand, M. (2016). Use of two aquatic macrophytes for the removal of heavy metals from synthetic medium. *Ecohydrology & Hydrobiology*, *16*(3), 194–200. doi:10.1016/j.ecohyd.2016.07.001

Pereira, M. R., Arroyo, P. A., de Barros, M. A. S. D., Sanches, V. M., da Silva, E. A., Fonseca, I. M., & Lovera, R. G. (2006). Chromium adsorption in olive stone activated carbon. *Adsorption*, *12*(2), 155–162. doi:10.1007/s10450-006-0377-z

Rahbar, A., Farjadfard, S., Leili, M., Kafaei, R., Haghshenas, V., & Ramavandi, B. (2016). Experimental data of biomaterial derived from Malva sylvestris and charcoal tablet powder for Hg 2+ removal from aqueous solutions. *Data in Brief*, *8*, 132–135. doi:10.1016/j.dib.2016.05.032 PMID:27294181

Rubio, J., & Kitchener, J. A. (1976). The mechanism of adsorption of poly (ethylene oxide) flocculant on silica. *Journal of Colloid and Interface Science*, *57*(1), 132–142. doi:10.1016/0021-9797(76)90182-X

Saha, R., Mukherjee, K., Saha, I., Ghosh, A., Ghosh, S. K., & Saha, B. (2013). Removal of hexavalent chromium from water by adsorption on mosambi (Citrus limetta) peel. *Research on Chemical Intermediates*, *39*(5), 2245–2257. doi:10.1007/s11164-012-0754-z

Santhy, K., & Selvapathy, P. (2004). Removal of heavy metals from wastewater by adsorption on coir pith activated carbon. *Separation Science and Technology*, *39*(14), 3331–3351. doi:10.1081/SS-200036561

Şencan, A., Karaboyacı, M., & Kılıç, M. (2015). Determination of lead (II) sorption capacity of hazelnut shell and activated carbon obtained from hazelnut shell activated with ZnCl2. *Environmental Science and Pollution Research International*, *22*(5), 3238–3248. doi:10.1007/s11356-014-2974-9 PMID:24801293

Singh, S. A., & Shukla, S. R. (2016). Adsorptive removal of cobalt ions on raw and alkali-treated lemon peels. *International Journal of Environmental Science and Technology*, *13*(1), 165–178. doi:10.1007/s13762-015-0801-6

Sobhanardakani, S., Parvizimosaed, H., & Olyaie, E. (2013). Heavy metals removal from wastewaters using organic solid waste—rice husk. *Environmental Science and Pollution Research International*, *20*(8), 5265–5271. doi:10.1007/s11356-013-1516-1 PMID:23381799

Song, S. T., Saman, N., Johari, K., & Mat, H. (2014). Surface chemistry modifications of rice husk toward enhancement of Hg (II) adsorption from aqueous solution. *Clean Technologies and Environmental Policy*, *16*(8), 1747–1755. doi:10.1007/s10098-014-0803-y

Srivastava, S., Agrawal, S. B., & Mondal, M. K. (2016). Characterization, isotherm and kinetic study of Phaseolus vulgaris husk as an innovative adsorbent for Cr (VI) removal. *Korean Journal of Chemical Engineering*, *33*(2), 567–575. doi:10.1007/s11814-015-0165-0

Sublet, R., Simonnot, M. O., Boireau, A., & Sardin, M. (2003). A kinetic study of 3-chlorophenol enhanced hydroxyl radical generation during ozonation. *Water Research*, *37*, 4904–4912. doi:10.1016/j.watres.2003.08.010 PMID:14604636

Sudha, R., & Srinivasan, K. (2015). Nickel (II) removal using modified Citrus limettioides peel. *International Journal of Environmental Science and Technology*, *12*(12), 3993–4004. doi:10.1007/s13762-015-0780-7

Sudha, R., Srinivasan, K., & Premkumar, P. (2016). Kinetic, mechanism and equilibrium studies on removal of Pb (II) using Citrus limettioides peel and seed carbon. *Research on Chemical Intermediates*, *42*(3), 1677–1697. doi:10.1007/s11164-015-2111-5

Thenmozhi, R., & Santhi, T. (2015). Kinetics and equilibrium studies of the adsorption of Cu (II) from aqueous solution using Acacia nilotica seed pods on $ZnCl_2$ activation. *Research on Chemical Intermediates*, *41*(3), 1327–1341. doi:10.1007/s11164-013-1276-z

Tinge, J. T., Mencke, K., & Drinkenburg, A. A. H. (1987). The absorption of propane and ethene in slurries of activated carbon in water—I. *Chemical Engineering Science*, *42*(8), 1899–1907. doi:10.1016/0009-2509(87)80136-7

Uçar, S., Erdem, M., Tay, T., & Karagöz, S. (2015). Removal of lead (II) and nickel (II) ions from aqueous solution using activated carbon prepared from rapeseed oil cake by Na2CO3 activation. *Clean Technologies and Environmental Policy*, *17*(3), 747–756. doi:10.1007/s10098-014-0830-8

UNESCAP. (1999). Integrating environmental consideration into economic policy making process: background readings: Vol. 1. Institional arrangement and mechanism at national level (ST/ESCAP/1944). UNESCAP.

UNESCAP. (2000*). Integrating environmental consideration into economic policy making: Institutional issues* (ST/ESCAP/ 1990). UNESCAP.

Vassileva, P., Tzvetkova, P., Lakov, L., & Peshev, O. (2008). Thiouracil modified activated carbon as a sorbent for some precious and heavy metal ions. *Journal of Porous Materials*, *15*(5), 593–599. doi:10.1007/s10934-007-9138-y

Venkatesan, G., Senthilnathan, U., & Rajam, S. (2014). Cadmium removal from aqueous solutions using hybrid eucalyptus wood based activated carbon: Adsorption batch studies. *Clean Technologies and Environmental Policy*, *16*(1), 195–200. doi:10.1007/s10098-013-0628-0

Wang, Y., Wang, F., Wan, T., Cheng, S., Xu, G., Cao, R., & Gao, M. (2013). Enhanced adsorption of Pb (II) ions from aqueous solution by persimmon tannin-activated carbon composites. *Journal of Wuhan University of Technology-Mater. Science Editor*, *28*(4), 650–657.

Wang, Z., Feng, Y., Hao, X., Huang, W., & Feng, X. (2014). A novel potential-responsive ion exchange film system for heavy metal removal. *Journal of Materials Chemistry. A, Materials for Energy and Sustainability*, *2*(26), 10263–10272. doi:10.1039/C4TA00782D

Wase, J., & Foster, C. (1997). *Biosorbent for metal Ions*. London: Taylor and Francis, Ltd.

Weng, C. H., Lin, Y. T., Hong, D. Y., Sharma, Y. C., Chen, S. C., & Tripathi, K. (2014). Effective removal of copper ions from aqueous solution using base treated black tea waste. *Ecological Engineering*, *67*, 127–133. doi:10.1016/j.ecoleng.2014.03.053

Wu, Y., Yilihan, P., Cao, J., & Jin, Y. (2013). Competitive adsorption of Cr (VI) and Ni (II) onto coconut shell activated carbon in single and binary systems. *Water, Air, and Soil Pollution*, *224*(9), 1662. doi:10.1007/s11270-013-1662-6

Yin, C. Y., Aroua, M. K., & Daud, W. M. A. W. (2008). Enhanced adsorption of metal ions onto polyethyleneimine-impregnated palm shell activated carbon: Equilibrium studies. *Water, Air, and Soil Pollution*, *192*(1-4), 337–348. doi:10.1007/s11270-008-9660-9

Zakhama, S., Dhaouadi, H., & M'henni, F. (2011). Nonlinear modelisation of heavy metal removal from aqueous solution using Ulva lactuca algae. *Bioresource Technology*, *102*(2), 786–796. doi:10.1016/j.biortech.2010.08.107 PMID:20855200

Zhu, L. N., Jun, M., & Yang, S. D. (2007). Removal of phenol by activated alumina bed in pulsed high-voltage electric field. *Journal of Environmental Sciences (China)*, *19*(4), 409–415. doi:10.1016/S1001-0742(07)60068-0 PMID:17915702

Zhuang, X., Wan, Y., Feng, C., Shen, Y., & Zhao, D. (2009). Highly efficient adsorption of bulky dye molecules in wastewater on ordered mesoporous carbons. *Chemistry of Materials*, *21*(4), 706–716. doi:10.1021/cm8028577

Zongo, I., Leclerc, J. P., Maïga, H. A., Wéthé, J., & Lapicque, F. (2009). Removal of hexavalent chromium from industrial wastewater by electrocoagulation: A comprehensive comparison of aluminium and iron electrodes. *Separation and Purification Technology*, *66*(1), 159–166. doi:10.1016/j.seppur.2008.11.012

ADDITIONAL READING

Alinnor, I. J. (2007). Adsorption of heavy metal ions from aqueous solution by fly ash. *Fuel*, *86*(5), 853–857. doi:10.1016/j.fuel.2006.08.019

Baccar, R., Bouzid, J., Feki, M., & Montiel, A. (2009). Preparation of activated carbon from Tunisian olive-waste cakes and its application for adsorption of heavy metal ions. *Journal of Hazardous Materials*, *162*(2), 1522–1529. doi:10.1016/j.jhazmat.2008.06.041 PMID:18653277

Cho, H., Oh, D., & Kim, K. (2005). A study on removal characteristics of heavy metals from aqueous solution by fly ash. *Journal of Hazardous Materials*, *127*(1), 187–195. doi:10.1016/j.jhazmat.2005.07.019 PMID:16125307

Fu, F., & Wang, Q. (2011). Removal of heavy metal ions from wastewaters: A review. *Journal of Environmental Management*, *92*(3), 407–418. doi:10.1016/j.jenvman.2010.11.011 PMID:21138785

Hua, M., Zhang, S., Pan, B., Zhang, W., Lv, L., & Zhang, Q. (2012). Heavy metal removal from water/wastewater by nanosized metal oxides: A review. *Journal of Hazardous Materials*, *211*, 317–331. doi:10.1016/j.jhazmat.2011.10.016 PMID:22018872

Karnitz, O. Jr, Gurgel, L. V. A., De Melo, J. C. P., Botaro, V. R., Melo, T. M. S., de Freitas Gil, R. P., & Gil, L. F. (2007). Adsorption of heavy metal ion from aqueous single metal solution by chemically modified sugarcane bagasse. *Bioresource Technology*, *98*(6), 1291–1297. doi:10.1016/j.biortech.2006.05.013 PMID:16843656

Kumar, U. (2006). Agricultural products and by-products as a low cost adsorbent for heavy metal removal from water and wastewater: A review. *Scientific research and essays*, *1*(2), 033-037.

Liu, A. M., Hidajat, K., Kawi, S., & Zhao, D. Y. (2000). A new class of hybrid mesoporous materials with functionalized organic monolayers for selective adsorption of heavy metal ions. *Chemical Communications*, (13): 1145–1146. doi:10.1039/b0026611

Liu, J. F., Zhao, Z. S., & Jiang, G. B. (2008). Coating Fe3O4 magnetic nanoparticles with humic acid for high efficient removal of heavy metals in water. *Environmental Science & Technology*, *42*(18), 6949–6954. doi:10.1021/es800924c PMID:18853814

Nascimento, M., Soares, P. S. M., & de Souza, V. P. (2009). Adsorption of heavy metal cations using coal fly ash modified by hydrothermal method. *Fuel*, *88*(9), 1714–1719. doi:10.1016/j.fuel.2009.01.007

Ray, P. Z., & Shipley, H. J. (2015). Inorganic nano-adsorbents for the removal of heavy metals and arsenic: A review. *RSC Advances*, *5*(38), 29885–29907. doi:10.1039/C5RA02714D

Stafiej, A., & Pyrzynska, K. (2007). Adsorption of heavy metal ions with carbon nanotubes. *Separation and Purification Technology*, *58*(1), 49–52. doi:10.1016/j.seppur.2007.07.008

Visa, M., Bogatu, C., & Duta, A. (2010). Simultaneous adsorption of dyes and heavy metals from multicomponent solutions using fly ash. *Applied Surface Science*, *256*(17), 5486–5491. doi:10.1016/j.apsusc.2009.12.145

Wang, S., Terdkiatburana, T., & Tadé, M. O. (2008). Single and co-adsorption of heavy metals and humic acid on fly ash. *Separation and Purification Technology*, *58*(3), 353–358. doi:10.1016/j.seppur.2007.05.009

KEY TERMS AND DEFINITIONS

Activated Carbon: This is a form of carbon processed to have small, low-volume pores that increase the surface area available for adsorption or chemical reactions. Activated carbon is usually derived from charcoal and is sometimes utilized as biochar. Those derived from coal and coke is referred as activated coal and activated coke, respectively.

Adsorption: This refers to the adhesion of atoms, ions, or molecules from a gas, liquid, or dissolved solid to a surface and creates a film of the adsorbate on the surface of the adsorbent. Adsorption is a surface-based process while absorption involves the whole volume of the material.

Bioaccumulation: This is the accumulation of substances, such as pesticides, or other chemicals in an organism. Bioaccumulation occurs when an organism absorbs a substance at a rate faster than that at which the substance is lost by catabolism and excretion.

Biomass: Biomass is the living matter used usually in industry for getting energy by burning wood and other organic matter.

Carbon Sequestration: The process involved in carbon capture and the long-term storage of atmospheric carbon dioxide is known as carbon sequestration. It involves long-term storage of carbon dioxide or other forms of carbon to mitigate or defer global warming.

Chemisorption: Chemisorption is a kind of adsorption that involves a chemical reaction between the surface and the adsorbate. New chemical bonds are generated at the adsorbant surface. The strong interaction between the adsorbate and the substrate surface creates new types of electronic bonds.

Environmental Remediation: The removal of pollution or contaminants from environmental media such as soil, groundwater, sediment, or surface water is known as environmental remediation.

Municipal Solid Waste: This refers to the waste consisting of everyday items that are discarded by the public.

APPENDIX

Table 1. Biobsorption efficiency and techniques used for removal of heavy metals

Bio-absorbent	Heavy metal	Instrumentation &Characterization	Region	Maximum Removal efficiency	Isotherm and Kinetics models	Reference
Palm shell	Nickel Copper Lead	Atomic Adsorption Spectrophotometer (Perkinelmer Hga900	Kuala Langat, Malaysia	55% 97% 100%	Langmuir and Fruendlich	(Onundi, 2010)
Palm shell	Cd(II) Pb(II)	XRD, SEM, PSA, N2 adsorption - desorption	Janguza barrack in Gwarzo L.G.A., Kano	100%	Langmuir and fruendlich, Pseudo first order, Enthalpy, free energy, Entropy.	(Gaya, Otenel, & Abdullah, 2015)
Cassava peels	Cu(II) Pb(II)	AAS	Ibusa, Oshimili North Local Government Area of Delta State, Nigeria	79% 73%	Langmuir and Fruendlich, Pseudo first order,, free Energy, Enthalpy, Entopy	(Owamah, 2014)
Hybrid eucalyptus	Cd(II)	Spectrophotometer	Tiruchirapalli, India	90.25%	Langmuir and Fruendlich	(Venkatesan, Sentthilnatham, & Rajan, 2014)
Coconut Shell	Cr(VI) Ni(II)	FTIR, XRD, SEM	--	94.5% 58.92%	--	(Wu et al., 2013)
Malva sylvestris	Hg^{2+}	FTIR/BET/AAS	Bushehr city, Iran.	--	-----------	(Rahbar et al., 2016)
Barley straw ash Cereal straw,	Ni(II) Cd(II) Cu(II)	BET/AAS/ICP-AES/ BJH/SEM	Fars, Iran	60%	Langmuir and Fruendlich, Pseudo first order and second order,	(Arshadi, Amiri, & Mousavi, 2014)
Hemp fibre, Hemp shives	NI(ii)	(FTIR, XRD, SEM/EDAX, crystallinity, and cellulose content)	HempFlax company (the Netherlands)/ Thessaloniki, Greece GREECE	66% 70% 62% 64%	Langmuir and Fruendlich, Pseudo first order and Second order, Elovich Model	(Kyzas et al., 2015)
Lettuce	Pd Cd Ni Cu	BET/FTIR/	Monastir (Tunisia) coastal zone	1.02 mg/g 1.66 mg/g 2.62 mg/g 3.04 mg/g	Langmuir and Fruendlich, Pseudo first order and second order,	(Zakhama, & Dhaouadi M'Henni, 2011)
Pine bark	Cd(II) Pd(II) Ni(II) Zn(II)	AAS/X-ray fluorescence	Geolia (Madrid, Spain).	84 84 99 75	Langmuir and Fruendlich	(Barreiro et al., 2014)
Pink shrimp (Solenocera melantho) shell	Cu(II) Co(II) NI(II)	SEM/FTIR/XRD/NMR/ TGA	Philadelphia, PA, USA	--	Langmuir and Fruendlich, Temkin isotherm model Pseudo first order and second order,	(Monier et al., 2010)
Banana peels	Cr(VI)	SEM	Khyber Pakhtunkhwa, Pakistan	98%	Langmuir and Fruendlich, First and second order, Intra particle diffusion	(Ali, Saeed, & Mabood, 2016)

continued on following page

Bio-absorbent	Heavy metal	Instrumentation &Characterization	Region	Maximum Removal efficiency	Isotherm and Kinetics models	Reference
Waste newspaper	Cr(VI)	FTIR	Tehran, Islamic Republic of Iran	--	Temkin, Langmuir and Freundlich,first order ad second order	(Dehghani, Sanaei, & Ali, 2016)
Corn cobs, wheat bran, bagasse, Peanut skin, Paddy husk, Human hair.	Zn, Pd,Fe, Mn Zn, Pd,Fe, Mn Zn, Pd,Fe, Mn Zn, Pd,Fe, Mn Zn, Pd,Fe, Mn Zn, Pd,Fe, Mn	AAS	Nigeria	62.5 - 98.5%	---------	(Asubiojo, & Ajelabi, 2009)
Coir pith	Cd Cu Zn	------	Anna University, Chennai	96.6% 99.5% 77.1%	Langmuir and Fruendlich	(Santhy, & Selvapathy, 2004)
Tea waste	Cr	NIR/PSD/XPS/SEM/ HPLC/ELS.	Xiamen University,china	------	----------	(Ng et al.,2013)
Prunus Domestica Jacaranda mimosifolia	Pd	LECO/SEM/EDX	Aguascalientes (Mexico	------	Langmuir and Fruendlich, Ships model	(Corderoa et al., 2013)
Banana, kiwi tangerine peels,	Cr Cd Zn	------	------	35%,35%,25%	Pseudo Second order, and Langmuir model	Al-Qahtani, (2016)
(Mango) Seed Shell	Cr	------	Gulbarga, Karnataka	60%	Langmuir and Fruendlich	(Shashikant, & Tripati, 2015)
Paper Plastic palm	Cu, Pb	------	Saudi Arabia	---	Pseudo second order, Enthalpy, Entropy, Free energy.	(Al-Othmanet al., 2016)
Olive stone	Cr(III)	Dubinin-Raduskhevich and BJH methods/AAS/ BET/Boehm titration methods	Brazil	----	Langmuir model	(Pereira et al., 2006)
Coconut shell	Cr (VI) Ni(II)_	SEM/XRD./FTIR	Hohai University, Nanjing, China	94.5% 58.92%	Langmuir and fruedlich, Dubinin-Radushkevich, Enthalpy, Entropy, Free energy, Elovich	(Wu et al., 2013)
hazelnut shell	Pb (II)	FFTIR/SEM/ BET/ ICP-OES	Ünye in Turkey	----------	Langmuir and Fruendlich, pseudo first and Second order.	(Şencan, Karaboyac, & Kılıç, 2015)
Palm Shell	Ni2+, Cd2+ Pb2	BET	Bravo Green Sdn Bhd (Malaysia	-------	Langmuir and Fruendlich, Redlich-Peterson	(Yin, Aurora, & Duad, 2008)

continued on following page

Table 1. Continued

Bio-absorbent	Heavy metal	Instrumentation &Characterization	Region	Maximum Removal efficiency	Isotherm and Kinetics models	Reference
Persimmon Tannin	Pb (II)	AAS/Uv-V spectophotometery/	------	------	Langmuir and Fruendlich, pseudo first and second order.	(Wang, et al., 2013)
African palm fruit	Cd Cu Ni Pb	AAS	Panteka Market Nigeria	93.0% 96.71% 92.0% 95.42%	------	(Abdulrazak, Hussaini, & Sani, 2016)
Olive stones	Co(II) Ni(II) Cu(II)	SEM/FTIR/ ICP—AES/ BOEHM TITRATION	Tunisia	------	Pseudo second order, Redlich-Peterson Model	(Bohli et al., 2016)
Coconut shell	Pb(II)	AAS/BET	Algeria	94.0%	Langmuir and Fruendlich, and Temkin, Pseudo first and second order.	(Boudrahem, Aissani-Benissad, & Soualah, 2013)
Ficus Nitida	Cr(VI)	flame atomic absorption photometer/BET	King Khalid University, Abha, Saudi Arabia	99.0%	Langmuir and Fruendlich, Enthalpy, Free energy, Entropy	(Ali, & Alrafi, 2016).
Telfairia Occidentalis (pumkin) seed shells	Pb	------	Nigeria	------	Langmuir and Fruendlich, pseudo first and second order, Batacharya-Venkocher, Elovich, Intra particle and film diffusion.	(Okoye, Ejikeme, & Onukwuli, 2010)
Date stones	Cu2+, Ni2+, Zn2	AAS	Tunisian	------	Langmuir and Fruendlich, Enthalpy, Entropy, Free energy.	(Bouhamed et al., 2016)
kenaf fiber	Cu	Autosorb/ SEM/BET/FTIR/AAS	Malaysia.	48.43%	Langmuir and Fruendlich, pseudo first and Second order, Elovich.	(Chowdhury et al., 2012)
Rapeseed oil cake	Pb (II) Ni(II)	Nano ZS potentiometer/ SEM/ Boehm's titration	Turkey	------	------	(Ucar et al., 2015)
Watermelon shell, Walnut shell	Pb (II) Zn(II)	Boehm titration/SEM/ FTIR	Bogota´, Colombia	------	------	(Barbosa et al., 2013)
Watermelon	Cu(II).	/FTIR/SEM/UV-VIS SPECTOPHOTOMETRY	MUNBAI INDIA	------	------	(Gupta, & Gogate, 2016)
Apricot stones	Ni Co Cd Pb Cu Cr(II) Cr(VI)	SEM/BET	Malatya in Turkey	98.51% 99.11% 99.68% 99.13% 97.48% 98.99% 99.99%	---------	(Kobya et al., 2005).
Bamboo husk	FE (III) Ni(II)	FAAS/FTIR/ BET	Taichung, Taiwan.	------	------	(Asberry et al., 2014).
Pomegranate peel	Cd	BET/FTIR/ASTM/AAS/ Centrifuge/UV–Vis spectrophotometer	Iran	------	Langmuir and Fruendlich, pseudo first and second order, Intra particle diffusion, diffusion,	(Abedi, Salmani, & Mozaffari, 2016)

continued on following page

Table 1. Continued

Bio-absorbent	Heavy metal	Instrumentation &Characterization	Region	Maximum Removal efficiency	Isotherm and Kinetics models	Reference
Lemon peels	Co (II)	FTIR/SEM/TGA/BET/ EDAX	Mumbai I ndia	------	Langmuir and Fruendlich, pseudo first and second order.	(Singh & Shukla 2016).
Andean Sacha inchi shell	Cu(ii) Pb(II)	SEM/AAS/FTIR/EDAX.	Ecudar / India	------	Langmuir and Fruendlich.	(Kumar et al., 2016)
Eupatorium adenophorum Buckwheat straw	Cr (VI).	Colorimetry	Sichuan Province, China	90%	Langmuir and Fruendlich, Enthalpy, Entropy, Fee Energy	(Chen et al., 2014).
Pomegranate seeds	Cr (VI).	Uv-vis spectrophotometer/ SEM	Yazd, Iran,	99.5%	Pseudo first and Second order, Elovich, Intra particle diffusion,and Boyd kinetics	(Ghaneian et al., 2016)
Eucalyptus bark saw dust, Mango bark saw dust, and Pineapple fruit peel	Zn(II)	AAS/ FTIR/SEM	IIT Roorkee, India.	84% 51.4% 22.9%	----------	(Mishra, Balomajumder, & Agarwal, 2010).
Rice husk and Rce straw	Cu (II) & Cd(II)	AAS/XRF?/ SEM/ FTIR	Shueng Shui in the New Territories, Hong Kong	97% & 95%, 96%.	Langmuir and Fruendlich, pseudo first and Second order.	(Li, Law, & Chan, 2015)
Phaseolus vulgaris	Cr(VI)	UV- Vis SPECT. /SEM/ FTIR	Varanasi, Uttar Pradesh, India	99.88%	Langmuir and Fruendlich, Pseudo first and Second order.	(Srivastava, Agrawal, & Mondal, 2016)
Ficus carcia	Cadmium (II) &lead(II)	FTIR/AAS/	Jordan	98.58%	Langmuir and Fruendlich, pseudo first and Second order.	(Farhan, Al-Dujaili, & Awaad, 2013).
Rice husk	Cr (III) Cu(II)	AAS/ Centrifuge	Hamedan Branch, Islamic Azad University, Hamedan, Iran	100%	-----------	(Sobhanardakan, Parvizimosaed, & Olyaie, 2013).
Citrus limettioides peel and seed	Pb(II)	SEM/EDX/FTIR/AAS/ Centrifuge	Timal Nadu < India	94.00% 99.30%	langmuir and Fruendlich, pseudo first and second order.	(Sudha, Srinivasan, & Premkumar 2016).
Acacia nilotica seed pods	Cu(II)	SEM, XRD, and FTIR/ UVS/ Centrifuge.	Coimbatore, Tamilnadu. India	85.96%	Langmuir and Fruendlich, Dubnin-Radukevich, Pseudo first and second order, Elovich, Intra particle kinetics, Enthalpy, Entropy, Free energy	(Thenmozhi, & Santhi, 2015)
Citrus limettioides peel	Nickel (II)	SEM, EDX and FT-IR/ AAS/Centrifuge	Timal Nadu < India	99.0%	Langmuir and Fruendlich, Enthalpy, Entropy, Free energy	(Sudha, & Srinivasan, 2015).

continued on following page

Table 1. Continued

Bio-absorbent	Heavy metal	Instrumentation &Characterization	Region	Maximum Removal efficiency	Isotherm and Kinetics models	Reference
Rice husk	Zn(II) and Se(IV)	AAS/Centrifuge	Al-Azhar University (Girls), Nasr City, Cairo, Egypt		Langmuir and Fruendlich,Temkin Model	(El-Said, Badawy, & Attyat, 2010).
Watermelon shell, Lemon, Banana peels	Cu(II)	AAS/FTIR/SEM.	Iraq	90% 78% 65%	langmuir and Fruendlich,	(Mohammed, & Ibrahim, 2016).
Rice husk	Cr(VI)	SEM/TEM/XRD/VSM/ XPS/FTIR/BET/UVS	Dalian, Liaoning Province CHINA	100%	Langmuir and Fruendlich, Pseudo first and second order	(Fan et al., 2016)
Pomegranate peel	Fe (II)	Spectophotometry	Iran		Langmuir and Fruendlich, Pseudo first and second order, Enthalpy, Entropy, Free Energy	(Moghadam, Dashti, & Babanezhad, 2013)
Barks of Acacia albida and leaves of Euclea schimperi	Cr(VI)	Double-beam UV–visible Spectrophotometer/FTIR	Ethiopia	95.63% 98.34%	Langmuir and Fruendlich, Pseudo first and second order	(Gebrehawaria, Hussem, & Rao, 2015)
(Citrus limetta) peel	Cr(VI)	SEM/FTIR? UV- VIS spect.	Mumbai India'		Langmuir and Fruendlich, Pseudo first and second order, Enthalpy, Entropy, Free Enrgy	(Saha et al., 2013)
Litchi pericarp and its leachate	Pb(II)	AAS/Centrifuge/	Tai'an, Shangdong, China.	98.0%	Langmuir and Fruendlich, Pseudo fdirst and second order	(Pan et al., 2016)
Sunflower stalk,	Pb Cd	AAS	Hamedan, Western Iran	97% 87%	Langmuir and Fruendlich, Pseudo second order	(Jalali, & Aboulghazi, 2013).
Rice husk	Hg(II)	SEM/FTIR/	Malaysia	50%	------------	(Song et al., 2014)
Rice husk ash	Mn Fe	Sears' Method/ XRF/ FTIR/ Spectrophotometrically/	Kwara State of Nigeria.	100% 70%	Langmuir and Fruendlich, Temkin, Pseudo first and second order, Free Energy, Enthalpy, Entropy	(Adekola, Hodonou, & Adegoke, 2016).

Chapter 4
Biosorption of Uranium Heavy Metals:
Technology and Methods

Ashok K. Rathoure
Independent Researcher, India

Sudhanshu Mishra
NIMS University, India

Sandeep Tripathi
NIMS University, India

ABSTRACT

Uranium is a seriously threatening heavy metal because of its high toxicity and radioactivity. Uranium contaminates surface and groundwater. Metal removed from aqueous solutions often leads to effective metal concentration. Apart from the slow natural process of metal mineralization, removal of heavy metals is attained when the metal becomes concentrated at a point that it is either returned to the process or resold. Physical adsorption takes place due to van-der Waals' forces. Conventional methods used for uranium removal are expensive and produce huge amount of sludge (consists of toxic substances) which blockade the membrane. In this chapter, uranium removal by biosorption method is discussed. Uranium removal is attained by the use of either living microorganisms (bacteria, algae, and fungi) or their dead biomasses.

DOI: 10.4018/978-1-5225-4162-2.ch004

Copyright © 2018, IGI Global. Copying or distributing in print or electronic forms without written permission of IGI Global is prohibited.

INTRODUCTION

The heavy metal pollution and their most serious environmental problems have drawn more attention. Due to rapid industrialization and economic development a number of industries like mining, surface finishing, energy and fuel producing, fertilizer, pesticide, metallurgy, electroplating, electrolysis, electro osmosis, leather, iron and steel photography, electric appliance manufacturing, metal surface treating, aerospace and atomic energy installations have been established. The wastes from these industries contain metals are directly or indirectly discharged into the environment which causing serious environmental pollution and even threatening human life. Apart from waste various metals including uranium also contributed from the natural sources. Uranium is naturally occurred heavy metal. It is found in various chemical forms in all types of soils, rocks, seas and oceans. It is also found in drinking water and some of the foods. Uranium (atomic number 92) is the heaviest known natural element with specific density of 18.7 g/cm^3 and is slightly radioactive. Uranium is about forty times more common than silver. Uranium occurs naturally in the +2, +3, +4, +5 and +6 valence states. The hexavalent is most common form of Uranium. World health organization (WHO, 2001) reports that approximately 90 micrograms an average of Uranium present in the human body through drinking water, food and air. It is supposed to be found in the skeleton (66%), in liver (16%), in kidneys (8%) and in other body tissues (10%). Apart from natural resources there are various other sources of exposure are mining and milling, metal, combustion, corrosion and nuclear waste. The atomic reactors are rich in the source of uranium and other radioactive materials. There major problem resides in their disposal. In lack of appropriately disposal, they can harm not only living organisms but also the human being.

ROUTE OF EXPOSURE

There are various route of exposure of uranium particularly inhalation, ingestion and dermal contact.

1. **Inhalation:** Inhalation is the most common route of route exposure take place when depleted uranium is re-suspended in the atmosphere through wind or other disturbances.
2. **Ingestion:** Ingestion may occur during the uranium contamination in drinking-water or food.
3. **Dermal contact:** Uranium is able to passes the skin and enter into the blood stream and further it could be deposited in the body organs.

REMOVAL OF URANIUM FROM THE WASTE

Several methods are being used for Uranium removal from the aqueous wastes. These may include chemical precipitation, ion exchange, electrochemical treatment, membrane technologies, adsorption on activated carbon, etc. The most common conventional methods for removal of heavy metals from aqueous streams (Rich and Cherry, 1987) are as follows:

- Chemical Precipitation
- Ultrafiltration
- Electrolysis
- Ion Exchange
- Reverse osmosis
- Phytoremediation

CONVENTIONAL METHODS OF HEAVY METALS REMOVAL AND THEIR DISADVANTAGES

1. Chemical Precipitation

The heavy metal ions are precipitated by mix up with coagulants (alum, iron salts, lime and organic polymers). Te main demerit of this methods s the production of huge amount of sludge consists of toxic substances.

2. Electro-Dialysis

The heavy metal ions are separated through ion-selective semi-permeable membranes and an electrical potential is applied between two electrodes that cause migration of cations and anions towards respective electrodes. The demerit of this method s that the metal hydroxide is formed during the process which blockade the membrane.

3. Ion-Exchange

The heavy metal ions in diluted solutions get exchanged with ions due to electrostatic forces on exchange resin. The demerit of this method is that it is more expensive and partial removal of some ions takes place.

4. Reverse Osmosis

The heavy metals separated through a semi-permeable membrane at a pressure greater than osmotic pressure which is due to dissolved solids in wastewater. The demerit of this method is that it very expensive method.

5. Phytoremediation

Some plants are used to remove heavy metals from soil and water contaminated with metals. The demerit of this method is that this method takes too much time for removal of metals.

6. Ultrafiltration

The heavy metals are removed by using porous membranes. It is a pressure driven membrane operations technique for removal of heavy metals. The demerit of the method is the production of excess amount of sludge.

The conventional methods used for removal of heavy metals are too expensive (Horsfall *et al.,* 2003). Thus it has become necessary to develop another method of removal of heavy metals that must be cost effective.

In 1990s, a new method known as "biosorption" has been developed which is not only cost effective but also accumulate heavy metals because of their metal-binding capacity. Thus the metal-binding capacity of any biological materials (living or dead) that cause accumulation of heavy metals from waste water (either by metabolic or physiochemical pathway) is known as "Biosorption". Some microorganisms (algae, bacteria, fungi and yeasts) are reported as potential metal sorbents (Wang *et al.,* 2017, Volesky, 1986). Biosorption by is not only used for toxic metal removal but also for precious metal recovery such as gold and silver (Volesky, 1986; Brierley *et al.,* 1986; Kuyucak *et al.,* 1988). The non-living and living biomass found to be effectively removed metals from environment (Horsfall *et al.*, 2003). The living microorganisms or their dead component which are capable of removing heavy metals are also referred as "biotrap" (Crusberg *et al,*. 2004). The merits of the biosorption are as follows:

- High Efficiency
- Low Cost
- Minimization of chemical or Biological sludge
- No additional Nutrient requirement
- Regeneration of biosorbent
- Possibility of metal recovery

BIOSORPTION

Uranium can be removed from aqueous solution only when they are appropriately immobilized. The procedures of removal from aqueous solutions lead to concentre the metal. The aspect of recovery of uranium heavy metal is easier and economically good. Biosorption is a physiochemical method which is occurred naturally in certain microbial biomasses that passively concentrate and bind metals onto its cellular structure.

- Chemical groups that act as biosorption of metals in Biomass
- Acetamido groups of chitin
- Polysaccharides of fungi
- Hydroxyls in Polysaccharides
- Amino and phosphate groups in Nucleic acid
- Amid, Amino Sulphhydryl and carboxyl groups in proteins

The process of biosorption comprises two phases:

1. Solid phase (sorvent or biosorbent; Any biological material)
2. Liquid phase (solvent, generally water; and a dissolved species to be sorbed (sorbate, metal ions).

The sorvent having the greater affinity for sorbate species, Thus sorbate species is attracted and bound to the solvent. This process continues till establishment of equilibrium between solid-bound sorbate and its parts remaining in the solution.

BIOSORBENT MATERIAL

The search for an economical and eco-friendly option has led to the utilization of biological materials (microbial and plant origin) as adsorbents. These biomaterials interact effectively with heavy metals. Biosorption is a term that describes the removal of heavy metals by dead biomass from an aqueous solution which is attributed mainly to the ligands present in their cell wall biomolecules. Various dead biomasses, such as, marine algae, bacteria, fungi, industrial wastes and several other materials have been successfully used to adsorb heavy metal ions from aqueous solution. Among these, the brown algal biomass has been shown to be highly effective, reliable, and predictable in the removal of heavy metals from aqueous solutions. Some biosorbents are effectively bound and collect majority of heavy metals having no specific activity while some are specific for certain metals. Now a days, industrial waste products such as waste mycelia from fermentation processes (Pagnanelli, *et al*., 2002), activated sludge from sewage treatment plants (Hammaini *et al.,* 2003) and oceans as seaweeds.

The appropriate choice of metals for biosorption dependent on the metals which can be classified in to four major categories (Raoa and Prabhakara, 2011) namely toxic heavy metals, strategic metals, precious metals and radio nuclides:

- Strategic metals
- Precious metals
- Toxic metals
- Radionuclides

BIOSORPTION MECHANISMS

The biosorption process can be classified as follows:

1. **Metabolism Dependent Biosorption:** The metal uptake is by transport of metal across the cell membrane. This process takes place only within viable cells.
2. **Non-Metabolism Dependent Biosorption:** The physico-chemical interaction is necessary between metal and functional groups (carboxyl, sulphate, phosphate and amino groups) present on microbial cell surface to uptake metal. This is based on physical adsorption, ion exchange and chemical sorption. This type of biosorption is occurs rapidly and can be reversible.

Transport Across Cell Membrane

The heavy metal is transported across microbial cell membranes in the same manner of transport of metabolically important ions such as potassium, magnesium and sodium. The biosorption by living organisms is two steps process:

1. The metals are bound to microbial cell walls (metabolism independent binding).
2. The metal ions transported across cell membrane (Metabolism dependent intracellular uptake) Nourbaksh *et.al.,* 1994.

Physical Adsorption

The physical adsorption achieved with help of van der Waals' forces. The biosorption of heavy metals (uranium, cobalts, cadmium, copper and zinc) was attained by using dead biomasses of microorganism (algae, fungi and yeasts) takes place through electrostatic force. The biosorption of thorium, uranium, nickel, zinc, cadmium and lead was achieved by *Rhizopus arrhizus* (Muraleedharan *et al.*, 1990).

Ion Exchange

The microbial cell wall possesses polysaccharides and bivalent metal ions which exchange with counter ions of the polysaccharides. The brown algal cell wall possess alginates that occurs as salts of ions such as (K^+, Na^+, Ca^{2+} and Mg^{2+}). The exchange of these ions can get exchanged with counter ions (CO^{2+}, Cu^{2+}, Cd^{2+} and Zn^{2+}) and leads to the biosorptive uptake of heavy metals (Kuyucak and Volesky 1988). The uptake of copper by fungi *Ganoderma lucidium* (Muraleedharan *et al.*, 1990) and *Aspergillus niger* (Zhou *et al.*, 1994) is one of the example of uptake by ion exchange mechanism.

Complexation

The metals are removed by complex formation on cell surface after interaction between the metal and active groups. The removal of Thorium and Uranium is achieved by biosorption by *Rhizopus arrhizus (*Tsezos *et al.,* 1981) is based on physical adsorption as well as by formation of complex. In this process, metals coordinate with nitrogen of chitin cell wall network. The accumulation of metal ions such as calcium, magnesium, cadmium, zinc, copper and mercury is achieved through the complexation mechanism by *Pseudomonas syringae* (Cabral, 1992).

Precipitation

The process of precipitation either dependent or independent on cellular metabolism. The precipitation process is consequence of the chemical interaction between the metal and the cell surface. For example; Uranium biosorption by *Rhizopus arrhizus (*Tsezos *et al.,* 1982) is terminated by the formation of the complex uranium-chitin which is followed by hydrolysis and precipitation of the hydrolysis end product (uranylhydroxide) in the cell wall.

Table 1. Citation of different biosorbent materials for specific metal ions

Biosorbent Material	Metal	References
Active sludge bacteria	Copper	(Aksu *et al.*, 1991)
	Zinc	(Norton *et al.* 2003)
Aapergillus niger	Cobalt	(Kuyucak *et al.*, 1988)
	Copper	(Venkobachar, 1990)
	Uranium	(Sakaguchi *et al.*, 1991)
	Gold	(Kuyucak *et al.*, 1988)
Chitin of Rhizopus arrhizus	Uranium	(Tsezos., 1983)
Chlorella regularis	Uranium	(Sakaguchi *et al* 1991)
Dried activated sludge	Copper	(Hammaini *et al.*, 2003)
	Cadmium	
	Zinc	
Pennicillium chrysogenum	Thorium	(Tsezos *et al.*, 1981)
Pennicillium digitatum	Lead	(Galu., 1987)
Pseudomonasfluorescens	Uranium	(Tsezos *et al.*, 1981)
Rhizopus arrhizus	Cobalt	(Sakaguchi *et al.*, 1991)
Saccharomyces cerevisiae	Copper	(Huang *et al.*, 1990)
	Thorium	(Norris *et al.*, 1988)
	Uranium	(Kuyucak *et al.*, 1988)40
	Zinc	(Volesky., 1992)
Streptomyces noursei	Silver	(Mattuschka *et al.*, 1993)
Streptomyces niveus	Uranium	(Tsezos *et al.*, 1981)
Streptomyces longwoodensis	Uranium	(Friis and Myers-Keith., 1986)
Sargassum natans	Cadmium	(Volesky., 1992)
	Lead	(Kuyucak ., 1988)
	Gold	(Kuyucak *et al.*, 1988)
Seaweed *Durvillea potatorum*	Zinc	(Aderhold *et al.* 1996)
Submerged aquatic plant *Myriophyllum spicatum.*	Lead	(Keskinkan *et al.* 2003)

FACTORS AFFECTING BIOSORPTION

There are a number of factors that can affect the biosorption process. These factors are given below.

1. Temperature range of 20-30 ^{0}C does not influence the biosorption (Aksu *et al.*, 1992).

2. pH is also an important parameter in the biosorptive process. The chemistry of the metals in solution and activity of the functional groups in the biomass can be affected by pH directly (Friis and Myers-Keith, 1986).
3. Biomass concentration in solution influences uptake of metal ions. Metal ion uptake is increased in lower biomass concentrations (Fourest and Roux, 1992) while decreased due to greater concentration of biomass that interferes between the binding sites (Gadd *et al.,* 1988).
4. Biosorption is more effective in the treatment of that wastewater which is contaminated with many different types of metal ions. The presence of more than one metals ion may or may not interfere in the removal of other ions. The manganese, cobalt, copper, cadmium, mercury and lead in solution not affect the Uranium uptake by microbial biomass (Sakaguchi and Nakajima, 1991) while presence of Fe^{2+} and Zn^{2+} influence uranium uptake by *Rhizopus arrhizus* (Tsezos and Volesky, 1982).

CONCLUSION

Uranium is highly toxic and radioactivity metal. Uranium exposure to the human take place through direct exposure or through drinking water, contaminated with Uranium which leads to health related complications such as kidney damage, neurological disorders and many more. So, it has become more important to remove the Uranium from drinking water. Many conventional methods such as chemical precipitation, electro-dialysis, ion-exchange, reverse osmosis, phytoremediation, ultra-filtration have been applied for Uranium removal from the aqueous solution but there are some demerits of using the conventional methods include production of huge amount of sludge and metal hydroxide that blockade the membrane, more expensive and take much time for removal of metals. It has become necessary to search for new technique for removal of Uranium. So, biosorption method is applied. In this technique, Uranium is removal by biosorption by either living microorganisms (algae, fungi and yeasts) or their dead biomasses through electrostatic interactions between the metal ions in solutions and cell walls of microbial cells.

REFERENCES

Aksu, Z., Kutsal, T., Gün, S., Haciosmanoglu, N., & Gholaminejad, M. (1991). Investigation of biosorption of Cu (It), Ni (II) and Cr (VI) ions to activated sludge bacteria. *Environmental Technology, 12*(10), 915–921. doi:10.1080/09593339109385086

Aksu. (1992). The biosorption of copper (II) by C. vulgaris and Zramigera. *Environ Technol, 13*, 579-586.

Cabral, J. P. S. (1992). Selective binding of metal ions to Pseudomonas syringae cells. *Microbios, 71*, 47–53.

Ercole, V. F., Toro, L., Ficara, G., & Lepidi, A. (1994). Immobilisation of microbial cells for metal adsorption and desorption. In Mineral Bioprocessing II. Snowboard.

Fourest, E., & Roux, J. C. (1992). Heavy metal biosorption by fungal mycelial by-products: Mechanism and influence of pH. *Applied Microbiology and Biotechnology, 37*(3), 399–403. doi:10.1007/BF00211001

Francesca, L. T. (2002). Olive mill solid residues as heavy metal sorbent material: A preliminary study. *Waste Management (New York, N.Y.), 22*(8), 901–907. doi:10.1016/S0956-053X(02)00086-7 PMID:12423052

Friis, N., & Myers-Keith, P. (1986). Biosorption of uranium and lead by Streptomyces longwoodensis. *Biotechnology and Bioengineering, 28*(1), 21–28. doi:10.1002/bit.260280105 PMID:18553838

Friis, N., & Myers-Keith, P. (1986). Biosorption of uranium and lead by Streptomyces longwoodensis. *Biotechnology and Bioengineering, 28*(1), 21–28. doi:10.1002/bit.260280105 PMID:18553838

Gadd, G. M., & De Rome, L. (1988). Biosorption of copper by fungal melanine. *Applied Microbiology and Biotechnology, 29*(6), 610–617. doi:10.1007/BF00260993

Gadd, G. M., White, C., & De Rome, L. (2003). Heavy metal and radionuclide uptake by fungi and yeasts. In: Hammaini, A., Simultaneous uptake of metals by activated sludge. *Minerals Engineering, 16*(8), 723–729. doi:10.1016/S0892-6875(03)00166-3

Galun, M., Galun, E., Siegel, B. Z., Keller, P., Lehr, H., & Siegel, S. M. (1987). Removal of metal ions from aqueous solutions by Penicillium biomass: Kinetic and uptake parameters. *Water, Air, and Soil Pollution, 33*(3-4), 359–371. doi:10.1007/BF00294204

Harnessing Biotechnology for the 21st Century. (n.d.). American Chemical Society.

Huang, C. P., Huang, C. P., & Morehart, A. L. (1990). The removal of Cu (II) from dilute aqueous solution by Saccharomyces cerevisiae. *Water Research, 24*(4), 433–439. doi:10.1016/0043-1354(90)90225-U

Kuyucak, N., & Volesky, B. (1988). Biosorbents for recovery of metals from industrial solutions. *Biotechnology Letters, 10*(2), 137–142. doi:10.1007/BF01024641

Mattuschka, B., & Straube, G. (1993). Biosorption of metals by a waste biomass. *Journal of Chemical Technology and Biotechnology (Oxford, Oxfordshire)*, *58*(1), 57–63. doi:10.1002/jctb.280580108

Muraleedharan, T. R., & Venkobachar, C. (1990). Mechanism of biosorption of copper (II) by Ganoderma lucidum. *Biotechnology and Bioengineering*, *35*(3), 320–325. doi:10.1002/bit.260350314 PMID:18592525

Nageswara, L. R., & Prabhakara, G. (2011). Removal of heavy metals by biosorption– an overall review J. *Engg. Res. Sci*, *II*(IV), 17–22.

Norris, P. R., & Kelly, D. P. (1988). *Biohydrometallurgy*. Chippenham, UK: A. Rowe.

Nourbakhsh, M., Sag, Y., Ozer, D. A., Aksu, Z., Kutsal, T., & Çağlar, A. (1994). Comparative study of various biosorbents for removal of Cr (VI) ions from industrial waste waters. *Process Biochemistry*, *29*(1), 1–5. doi:10.1016/0032-9592(94)80052-9

Sakaguchi, T., & Nakajima, A. (1991). Accumulation of heavy metals such as uranium and thorium by microorganisms. In R. W. Smith & M. Misra (Eds.), *Mineral Bioprocessing*. The Minerals, Metals and Materials Society.

Smith, R. W., & Misra, M. (1991). *Mineral Bioprocessing*. The Minerals, Metals and Materials Society.

Tsezos, M. (1983). The role of chitin in uranium adsorption by R. arrhizus. *Biotechnology and Bioengineering*, *25*(8), 2025–2040. doi:10.1002/bit.260250812 PMID:18551548

Tsezos, M., & Volesky, B. (1981). Biosorption of uranium and thorium. *Biotechnology and Bioengineering*, *23*(3), 583–604. doi:10.1002/bit.260230309

Tsezos, M., & Volesky, B. (1982). The mechanism of uranium biosorption by Rhizopus arrhizus. *Biotechnology and Bioengineering*, *24*(2), 385–401. doi:10.1002/bit.260240211 PMID:18546310

Venkobachar, C. (1990). Metal removal by waste biomass to upgrade wastewater treatment plants. *Water Science and Technology*, *22*, 319–320.

Volesky, B. (1992). Removal of heavy metals by biosorption. In M. R. Ladisch & A. Bose (Eds.), *Harnessing Biotechnology for the 21st Century*. Washington, DC: American Chemical Society.

Wang, X., Wang, T., Zheng, X., Shen, Y., & Lu, X. (2017). Isotherms, thermodynamic and mechanism studies of removal of low concentration uranium (VI) by Aspergillus niger. *Water Science and Technology*, *75*(12), 2727–2736. doi:10.2166/wst.2017.055 PMID:28659512

Zhou, J. L., & Kiff, R. J. (1991). The uptake of copper from aqueous solution by immobilized fungal biomass. *Journal of Chemical Technology and Biotechnology (Oxford, Oxfordshire)*, *52*(3), 317–330. doi:10.1002/jctb.280520305

KEY TERMS AND DEFINITIONS

Activated Carbon: Charcoal that has been heated or otherwise treated to increase its adsorptive power.

Anion: A negatively charged ion which get attracted to the anode in electrolysis.

Aqueous: Containing water (example: an aqueous solution of Sodium chloride).

Biomass: Total quantity or weight of organisms in a given area or volume.

Cation: A positively charged ion, which get attracted to the cathode in electrolysis.

Cellulose: A polysaccharide; an insoluble substance; main constituent of plant cell walls.

Chitin: A fibrous substance consisting of polysaccharides which is the major constituent in the exoskeleton of arthropods and the cell walls of fungi.

Equilibrium: State in which opposing forces or influences are balanced.

Fermentation: Chemical breakdown of a substance by bacteria, yeasts, or other microorganisms.

Hydrolysis: Chemical breakdown of a compound due to reaction with water.

Mycelia: Network of hyphae that form vegetative part of a fungus.

Osmosis: A process by which molecules of a solvent tend to pass through an impermeable membrane from a less concentrated solution into a more concentrated one.

Phytoremediation: Use of living plants to remove hazardous chemicals and heavy metals from air, water, and soil.

Polysaccharide: A type of carbohydrate (e.g., starch, cellulose, or glycogen).

Precipitation: Action or process of precipitating a substance from a solution.

Chapter 5
Demethylation of Arsenic and Nickel From Tannery Effluent

Shankar Mukundrao Khade
Indian Institute of Technology (BHU), India

Suresh Kumar Srivastava
Indian Institute of Technology (BHU), India

ABSTRACT

The tanning Industry is considered to be a major source of pollution and tannery wastewater in particular is a potential environmental concern. Wastewaters produced by tanneries contain high concentration of inorganic and organic pollutant. The pollutants of concern within the tanning industry include Azodyes, Cadmium compounds, Cobalt, Copper, Antimony, Barium, Lead, Selenium, Mercury, Zinc, Arsenic, Polychlorinated Biphyenls (PCB), Nickel, Formaldehyde resins, and Pesticides residues. Because tannery wastewater contains a complexity of pollutants including chromium and chlorinated phenols as indicated earlier, it is vital to dissect the toxic nature of such wastewater both to understand its environmental impacts and identify potential remediation strategies. Furthermore, there are strict regulations imposed for the environmental control of pollutants such as heavy metals and persistent organic pollutants. In this chapter, the authors discuss the demethylation process of Arsenic and Nickel from Tannery Effluent.

DOI: 10.4018/978-1-5225-4162-2.ch005

Copyright © 2018, IGI Global. Copying or distributing in print or electronic forms without written permission of IGI Global is prohibited.

INTRODUCTION

The setback of environmental pollution on report of vital industrial growth in practical requisites is the problem of disposal of industrial wastes in the form of solid, liquid, or gas. These forms are responsible for ultimately polluting water. The polluted water indeed affects the soil of industrial areas along with the fertile agricultural soils. In addition to these effects, secondary pollution is also prominent by mixing with the ground water. Industries are the one of the main cause of environmental pollution. The major of quantity and type of pollutants released into the environment depends on the type of industry. The effluent discharged from the industries cause severe metal pollution in the environmental. Most of the metals released into the environment cause severe risk to the human and animal's health. Many metals including arsenic, copper, mercury, chromium, cadmium, aluminium, nickel, zinc, and uranium affects the environment. These metals impose severe concerns to the organisms due to their persistence, biomagnification and bioaccumulation in food chain. So to avoid these effects, prior treatment of industrial waste is necessary. The exposure to these metals affects the health of an organism by retarding the development and growth, organ damage, neurological damages, cancer and at higher concentration may leads to death. The exposure to mercury leads to minamata syndrome. In addition to this, mercury and lead causes autoimmune disease along with kidneys, circulatory system (Barregård et al., 1999; Barsoum, 2006; Soderland et al., 2010).

The release of industrial effluents containing soluble or insoluble compounds into the ground water, streams, ponds, rivers, and oceans leads to cause primary and secondary water pollution (Akcil & Koldas, 2006; Gupta et al., 2012). However, the released pollutants are either removed or transformed to the less harmful compound by microbial approach termed as 'bioremediation'. Biotransformation is an enzymatic process which includes oxidation, reduction, methylation or demethylation of highly toxic pollutants. The microbes involved in maintaining the biogeochemical cycle of arsenic by demethylation process include *Candida humicola, Mycobacterium neoaurum, Bacillus sp.* etc. Most of the water pollution in the developing countries occurs because of the poor management of wastes generated in huge quantity due to many anthropogenic activities (Giusti, 2009; Goel, 2006).

ENVIRONMENTAL POLLUTION AND ITS EFFECTS

Globally environmental pollution is now days a big challenge, due to the usage of huge amount of chemicals and discharge of untreated industrial effluents into the water bodies. Heavy metals which are used in the various industries or produced as by-products like mining industries becoming the major challenge due to their persistent nature in the environment (Duruibe et al., 2007; Khan et al., 2008).

Water Pollution and Toxic Heavy Metals

Effluents from chromium-based industries are mostly becoming source of pollution of ground water and surface water creating the scarcity of safe drinking water in most of the countries (Gadgil, 1998; Urbansky & Schock, 1999). In the developing countries like India, most of the rural and urban areas utilize ground water as the first priority. However, since 2-3 decades, the safety of drinking water has been affected due to the untreated discharge of wastewater containing toxic heavy metals from the industries (Chatterjee et al., 1995; Chowdhury et al., 2000). Globally, effluents from tannery and metal plating industries containing chromium metal pose severe cause of pollution of water (Akbal & Camcı, 2011; Owlad et al., 2009). Wastewaters containing heavy metals are hazardous to the nature which indeed affects organism's habitats. Bioremediation treatment of wastewaters by using microorganism or in consortium can be applied to remove the hazardous metals. Bioremediation approach can be used without affecting environment over the conventional process. Bioremediation of chromium is vital, because it is released from many industries and also it is present in various forms of oxidation states. Cr (III) is naturally available in the environment which acts as a vital nutrient element to organisms; however Cr (VI) acts as a carcinogen to humans in addition to kidney and liver damages. So to prevent these effects, Cr (VI) must be significantly converted to Cr (III) (Lytle et al., 1998; Zhitkovich, 2005).

Leather Tanning Industry

Tanning industry in India is among the fully flourished industries. Even though the developed sector of tanning industries, tannery effluents are released at a high rate among the pollutants from the other industrial wastes (Belay, 2010; Bosnic et al., 2000; Rajaram & Das, 2008).

Tanning is the chemically driven process or technology in which partly soluble collagen protein is transformed to tough, flexible, insoluble leather, requiring large amount of water (Andualem, 2014; Muralidhara et al., 1982). Here animal skins are used to make leather because of their ability to absorb tannic acid and chemical compounds that avoids biological decaying, make withstand in wet. Mostly Cr and extracted tree bark vegetable tannins are essentially used as tanning agents. Most of the tanneries use Cr (III) as tanning agent, but it also get converted into carcinogenic Cr (VI) after reacting with the oxygen at high temperatures. Further Cr (VI) adversely affects the environment as well as living organisms.

The leather processing follows the following stages as represented in flow chart of Figure 1 (Rivela et al., 2004).

During the various processing stages of leather, many chemicals are being used like sodium sulphide, calcium oxide, sulphuric acid, formic acid, aldehyde, sulphur, acid dye, neutralizing agents, formalin, chromium etc. These chemicals after the processing of leather are being released into the river, pond, or open environment. Hence tannery industries are considered to be one of the highly pollution creating sector affecting the environment rigourously. Large amount of chromium is released with the effluent during the chrome tanning process.Around 80 o 90% of leather processing follows the chrome tanning process. In India leather processing is the highest pollution creating sector among the all other industries. In India, the permissible discharge of chromium in water bodies is 2mg/l, but the tanneries are releasing 2000 to 5000 mg/l affecting the environment adversly (Mandal et al., 2011; Yayintaş et al., 2007).

Figure 1. Stages of leather processing

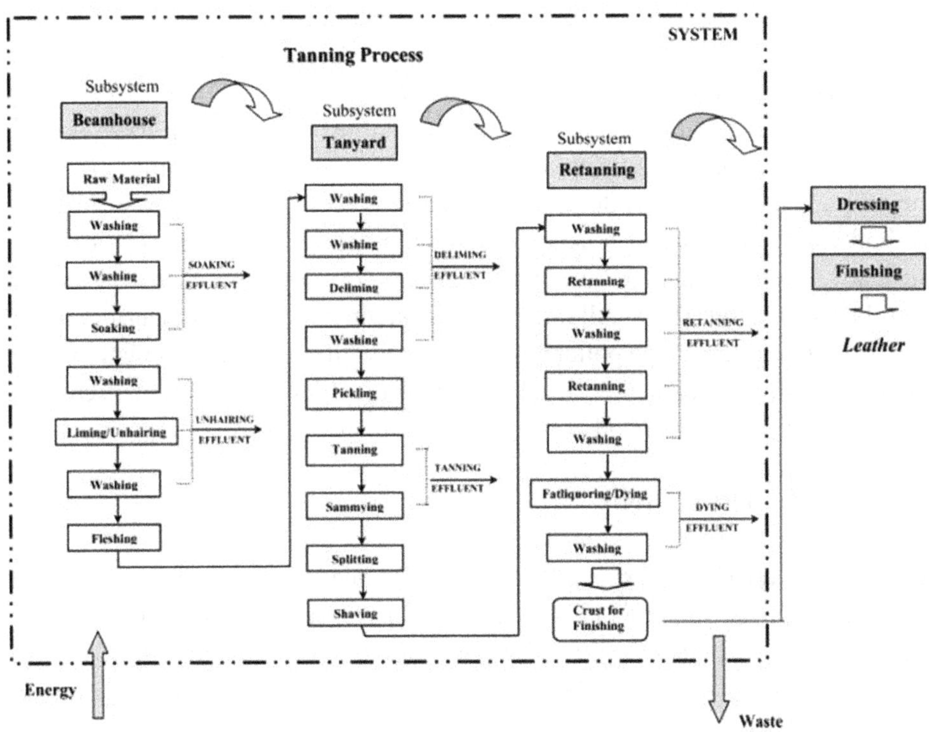

BIOREMEDIATION AND PHYTOREMEDIATION MECHANISMS

Plants based removal of hazardous metals from the polluted sites is defined as phytoremediation. Biological means of removal or transformation of toxic chemicals is defined as bioremediation which significantly removes the pollutants than costlier chemical or physical technology (Bamforth & Singleton, 2005; Robinson et al., 2001). Mainly *Brassicaceae* family is involved in the removal of metals. Among the family, *Brassica juncea* has significant ability to grow at the highly metal contaminated site and remove them by accumulating in the root and stem tissues.

Bioremediation Mechanism

The significant removal of heavy metals from the sewage or industrial wastes can be achieved using the microbes. Microbes have ability to scavenge the metals or transform into less toxic forms at the polluted site without any adverse effects. It has been reported that *Pseudomonas aeruginosa* is actively participates in the reduction of hexavalent chromium Cr (VI) which is usually carcinogenic to the organisms (Ganguli & Tripathi, 2002). The chromate reductase enzyme is responsible for the reduction of carcinogenic chromium. In this method, reduction of chromium is achieved in chrome-plating effluent in batch culture by entrapping the cells in dialysis sac as shown in Figure 2 (Ganguli & Tripathi, 2002).

The bioremediation of hexavalent chromium detoxification using *Aspergillus niger* facilitates the removal of toxic chromium from the contaminated site (Srivastava & Thakur, 2006). In addition to these microbes, *Bacillus, Enterobacter, Arthrobacter Escherichia, Pseudomonas*, yeasts and fungi are efficiently transforms the hexavalent chromium into nontoxic trivalent chromium. Bioremediation process is efficient at the optimum conditions such as microbial population, temperature, pH, oxygen, nutrients. At some point of polluted area, nutrients are externally added to enrich the microbial population. This is called as 'Biostimulation' or at some moment, one needs to add microbial flora at the polluted site to increase the microbial population. This is referred as 'Bioaugmentation'. In addition, with the chromium, effluent discharged from the tanneries contains other metals including nickel, arsenic, copper, etc.

Case Study: Nickel (Ni)

According to WHO the permissible nickel (Ni) concentration in water is 0.02 to 0.1mg/l. The nickel concentration in Challawa River has observed to be in huge amount and considered as highly polluted with nickel. This was due to the discharge of tannery effluents (J.C et al., 2009). Nickel in small amount is essentially required for various metabolic activities of human, but however in large quantity affects in

Figure 2. Bioremediation in Dialysis bioreactor

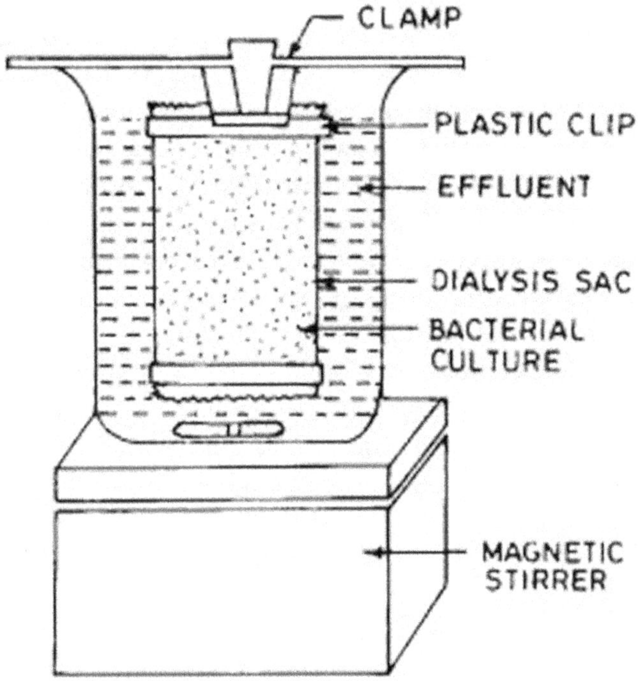

many ways. Nickel at higher concentrations causes carcinogenic, hypersensitive, teratogenic, respiratory, asthma and heart defects (Li et al., 2015). Higher nickel concentration accumulates in plants and affects the growth. Also nickel in water bodies gets accumulated in algae inhibiting their further growth and development.

Case Study: Arsenic (As)

Arsenic is commonly present in nature in the form of As^{-3}, As^{+3}, and As^{+5} oxidation states. Arsenic contamination in the drinking water is globally a common problem affecting the organism health. India and Bangladesh are frequently getting As pollution as discharge of wastes into the environment without any prior treatment (Alam et al., 2002). Arsenic is one of the metal disposed from the tannery wastes causing pollution in the environment. At the tannery industry, arsenic is used as a pesticide in the form of sodium arsenite. In India, heavy contamination of arsenic in drinking water was reported in West Bengal in1983, when spotted with people suffering from arsenicosis (Saha, 2003). Prolonged consumption of food or water leads to pigmentation, keratosis and in some cases causes cancer (Mazumder et al., 2010).

Demethylation of Ni and As

The metals are essentially required for metabolic activities of organisms, but in little quantity. Most of the metabolic enzymes require metal as a cofactor for their normal reactions. The higher concentrations of any metal results into toxicity to that organism. So in order to withstand or show resistance at higher concentration of metals, eukaryotic organisms render intracellular complexation of the hazardous metals (Stavreva Veselinovska, 2011). This would leads to resist the higher concentration of toxic metals. Due to these mechanisms, toxicity to that organism may be minimized. However, the prokaryotes are having specialized different defense mechanisms to the higher concentration of toxic metal ions. The toxic metals are enzymatically transformed by oxidation, reduction, addition of methyl group (methylation) or removal of methyl group (demethylation) (Williams & Silver, 1984). This would result into survival even at higher concentration of metals. The demethylation of arsenic can be understood by following reaction mechanism:

$CH_3AsO(OH)_2 + 1.5\ O_2 \leftrightarrow H_3AsO_3 + CO_2 + H_2O$ (Davis et al., 1994)

The relatively mobilization of arsenic by bioremediation using plants and microbes is represented in the following Figure 3 (Singh et al., 2014).

Figure 3. Mobilization of arsenic by bioremediation using plants and microbes

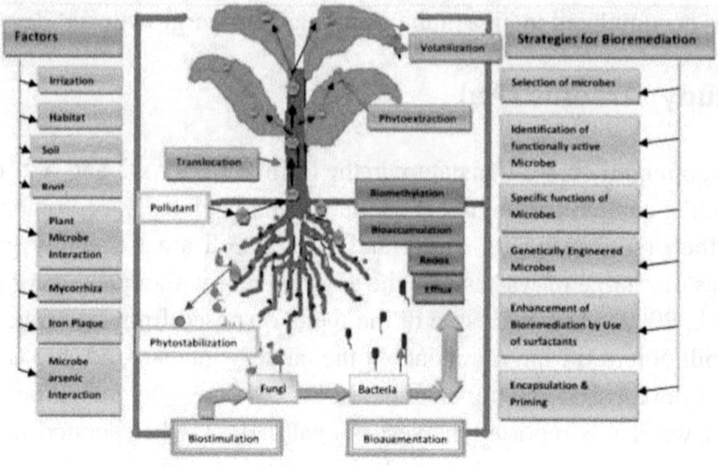

DISADVANTAGES OF BIOREMEDIATION BY DEMETHYLATION

In contrast to the advantages of bioremediation of metal removal by transformation, there are several disadvantages by this approach mentioned as:

- The transformation of metals using microbial approach indeed needs to add microbial community at the site which again increase the microbial load.
- Even though the removal of metallic pollution by bioremediation is efficient, the process is slow.
- For bioremediation the physical parameters like pH, temperature need to be optimum according to the microbial activity.

CONCLUSION

The effluent discharged from the tannery industries is responsible for polluting the environment. It consists of acids, bases, or heavy metals contaminate the water bodies drastically. Arsenic is used as pesticide in the form of sodium arsenate which is further discharged along with the waste responsible for pollution. The nickel and arsenic affects the organism in several ways by causing cancer, birth defects, asthma, heart defects etc. The relative use of arsenic in tannery industry should be replaced with other pesticide in order to control arsenic pollution. Further removal of metal pollution can be achieved by means of cost effective bioremediation technology instead of the conventional physical and chemical methods. The microbial bioremediation is essentially effective due to the presence of transforming ability of metals either by oxidation, reduction, methylation and demethylation approach.

REFERENCES

Akbal, F., & Camcı, S. (2011). Copper, chromium and nickel removal from metal plating wastewater by electrocoagulation. *Desalination*, *269*(1), 214–222. doi:10.1016/j.desal.2010.11.001

Akcil, A., & Koldas, S. (2006). Acid Mine Drainage (AMD): Causes, treatment and case studies. *Journal of Cleaner Production*, *14*(12), 1139–1145. doi:10.1016/j.jclepro.2004.09.006

Alam, M., Allinson, G., Stagnitti, F., Tanaka, A., & Westbrooke, M. (2002). Arsenic contamination in Bangladesh groundwater: A major environmental and social disaster. *International Journal of Environmental Health Research*, *12*(3), 235–253. doi:10.1080/09603120210000000998 PMID:12396524

Bamforth, S. M., & Singleton, I. (2005). Bioremediation of polycyclic aromatic hydrocarbons: Current knowledge and future directions. *Journal of Chemical Technology and Biotechnology (Oxford, Oxfordshire)*, *80*(7), 723–736. doi:10.1002/jctb.1276

Barregård, L., Svalander, C., Schütz, A., Westberg, G., Sällsten, G., Blohmé, I., ... Haglind, P. (1999). Cadmium, mercury, and lead in kidney cortex of the general Swedish population: A study of biopsies from living kidney donors. *Environmental Health Perspectives*, *107*(11), 867–871. doi:10.1289/ehp.99107867 PMID:10544153

Barsoum, R. S. (2006). Chronic kidney disease in the developing world. *The New England Journal of Medicine*, *354*(10), 997–999. doi:10.1056/NEJMp058318 PMID:16525136

Belay, A. A. (2010). Impacts of chromium from tannery effluent and evaluation of alternative treatment options. *Journal of Environmental Protection*, *1*(01), 53–58. doi:10.4236/jep.2010.11007

Bosnic, M., Buljan, J., & Daniels, R. (2000). Pollutants in tannery effluents. *UNIDO, Vienna, Rev*.

Chatterjee, A., Das, D., Mandal, B. K., Chowdhury, T. R., Samanta, G., & Chakraborti, D. (1995). Arsenic in ground water in six districts of West Bengal, India: The biggest arsenic calamity in the world. I: Arsenic species in drinking water and urine of the affected people. *Analyst (London)*, *120*(3), 643–650. doi:10.1039/an9952000643

Chowdhury, U. K., Biswas, B. K., Chowdhury, T. R., Samanta, G., Mandal, B. K., Basu, G. C., ... Mukherjee, S. K. (2000). Groundwater arsenic contamination in Bangladesh and West Bengal, India. *Environmental Health Perspectives*, *108*(5), 393–397. doi:10.1289/ehp.00108393 PMID:10811564

Davis, A., Kempton, J. H., Nicholson, A., & Yare, B. (1994). Groundwater transport of arsenic and chromium at a historical tannery, Woburn, Massachusetts, USA. *Applied Geochemistry*, *9*(5), 569–582. doi:10.1016/0883-2927(94)90019-1

Duruibe, J., Ogwuegbu, M., & Egwurugwu, J. (2007). Heavy metal pollution and human biotoxic effects. *International Journal of Physical Sciences*, *2*(5), 112–118.

Gadgil, A. (1998). Drinking water in developing countries. *Annual Review of Energy and the Environment*, *23*(1), 253–286. doi:10.1146/annurev.energy.23.1.253

Ganguli, A., & Tripathi, A. (2002). Bioremediation of toxic chromium from electroplating effluent by chromate-reducing Pseudomonas aeruginosa A2Chr in two bioreactors. *Applied Microbiology and Biotechnology*, *58*(3), 416–420. doi:10.1007/s00253-001-0871-x PMID:11935196

Giusti, L. (2009). A review of waste management practices and their impact on human health. *Waste Management (New York, N.Y.)*, *29*(8), 2227–2239. doi:10.1016/j.wasman.2009.03.028 PMID:19401266

Goel, P. (2006). *Water pollution: causes, effects and control*. New Age International.

Gupta, V. K., Ali, I., Saleh, T. A., Nayak, A., & Agarwal, S. (2012). Chemical treatment technologies for waste-water recycling—an overview. *RSC Advances*, *2*(16), 6380–6388. doi:10.1039/c2ra20340e

Khan, S., Cao, Q., Zheng, Y., Huang, Y., & Zhu, Y. (2008). Health risks of heavy metals in contaminated soils and food crops irrigated with wastewater in Beijing, China. *Environmental Pollution*, *152*(3), 686–692. doi:10.1016/j.envpol.2007.06.056 PMID:17720286

Li, Q., Liu, H., Alattar, M., Jiang, S., Han, J., Ma, Y., & Jiang, C. (2015). The preferential accumulation of heavy metals in different tissues following frequent respiratory exposure to PM2. 5 in rats. *Scientific Reports*, 5. PMID:26582271

Lytle, C. M., Lytle, F. W., Yang, N., Qian, J.-H., Hansen, D., Zayed, A., & Terry, N. (1998). Reduction of Cr (VI) to Cr (III) by wetland plants: Potential for in situ heavy metal detoxification. *Environmental Science & Technology*, *32*(20), 3087–3093. doi:10.1021/es980089x

Mandal, B. K., Vankayala, R., & Uday Kumar, L. (2011). Speciation of chromium in soil and sludge in the surrounding tannery region, Ranipet, Tamil Nadu. *ISRN Toxicology*. PMID:23724287

Mazumder, D. N. G., Ghosh, A., Majumdar, K. K., Ghosh, N., Saha, C., & Mazumder, R. N. G. (2010). Arsenic Contamination of Ground Water and its Health Impact on Population of District of Nadia, West Bengal, India. *Indian Journal of Community Medicine: Official Publication of Indian Association of Preventive & Social Medicine*, *35*(2), 331–338. doi:10.4103/0970-0218.66897 PMID:20922118

Muralidhara, H., Maggin, B., & Phipps, H. Jr. (1982). Conversion of tannery waste to useful products. *Resources and Conservation*, *8*(1), 43–59. doi:10.1016/0166-3097(82)90052-9

Owlad, M., Aroua, M. K., Daud, W. A. W., & Baroutian, S. (2009). Removal of hexavalent chromium-contaminated water and wastewater: A review. *Water, Air, and Soil Pollution*, *200*(1-4), 59–77. doi:10.1007/s11270-008-9893-7

Rajaram, T., & Das, A. (2008). Water pollution by industrial effluents in India: Discharge scenarios and case for participatory ecosystem specific local regulation. *Futures*, *40*(1), 56–69. doi:10.1016/j.futures.2007.06.002

Rivela, B., Moreira, M., Bornhardt, C., Méndez, R., & Feijoo, G. (2004). Life cycle assessment as a tool for the environmental improvement of the tannery industry in developing countries. *Environmental Science & Technology*, *38*(6), 1901–1909. doi:10.1021/es034316t PMID:15074705

Robinson, T., McMullan, G., Marchant, R., & Nigam, P. (2001). Remediation of dyes in textile effluent: A critical review on current treatment technologies with a proposed alternative. *Bioresource Technology*, *77*(3), 247–255. doi:10.1016/S0960-8524(00)00080-8 PMID:11272011

Saha, K. (2003). Review of arsenicosis in west Bengal, India—a clinical perspective. *Critical Reviews in Environmental Science and Technology*, *33*(2), 127–163. doi:10.1080/10643380390814514

Singh, N., Srivastava, P.K., Tripathi, R.D., Srivastava, S., & Vaish, A. (2014). Microbial in-situ mitigation of arsenic contamination in plants and soils. *In-Situ Remediation of Arsenic-Contaminated Sites*, 115.

Soderland, P., Lovekar, S., Weiner, D. E., Brooks, D. R., & Kaufman, J. S. (2010). Chronic kidney disease associated with environmental toxins and exposures. *Advances in Chronic Kidney Disease*, *17*(3), 254–264. doi:10.1053/j.ackd.2010.03.011 PMID:20439094

Srivastava, S., & Thakur, I. S. (2006). Evaluation of bioremediation and detoxification potentiality of Aspergillus niger for removal of hexavalent chromium in soil microcosm. *Soil Biology & Biochemistry*, *38*(7), 1904–1911. doi:10.1016/j.soilbio.2005.12.016

Stavreva Veselinovska, S. (2011). Microorganisms-indicators of the level of soil pollution with lead. *1st National Agriculture Congress and Exposition on behalf of Ali Numan Kıraç with International Participation Proceedings*, 279-287.

Urbansky, E. T., & Schock, M. (1999). Issues in managing the risks associated with perchlorate in drinking water. *Journal of Environmental Management*, *56*(2), 79–95. doi:10.1006/jema.1999.0274

Williams, J. W., & Silver, S. (1984). Bacterial resistance and detoxification of heavy metals. *Enzyme and Microbial Technology*, *6*(12), 530–537. doi:10.1016/0141-0229(84)90081-4

Yayintaş, Ö. T., Yilmaz, S., Türkoĝlu, M., Şolakoĝlu, F. A., & Şakir, F. (2007). Seasonal variation of some heavy metal pollution with environmental and microbiological parameters in sub-basin of Kocabas Stream (Biga, Canakkale, Turkey) by ICP-AES. *Environmental Monitoring and Assessment*, *134*(1), 321–331. doi:10.1007/s10661-007-9623-4 PMID:17345011

Zhitkovich, A. (2005). Importance of chromium– DNA adducts in mutagenicity and toxicity of chromium (VI). *Chemical Research in Toxicology*, *18*(1), 3–11. doi:10.1021/tx049774+ PMID:15651842

Chapter 6 Role of Plant Growth Promoting Bacteria (PGPB) for Bioremediation of Heavy Metals: An Overview

Khushboo Chaudhary
Banasthali University, India

Suphiya Khan
Banasthali University, India

ABSTRACT

The heavy metal pollution problem is all over the world. Plant growth promoting bacteria (PGPB) has transformed heavy metals present in the soil, which removes and minimizes their toxic effects. This chapter highlights the role of PGPB for remediation of heavy metals, their mechanism of action, and their applications approach of hyperaccumulation. Further, it also highlights the role of uptake and detoxification of metals by cellular mechanisms which facilitate the bioremediation of heavy metals from contaminated areas. Bacteria may also enhance nutrient uptake, increasing plant growth and defenses while diminish heavy metals intake and their toxic effects. Therefore, this chapter focuses on the mechanisms by which microorganisms can mobilize or immobilize metals in soils and the bioremediation strategies are addressed for the improvement of phytoextraction as an innovative process for enhancement of heavy metals removal from soil.

DOI: 10.4018/978-1-5225-4162-2.ch006

Copyright © 2018, IGI Global. Copying or distributing in print or electronic forms without written permission of IGI Global is prohibited.

INTRODUCTION

The heavy metal pollution problem is all over the world. The chemicals released into the soil in the form of cadmium (Cd), cobalt (Co), copper (Cu), lead (Pb), zinc (Zn), mercury (Hg), chromium (Cr), antimony (Sb) and nickel (Ni) elements. Some of these elements are essential for many physiological functions in living beings while others have no such function in biological function in required level (Fassler et al., 2010; Dubey et al., 2014). These elements are commonly found in fungicides, animal waste, fertilizers and sewage sludge in soil. Deposition of these industrial wastes can increase the concentration of the elements at toxic level (Fassler et al., 2010). The use of a lot of fertilizers has been released heavy metals which dramatically affects to the agriculture (Saba et al., 2015). Lead (Pb) is the most toxic heavy metal found in several fertilizers, which is further translocated into plants and easily enter into the food chain. Heavy metal contaminants movement from soil to the water (ground level) is very slow because of the less mobility of them and it is the major cause of their absorption by plants (Rodriguez et al., 2011; Lori et al., 2015).

Plants under heavy metal stress produce reactive oxygen species (ROS). ROS includes hydrogen peroxide (H_2O_2), superoxide radicals (O_2), hydroxyl radicals (OH) and catalase which result in damages to plant cell or tissue (Migocka et al., 2014; Wang et al., 2015,). ROS produced continuously in different compartments of the plant with a number of antioxidant molecule as by-products (Kwankua et al., 2012; Palma et al., 2013; Dubey et al., 2014). However, the critical imbalance and an overload amount of production of ROS and antioxidant molecule in plant creates disorder and effect the plant enzymatic activity (Gratao et al., 2005; Qiao et al., 2015). There are number of remediation technologies are used such as landfill, excavation, thermal treatment, recovery by means of electricity and leaching of acids but those are not suitable due to their high cost and very less successful for specific metal contamination site. Therefore, it is important to develop economically practical and more effective method to decontaminate soils from heavy metal contamination.

The advantage of phytoremediation, it is a low-cost technology (Abdul and Schroder, 2009). It is about 1000-fold cheaper in compare to other conventional methods such as (flotation-filtration, evaporation, ion exchange, electrodialysis, and ultrafiltration). For example, to clean about 1 acre of sandy soil (55 cm depth) will cost around 60,000-100,000 \$ as compare to 400,000 \$ for conventional methods (Ali et al., 2013). Phytoremediation technique involves several techniques such as Phyto filtration, rhizofiltration, phytoextraction, phytostabilization, Phyto immobilization and phytodegradation, and rhizodegradation (Figure 1) (Abdul and Schroder, 2009; Ali et al., 2013). Phytoextraction is the most efficient and useful technique among them but it is much more difficult than others. Hyperaccumulator plants are those which uptake and tolerate about hundred times greater contaminants

Figure 1. Diagramatic representation of phytoremediation technique

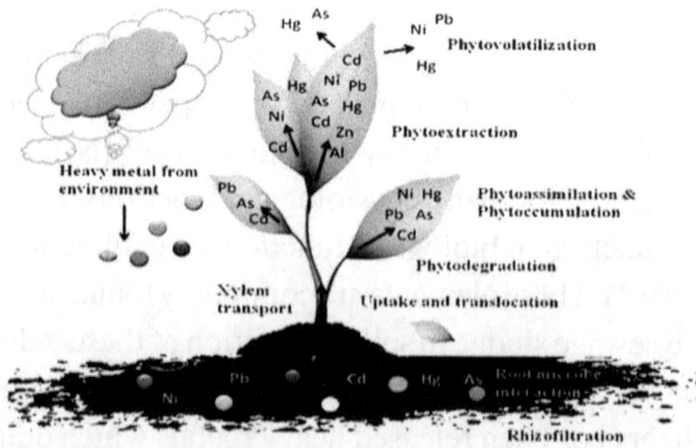

than normal plants without any visible symptoms (Memon and Schroder, 2009). Some different 450-500 plants have been identified as hyperaccumulator include *Thlaspi caerulescens* that accumulate (Pb, Ni, Cd and Zn), *Arabidopsis halleri* that can accumulate high levels of heavy metals (Cd and Zn but not Pb), *Alyssum bertolonii* can uptake (Ni and Co) (Lodewyckx et al., 2002; Padmavathiamma and Li 2007; Maestri et al., 2010; Park et al., 2011; Zhang et al., 2011). Bacteria have ability to tolerance of metals using different mechanism which maintain metal homeostasis with keeping concentration of essential metals (Nies, 2003). It involves both actively and passively for metal uptake and remaining. Soil has number of microbial inhabitants such as fungi, protozoa, and algae, with the most prominent being bacteria (Curl and Truelove, 1986). The rhizosphere is enormous with some representative genera *Agrobacterium, Asorhizobium, Achromobacter, Bacillus, Burkholderia, Pseudomonas and Cytophaga*, (Campbell and Greaves, 1990; Glick et al., 1999). Bacteria that give benefits to plants may be symbiotic bacteria, which are closely associated with plants such as *Rhizobia*.

Under heavy metals toxicity bacteria produced siderophores for protection, enhancing plant growth for providing iron at sufficient amount and pass them to rise above the toxic condition. Another factor for plant growth is ACC deaminase has been studied in respect to heavy metals stress, produced ethylene as an intermediate in same favored reaction. Ali et al. (2013) reported that ACC deaminase can produce by bacteria at lower levels of ethylene in plants for promoting plant growth. Both plants and bacteria produce the auxin IAA, which increasing plant biomass and metal accumulation (Sing et al., 2014). Cr-resistant bacteria were enhancing plant growth by re-inoculation and less accumulation of Cr compare to un-inoculated sunflower plants (Faisal and Hasnain, 2005). Plant growth promoting rhizobacteria play role to stimulate root growth, increased soil nutrients for obtainable to the plant, fix nitrogen, improve soil fertility, suppress soil-borne pathogens, hormone production, nodule formation, nutrients uptake, siderophore production N-fixation and protect from different diseases to important role in crop yield under various environmental stresses like heavy metals, drought, salinity and temperature (Naik et al., 2012). Bacteria have ability to produce some enzymatic activities which allowing to covert hazardous compounds to hamless froms to survive extreme heat, cold, and heavy metals stress conditions (Ali et al., 2013; Singh et al., 2014). Nickel (Ni)-resistant bacteria have been found in the rhizosphere of the N-hyperaccumulator Thlaspi goesingense and Alyssum serpyllifolium subsp. (Idris et al., 2004; Becerra-Castro et al., 2009). Bacterial species *Arthrobacter spp.* and *Streptomyces spp.* were dominant and could tolerate as much as 10 mM Ni in plate assays. *Pseudomonas aeruginosa* strain can tolerate up to 0.6 mM lead nitrate (Naik et al., 2012). The purpose of this chapter is to examine how microorganisms may be assisting phytoextraction and discuss the outcomes of microbial phytoextraction assess broadly The microbial metal mobilization and plant growth promoting bacteria (PGPB) are improve phytoextraction, and highlight the role of specific microbial processes in improving phytoextraction in the meta-analysis.

PHYTOREMEDIATION TECHNIQUES

Phytoremediation is a technology in which plants are used to remediate contaminated soil and water; it is a low-cost effective technique (Salla et al., 2011; Yao et al., 2012; Dixit et al., 2015. Phytoremediation involves a number of biological mechanisms including direct uptake, the release of exudates into the rhizosphere (to enhance bacterial and fungal processes), and metabolic processes within the root and shoots cells (Singh et al., 2003). Selected plants are grown on the contaminated site, where they draw up pollutants and concentrate them within various tissues. The plants are then harvested and may be further treated by burning them in a controlled system. The residue of the plants would then be very rich in metals and could be recycled or placed in landfills (Ma et al., 2011). Designing a phytoremediation system varies according to a wide range of factors including the contaminant conditions at the site, the level of clean up required and the plant species to be used. All these factors need to be considered for effective phytoremediation (John et al., 2009).

Phytoremediation is a broad term which involves several different techniques such as, Phyto filtration, phytovolatilization, phytodegradation, phytostabilization and phytoextraction (Figure 1) (Ali et al., 2013). Phytofiltration is used as a primary term for phytoremediation technique. It is mainly based on the adsorption of heavy metal contaminants from soil or water with the help of plant organs such as root, shoot, and leaves (Mukhopadhyay and Maiti, 2010). The second form of phytoremediation involves phytovolatilization, in this process, there is the conversion of the toxic pollutant compounds into a volatile form which makes them direct release from the soil into the atmosphere (Prasad and Freitas, 2003). Next one is phytodegradation, in this process, the toxic metal pollutant is degraded into the small particles and which can easily uptake by the roots (Dixit et al., 2015). Phytostabilization term is used when metal pollutants are restricted to the particular soil zone mainly near the roots (Figure 1) (Salt et al., 1998; Yao et al., 2012). One another technique is phytoextraction, it is very useful process than others. In this technique contaminant in soils which uptake large amounts through the roots and translocation into the aerial parts of the plant. Nature has not a tendency that all plant accumulates in equal proportion. Some plants have accumulated in large quantity and these hyperaccumulator plants are the base of phytoremediation technology.

HYPERACCUMULATOR PLANTS: A BASE FOR PHYTOREMEDIATION TECHNOLOGY

Hyperaccumulator plants are those which uptake and tolerate more metal ion contaminants without any visible symptoms (Memon and Schroder, 2009). Some 450-500 different plants have been identified as hyperaccumulator include *Thlaspi caerulescens* that accumulate (Pb, Ni, Cd and Zn), *Arabidopsis halleri* accumulate (Cd and Zn but not Pb), *Alyssum bertolonii* can uptake (Ni and Co) and some other plants which belong to different family can also participate to accumulate heavy metals such as *Caryophyllaceae, Fabaceae, Poaceae, Lamiaceae, Asteraceae, Cunoniaceae and Cyperaceae* and many others plants (Maestri et al., 2010).

Plants have specific properties that give us some specific advantages to remediate environment (Meagher, 2000; Meagher et al., 2000). Plants absorb metal particle through roots and root hairs that generate forces (cohesive and adhesive) through which pollutants can be extracted from contaminated soil and water. Plants are autotrophs, it takes up nutrients directly from the environment in gaseous form with the help of photosynthesis process. Different crop plants like *Ipomea alpine, Centella asiatica, Eichlarnia crassipes, Euphorbia maeleocloela, Berkheya coddii, Alyssum and Thlaspi, Zea mays, Pteris vittata, Arabidopsis bisulcatus, Sesbaniia drumnondia, Sedum alfredei, Euphorbia marcoleida Phragnites australis Phytolacca Americana, Astragolosus, craenosus, Cardanine hupingshas and Iberis intermedia* reported as hyperaccumulators for phytoextraction of particularly Cu, Ni, Cd, Zn, Cr, As, Mg, Se and Ti in their shoots (Sharma et al., 2003; Meers et al., 2005; Dong, 2005; Israr and Sahi, 2006 Colheiros et al., 2008; Jin et al., 2008; Pollard et al., 2009; Bani et al., 2010; Mehdami et al., 2011; Mokhtar et al., 2011; Namatian & Kazemein 2013; Tong et al., 2014).

The five willow tree species *Salix fragilis* (Belgisch Rood), *Salix schwerinii* (Christina) *Salix purpurea–Salix daphnoides* (Bleu), *Salix triandra* (Noir de villaines), *and Salix dasyclados* (Loden) accumulate Zn and Cd at a higher concentration in their shoots (Meers et al., 2007). Recently a plant named *Lolium multiflorum* is also found which is used for phytoremediation of Mn, Cu, Pb and Zn (Mugica-Alvarez et al., 2015). The aquatic plant *Hydrilla verticillata* also have a great capacity of heavy metal uptake and also consider as a good accumulator for Cr and Cd from ponds (Phunkan et al., 2015). *Sauropus androgynus* is a perennial plant that has the ability to phytoremeate soil and this plant can be used commercially due to its high capacity of toxic metal accumulation (Xia et al., 2013). One other plant named *Pannicum virgatum* (Switchgrass) reported as a heavy metal accumulator for Zn, Cr, and Cd (Chen et al., 2012).

HEAVY METAL TOXICITY EFFECT ON PLANT NUTRIENTS

Some heavy metals such as Cu, Zn, Al, Cd, Ni, As and Hg has negative effects on macronutrient and micronutrients (Mariano and Keltjens, 2005; Baligar et al., 1993). Calcium (Ca) ions worked as a "secondary messenger in plant signaling and uptake from roots and translocated into meristematic zones of young tissues. Ca is unable to recycle after deposited in the leaves (Hanger, 1979). Heavy metals bind to all Ca binding sites on the surface of the cell, at low pH (4.5) it work with Ca absorption and uptake by roots (Roy et al., 1988, Hossain et al., 2014). The high amount of heavy metals exposes to the inhibition (Ribeiro et al., 2013) of Ca ions (100 mM (Al) (Nichol et al., 1993). Al ions interfere with the action of Guanosine 5' triphosphate (GTP) binding protein as well as inhibit Ca ion uptake by binding the specific channels (verapamils) (Rengel and Elliott, 1992). In beech plants, there is a decrease in Ca ions uptake due to the combination of high concentration of nitrogen and aluminum (Bengtsson et al., 1994).

Potassium (K) channel influx inhibits due to the toxicity of heavy metals like Al. Active pathway involvement of uptake of K is also inhibited by high concentration of Al (Hossain et al., 2014). At the low pH (4.5) concentration of K decreases, when they are treated with heavy metals (Moustakes et al., 1995). K ions decrease in the roots cell as well as guard cell due to the Al toxicity which causes by blocking the pathway channels at cytoplasmic sites of the plasma membrane (Liu and Luan, 2001). Somewhat significantly increase K ion from barley roots under stress of Al ions (Kasai et al., 1992). The concentration of K ions increased in root and shoots of pine trees with an increase in Al concentration (Huang and Bachelard, 1993). Durum wheat is also very resistant to Al toxicity and it shows a decrease in K ion concentration with winter wheat (Zsoldos et al., 2000).

Magnesium (Mg) ions are also affected by heavy metals stress more than other nutrients present in the plant (Wheeler and Follett, 1991). Plant treated with different concentration of heavy metals in acidic medium (pH 4.5) to check the decreased level of Mg ions (Huang and Bachelard, 1993) The uptake of Mg ions through roots is much stronger than Ca ion uptake (Bose et al., 2011; Bose et al., 2013). Iron (Fe) concentration is also getting decreased at acidic pH of 4.5 when they are treated with different concentration of heavy metals and also significantly much-affected root growth (Moustakas et al., 1995). Some researchers focus on nutrition level value which really affected by heavy metals. The accumulated nitrogen concentration in roots gets decreased as a result of high exposure Al ions as well as it get translocated to aerial parts of plants (Gomes et al., 1985; Purcino et al., 2003).

Under the heavy stress of toxic metal, there is a decrease in concentration of lipoxygenase enzyme, antioxidant enzyme and ROS (Wang and Yang, 2005). Soils have a high amount of heavy metals stress than it has a low amount of phosphorus and positively correlated with aluminum (Liao et al., 2006). The concentration of metals is present in low amount than it has been increased such micronutrients and also macronutrients but if it is present more than the requirement of plant result to cause toxicity and also decreases the level of phosphorus (Cumming et al., 1986; Nichol et al., 1993). Hydrogen peroxidation level increases the plant tolerance to the heavy metals stress (Hameed et al., 2016). Nevertheless with the treatment given at 100 mM H_2O_2 for 1 day lower down the Cd stress by inducing some antioxidant enzymes such as CAT, SOD, GST, APX and GPX and increase the concentration of GSH and AsA. The activity of POD, SOD, GSH, GPX, APX, AsA, MDHAR and DHAR" content and their redox state gets increased in the response to the Al stress. Therefore H_2O_2 pretreatment in Cd and Al stress makes the plant more tolerant and inducing GSH, AsA, NPT and PCs contents as well as the GST activity in root tissues (Cho and Seo, 2005; Metwally et al., 2005; Bai et al., 2011, Kumar and Trivedi, 2016; Hameed et al., 2016). Al toxicity is also disturbed the cellular redox homeostasis and causes the oxidative burst of mitochondria and programmed cell death (PCD) in plants Figure 2 (Li and Xing, 2010).

HEAVY METAL TRANSPORTATION PATHWAY

Plants have ability to tolerate high concentration of toxic heavy metals. The uptake of heavy metals from the soil into roots, from where there are "translocated" through different transporters to aerial parts of the plant is called as phytoremediation (Figure 1) (Mahmood, 2010). This process involves to uptake of heavy metals from soil by roots cell and translocation from roots to shoots throughout the xylem channels which is known as xylem loading and the sequestration in leaves predominantly in

Figure 2. Reactive oxygen species (ROS) signaling activate under heavy metal stress

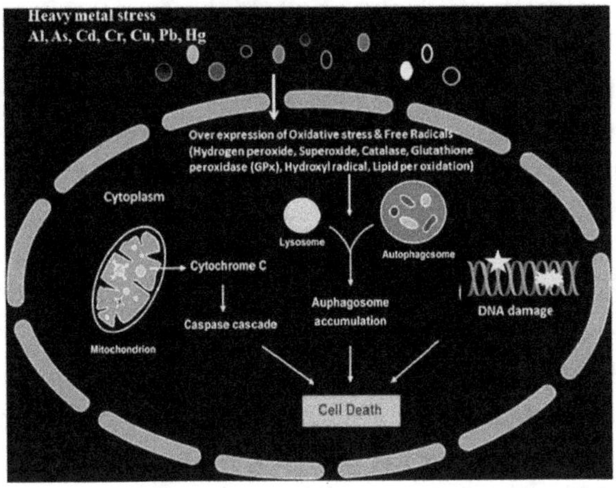

vacuoles (Bhargava et al., 2012). There are two pathways by which they can enter into the roots. First is an apoplastic pathway and the second is symplastic. The apoplastic pathway allows the soluble metals to travel without entering the cells and also through intracellular spaces. On the other hand, symplast pathway allows the movement of nonessential metals like Ni, Cd, and Pb through the cytoplasm by consuming energy (Lombi et al., 2002; Lu et al., 2009). It is transferred through symplast pathway where they are an energy consuming process that allows all non-essential heavy metals. This is due to normal transportation intracellular movement pathway block by casparian strip in the endodermis. Hence the movement of heavy metals through apoplast pathway gets blocked which means that the only way to enter xylem vessels is by taking the symplast pathway (Mahmood, 2010).

PLANT GROWTH PROMOTING BACTERIA (PGPB)

PGPR have the ability to reduce plant stress level directly by providing the plant with a compound that is made by the bacterium or facilitating uptake of nutrients. Microorganisms involve both actively and passively for metal uptake, remaining or sequestering. Soil has numerous microbial inhabitants including fungi, protozoa, and algae, with the most pronounced being bacteria. The bacterial consortium occupying the rhizosphere is enormous with some representative genera *Agrobacterium*, *Achromobacter*, *Asorhizobium*, *Bacillus*, *Burkholderia*, *Cytophaga* and *Pseudomonas*. Some of these can be beneficial to plant growth, directly or indirectly (by preventing damage to plants or providing plants with a synthesized compound or facilitating nutrient uptake). Bacteria that provide benefits to plants may be symbiotic bacteria, which are closely associated with plants and also bacteria that are free-living in the soil. This chapter highlights the role of PGPB for remediation of heavy metals, their mechanism of action, structure and their applications approach of hyperaccumulation. Further, it is also highlights the role of uptake and detoxification of metals by cellular mechanisms which facilitate the bioremediation of heavy metals from contaminated areas.

Many PGPR strains can synthesize a "phytohormone, indoleacetic acid (IAA) that acts to increase the various stages of plant growth (Glick et al., 1998). IAA is taken up by the plant, where it can encourage plant cell proliferation and cell elongation. It can also stimulate the activity of the enzyme 1-aminocycloproane-1-carboxylic acid (ACC) synthase, which is responsible for conversion of S-adenosylmethionine (SAM) to ACC, a precursor for the plant stress hormone ethylene (Glick et al., 1998). Bacterial strains that possess ACC deaminase can use ACC as a source of nitrogen, effectively lowering the outside concentration of ACC (Glick, 2003). To maintain the concentration gradient of internal and external ACC and to compensate for bacterial utilization of this compound, the plant must increase ACC exudation. When the internal ACC concentration is lowered, the downstream effect is decreased production of ethylene and lowering of the stress responses induced by this molecule in the plant. When added as a seed inoculant, the two *Pseudomonas* strains used throughout the study, UW3 and UW4 promote plant growth leading to a larger, healthier, richer rhizosphere, thus increasing the potential for remediation.

Mechanism of phytoremediation technology and metal accumulation somehow depends on the chelating molecules synthesized by bacteria. The combination of bacteria interaction with some chelating compounds in specific plants may give more benefits to the plants for better accumulation of contaminants from the soil with different process (bioaccumulation, bioleaching, biotransformation, biodegradation and biosorption) Figure 3. Inoculation of *Pseudomonas aeruginosa* was found to enhanced antioxidative enzyme such as peroxidase (180%), Catalase (83%), Superoxidase dismutase 9161%) and also the non-enzymatic components such as total phenolics (10%) and ascorbic acid (65%) compared with treated plants under Zn stress. Phytoremediation process successfully depends not only on the plants roots interaction with bacteria but also depends on the bioavailability of metals in soils. Plants roots formed some enzymatic activities to simply absorb contaminants of metals and mobilize metals ions for the bioavailable fraction so plants cannot simply absorb metals without help of enzymes produce around the roots" (Li et al., 2009). *Pseudomonas koreensis* AGB-1 strain increased SOD and CAT activities by 33% and 42% respectively (Babu et al., 2015).

Figure 3. Microbial cell-interactions with metal and process of (bioaccumulation, bioleaching, biotransformation, biodegradation and biosorption; M^{2+}- metal ions) in soil with plant root rhizosphere surface

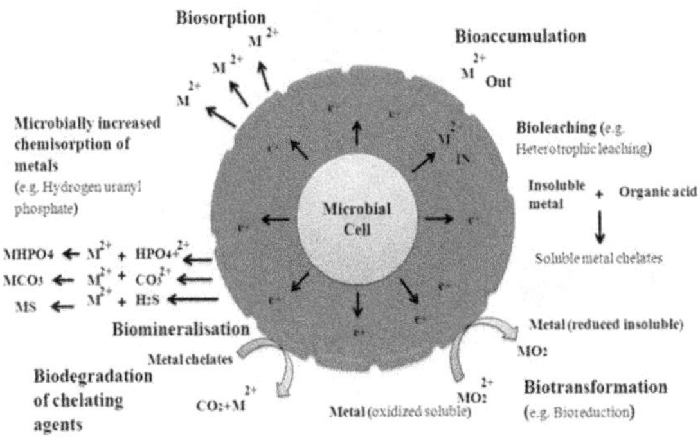

Mechanisms Involved in Heavy Metal Absorption

Microorganisms can enhance the mobility of metals in soil. There are three mechanisms which concerned complexolysis, redoxolysis and acidolysis which related to the ability of microbes to oxidize the structural elements of minerals, to decrease soil pH and to increase the concentration of metal chelators, respectively (Brandl and Faramarzi, 2006). In complexolysis also known as ligand-induced metal solubilization, complexes are formed at the surface of metal-bearing phases between metals and microbial chelators (Grybos et al., 2011; Ferret et al., 2014; Hazotte et al., 2016). The redoxolysis involves bacteria that obtain energy from solid-phase minerals. This mechanism has been widely studied for bioleaching purposes which generate strong acidic conditions arising from the bacterial production (Vestola et al., 2010). Therefore acidolysis, also called proton-induced metal solubilization, protons secreted by microbes and to bind with mineral surface which resulting in the release of metal ions from the solid surface (Brandl and Faramarzi, 2006). Recently studies were found about the completely nitrifying bacterium from the genus *Nitrospira* Figure 4 (Hinsinger et al., 2003; Daims et al., 2015).

There are number of pollutants all over the world, so recover metallic elements by using various methods derived from mining activities. Recent studies refer to applications of bioleaching for the recovery of Cu (Potysz et al., 2016). Chemolithotrophic bacteria have been tested for their ability to mobilize metals by acidolysis and redoxolysis such as "*Acidithiobacillus ferrooxidans, Acidithiobacillus caldus Acidithiobacillus thiooxidans, Leptospirillum ferroox- idans* and *Sulfobacillus thermotolerans* was able to solubilize 62% in 29 days of the Cu from Cu smelting industry slag waste (Kaksonen et al., 2011).

Figure 4. Mechanism of bioremediation

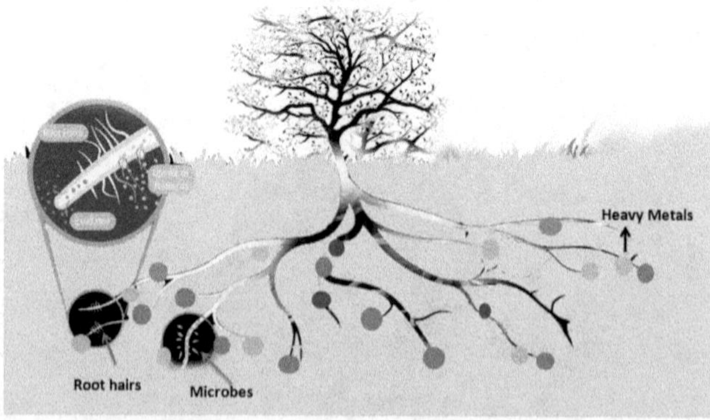

The plant and bacteria interaction can be symbiotic nature in which both of them are benefited from each other. The roots of plants are mainly in symbiosis relationship with bacteria (Glick, 1995). These bacteria can help plants to sustain its nutrient supply by recycling. The soil fertility is also improved by rhizobacteria. The harmful chemicals released by metals also detoxified by rhizobacteria. The host plant can also become pest resistant through symbiosis thus plant overall growth is enhanced (Giller et al., 1998; Elsgaard et al., 2001; Filip, 2002). Host plant help bacteria by providing it with a carbon source that in turn will help to decrease the phytotoxicity. Soil played an important role in the type of interaction with bacteria. The interactions can be specific and non-specific. The utilization of heavy metal contaminants also dictates terms of interactions of plant with bacteria. The metal toxicity is one part that inhibits plant growth but there are some other factors such as low soil fertility, water and nutrients low bioavailability" and dry and harsh conditions can also inhibit plant growth. The PGPR interacts with heavy metals that it can change the availability of them in soil and following features which bacteria possess (1) release of chelating substances (2) acidifying the microenvironment (3) inducing modifications in redox potential (Lasat, 2002; Whiting et al., 2001). The lower the interaction of bacteria with plant due to poor water supply, lack of soil fertility, harsh conditions and impossible nutrient exchanges.

SIGNIFICANCE OF PGPB IN PHYTOREMEDIATION

High amount of heavy metals decrease the microbial activity in the soil and affect the crop production by getting accumulated in plant organs. Metal ions present in the soil are first absorbed by roots and then transported to different organs. The plant proteins and enzymes present in cells have affinity for heavy metals which render them inactive and make them lose their function. Once they interact with heavy metals, a change in protein structure occurs, which ultimately affects the plant growth by inactivation of the photo-system. The resultant oxidative effect produced by free radicals has an adverse effect on the plant's biochemical and physiological processes. There is a huge amount of proteins which can bind with different heavy metals with greater affinity. These metal binding proteins are at the outer membranes in the plants and microorganisms where they interact with the metal ions present in the environment, thus ensuring the transport of these metals to the cytosol, which are then transferred to suitable receptor protein by the metallochaperones. Plants respond to the higher levels of the heavy metals by synthesizing: (1) metallothioneins and (2) phytochelatins under heavy metal stress.

CONCLUSION AND FUTURE PROSPECTS

This study demonstrates the ability of bacteria to mobilize or immobilize heavy metals in soil. The microorganisms can improve phytoextraction by increasing plant biomass or by enhancing the availability of heavy metals to the plant. However, use of microorganisms to improve plant biomass via improved N or P nutrition can have a significant positive impact on phytoextraction. Those soils, which are contaminated and possess the higher amount of toxic metal than plant tolerance, then at such sites phytoremediation through rhizobacteria is preferred and carried out. This method not only increases the plant tolerance towards metals but also stabilizes the growth of plants involving the remediation of metal polluted soil. The soil fertility which is being lost due to heavy metal accumulation has been restored through this method. As the role of plant growth-promoting rhizobacteria is known however a lot remains there to understand regarding the mechanism of phytoremediation. While on the other hand, bioremediation through rhizobacteria is an economic and ecological friendly process.

REFERENCES

Abdul, M. R., & Schroder, P. (2009). Implications of metal accumulation mechanisms to phytoremediation. *Environ. Sci. Pollut. Resear.*, *16*(2), 162–175. doi:10.1007/s11356-008-0079-z PMID:19067014

Ali, H., Khan, E., & Sajad, M. A. (2013). Phytoremediation of heavy metals-concepts and applications. *Chemosphere*, *91*(7), 869–881. doi:10.1016/j.chemosphere.2013.01.075 PMID:23466085

Bai, X. J., Liu, L. J., Zhang, C. H., Ge, Y., & Cheng, W. (2011). Effect of H_2O_2 pretreament on Cd tolerance of different rice cultivars. *Rice Science*, *18*(1), 29–35. doi:10.1016/S1672-6308(11)60005-X

Baligar, V. C., Shaffert, R. E., Dos Santos, H. L., Pitta, G. V. E., Bahia Filho, A. F., & De, C. (1993). Soil aluminium effects on uptake influx and transport of nutrients in sorghum genotypes. *Plant and Soil*, *150*(2), 271–277. doi:10.1007/BF00013024

Bani, A., Pavlova, D., Echevarria, G., Mullaj, A., Reeves, R. D., Morel, J. L., & Sulce, S. (2010). Nickel hyperaccumulation by the species of *Alyssum and Thlaspi* (*Brassicaceae*) from the ultramafic soils of the Balkans. *Botanica Serbica*, *34*(1), 3–14.

Bendtsson, B., Asp, H., & Jensn, P. (1994). Uptake and distribution of calcium and phosphorus in beech (*Fagus sylvatica*) as influences by aluminum and nitrogen. *Tree Physiology*, *14*(1), 63–73. doi:10.1093/treephys/14.1.63 PMID:14967634

Bhargava, A., Carmona, F. F., Bhargava, M., & Srivastava, S. (2012). Approaches for enhanced phytoextraction of heavy metals. *Journal of Environmental Management*, *105*, 103–120. doi:10.1016/j.jenvman.2012.04.002 PMID:22542973

Bose, J., Babourina, O., & Rengel, Z. (2011). Role of magnesium in alleviation of aluminium toxicity in plants. *Journal of Experimental Botany*, *62*(7), 2251–2264. doi:10.1093/jxb/erq456 PMID:21273333

Bose, J., Babourina, O., Shabala, S., & Rengel, Z. (2013). Low-pH and aluminium resistance in *Arabidopsis* correlates with high cytosolic magnesium content and increase magnesium uptake by plant roots. *Plant & Cell Physiology*, *54*(7), 1093–1104. doi:10.1093/pcp/pct064 PMID:23620479

Brandl, H., & Faramarzi, M. A. (2006). Microbe metal-interactions for the biotechnological treatment of metal-containing solid waste. *Chin Particuol*, *4*(2), 93–97. doi:10.1016/S1672-2515(07)60244-9

Calheiros, C. S. C., Rangel, A. O. S. S., & Castro, P. M. L. (2008). The effects of tannery waste water on the development of different plant species and chromium accumulation in *Phragmites australis*. *Archives of Environmental Contamination and Toxicology*, *55*(3), 404–414. doi:10.1007/s00244-007-9087-0 PMID:18214580

Chaudhary, K., Sumira, J., & Khan, S. (2016). Heavy metal ATPase (*HMA2, HMA3 & HMA4*) genes in hyperaccumulation mechanism of heavy metals. *Plant metal interaction (Emerging Remediation Techniques),* 545-552.

Chen, B. C., Lai, H. Y., & Juang, K. W. (2012). Model evaluation of plant metal content and biomass yield for the phytoextraction of heavy metals by switchgrass. *Ecotoxicology and Environmental Safety*, *80*, 393–400. doi:10.1016/j.ecoenv.2012.04.011 PMID:22541831

Cho, U. H., & Seo, N. H. (2005). Oxidative stress in *Arabidopsis thaliana* exposed to cadmium is due to hydrogen peroxide accumulation. *Plant Science*, *168*(1), 113–120. doi:10.1016/j.plantsci.2004.07.021

Cornu, J. Y., Huguenot, D., Jezequel, K., Lollier, M., & Lebeau, T. (2017). Bioremediation of copper contaminated soils by bacteria. *World Journal of Microbiology & Biotechnology*, *33*(2), 26. doi:10.1007/s11274-016-2191-4 PMID:28044274

Cumming, J. R., Eckert, R. T., & Evans, L. S. (1986). Effect of aluminium on ^{32}P uptake and translocation by red spruce seedlings. *Canadian Journal of Forest Research*, *16*(4), 864–867. doi:10.1139/x86-152

Daims, H., Lebedeva, E. V., Pjevac, P., Han, P., Herbold, C., Albertsen, M., ... Wagner, M. (2015). Complete nitrification by Nitrospira bacteria. *Nature*, *528*, 504–509. PMID:26610024

Dixit, R., Malaviya, D., Pandiyan, K., Singh, U. B., & Sahu, A. (2015). Bioremediation of heavy metals from soil and aquatic environment: An overview of principles and criteria of fundamental processes. *Sustainability*, *7*(2), 2189–2212. doi:10.3390/su7022189

Dubey, S., Shri, M., Misra, P., Lakhwani, D., Bag, S. K., Asif, M. H., & Chakrabarty, D. (2014). Heavy metals induce oxidative stress and genome-wide modulation in transcriptome of rice root. *Functional & Integrative Genomics*, *14*(2), 401–417. doi:10.1007/s10142-014-0361-8 PMID:24553786

Elsgaard, L., Petersen, S. O., & Debosz, K. (2001). Effects and risk assessment of linear alkylbenzene sulfonates in agricultural soil. 1. Short-term effects on soil microbiology. *Environmental Toxicology and Chemistry*, *20*(8), 1656–1663. doi:10.1002/etc.5620200806 PMID:11491546

Fassler, E., Robinson, B. H., Gupta, S. K., & Schulin, R. (2010). Uptake and allocation of plant nutrients and Cd in maize, sunflower and tobacco growing on contaminated soil and the effect of soil conditioners under field conditions. *Nut. Cycle Agro.*, *87*(3), 339–352. doi:10.1007/s10705-009-9342-z

Felestrino, E.B., Santigo, I.F., Freitos, L.S., Rosa, L.H., Ribeiro, S.P., & Moreira, L.M. (2017). Plant growth promoting bacteria associated with *Longsdorffia hypogae* Rhizosphere host biological interfere a neglected model of bacterial prospection. *Frontiers in Microbiol.* doi 10.3389ifmicb.2017.00172

Ferret, C., Cornu, J. Y., Elhabiri, M., Sterckeman, T., Braud, A., Jézéquel, K., ... Geoffroy, V. A. (2015). Effect of pyoverdine supply on cadmium and nickel complexation and phytoavailability in hydroponics. *Environ. Sci. Pollut.*, *22*(3), 2106–2116. doi:10.1007/s11356-014-3487-2 PMID:25167822

Ferret, C., Sterckeman, T., Cornu, J. Y., Gangloff, S., Schalk, I. J., & Geoffroy, V. A. (2014). Siderophore-promoted dissolution of smectite by fluorescent *Pseudomonas*. *Environmental Microbiology Reports*, *6*(5), 459–467. doi:10.1111/1758-2229.12146 PMID:25646536

Filip, Z. (2002). International approach to assessing soil quality by ecologically related biological parameters. *Agriculture, Ecosystems & Environment*, *88*(2), 169–174. doi:10.1016/S0167-8809(01)00254-7

Gamalero, E., & Glick, B. R. (2011). Mechanisms used by plant growth-promoting bacteria. In M. K. Maheshwari (Ed.), *Bacteria in Agrobiology: Plant Nutrient Management* (pp. 17–46). Berlin, Heidelberg, Germany: Springer-Verlag. doi:10.1007/978-3-642-21061-7_2

Giller, K. E., Witter, E., & Mcgrath, S. P. (1998). Toxicity of heavy metals to microorganisms and microbial processes in agricultural soils: A review. *Soil Biology & Biochemistry*, *30*(10-11), 1389–1414. doi:10.1016/S0038-0717(97)00270-8

Girisha, S. T., & Ragavendra, V. B. (2009). Accumulation of heavy metals in leafy vegetables grown in urban areas by using sewage water and its effect. *Arc. Phytopathol. Plant Protect.*, *42*(10), 956–959. doi:10.1080/03235400701543806

Gomes, M. M. S., Cambraia, J., Sant'anna, R., & Estevão, M. M. (1985). Aluminium effects on uptake and translocation of nitrogen in sorghum (*Sorghum bicolor, L. Moench*). *Journal of Plant Nutrition*, *8*(6), 457–465. doi:10.1080/01904168509363360

Gratao, P. L., Polle, A., Lea, P. J., & Azevedo, R. A. (2005). Making the life of heavy metal-stressed plant a little easier. *Functional Plant Biology*, *32*(6), 481–494. doi:10.1071/FP05016

Grison, C. (2015). Combining phytoextraction and ecocatalysis: A novel concept for greener chemistry, an opportunity for remediation. *Environ. Sci. Pollut. R*, *22*(8), 5589–5591. doi:10.1007/s11356-014-3169-0 PMID:24946705

Grybos, M., Billard, P., Desobry-Banon, S., Michot, L. J., Lenain, J. F., & Mustin, C. (2011). Bio-dissolution of colloidal-size clay minerals entrapped in microporous silica gels. *Journal of Colloid and Interface Science*, *362*(2), 317–324. doi:10.1016/j.jcis.2011.07.031 PMID:21807370

Hameed, A., Rasool, S., Azooz, M.M., Hossain, M.A., Ahanger, M.A, & Ahmad, P. (2016). Heavy metal stress: Plant responses and signalling. In *Plant metal interaction*. Academic Press.

Hanger, B. C. (1979). The movement of calcium in plants. *Communications in Soil Science and Plant Analysis*, *10*(1-2), 171–193. doi:10.1080/00103627909366887

Hazotte, A. A., Peron, O., Abdelouas, A., Montavon, G., & Lebeau, T. (2016). Microbial mobilization of cesium from illite: The role of organic acids and siderophores. *Chemical Geology*, *428*, 8–14. doi:10.1016/j.chemgeo.2016.02.024

Hinsinger, P., Plassard, C., Tang, C. X., & Jaillard, B. (2003). Origins of root mediated pH changes in the rhizosphere and their responses to environmental constraints: A review. *Plant and Soil*, *248*(1/2), 43–59. doi:10.1023/A:1022371130939

Hossain, M. A., Ashrafuzzaman, M., Hossain, A. K. M. Z., Ismail, M. R., & Koyama, H. (2014). Role of accumulated calcium in alleviating aluminum injury in wheat plants. *The Scientific World Journal*. PMID:24701169

Huang, J., & Bachelard, E. P. (1993). Effects of aluminium on growth and cation uptake in seedlings of *Eucalyptus mannifera and Pinus radiate. Plant and Soil*, *149*(1), 121–127. doi:10.1007/BF00010769

Indriolo, E., Na, G., Ellis, D., Salt, D. E., & Banks, J. A. (2010). A vacuolar arsenite transporter necessary for arsenic tolerance in the arsenic hyperaccumulating fern *pteris vittta* is missing in flowering plants. *The Plant Cell*, *22*(6), 2045–2057. doi:10.1105/tpc.109.069773 PMID:20530755

Israr, M., Sahi, S. V., & Jain, J. (2006). Cadmium accumulation and antioxidative responses in the *Sesbania drummondii* callus. *Archives of Environmental Contamination and Toxicology*, *50*(1), 121–127. doi:10.1007/s00244-005-5029-x PMID:16247566

Jin, X., Yang, X., Islam, E., Liu, D., & Mahmood, Q. (2008). Effects of cadmium on ultrastructure and antioxidative defense system in hyperaccumulator and nonhyperaccumulator ecotypes of *Sedum alfredii* Hance. *Journal of Hazardous Materials*, *156*(1-3), 387–397. doi:10.1016/j.jhazmat.2007.12.064 PMID:18242844

Kaksonen, A. H., Lavonen, L., Kuusenaho, M. K., Kolli, A., Närhi, H. M., Vestola, E. A., ... Tuovinen, O. H. (2011). Bioleaching and recovery of metals from final slag waste of the copper smelting industry. *Minerals Engineering*, *24*(11), 1113–1121. doi:10.1016/j.mineng.2011.02.011

Kasai, M., Sasaki, M., Yamamoto, Y., & Matsumoto, H. (1992). Aluminum stress increases Kp efflux and activities of ATP- and PPi dependent Hp pumps of tonoplast-enriched membrane vesicles from barley roots. *Plant & Cell Physiology*, *33*(7), 1035–1039.

Kumar, S., & Trivedi, P.K. (2016). Heavy metal stress signalling in plants. In *Plant metal interaction*. Academic Press.

Kwankua, W., Sengsai, S., Muangphra, P., & Euawong, N. (2012). Screening for plants sensitive to heavy metals using cytotoxic and genotoxic biomarkers. *Witthayasan Kasetsat Witthayasat*, *46*, 10–23.

Lasat, M. M. (2002). Phytoextraction of toxic metals. *Journal of Environmental Quality*, *31*(1), 109–120. doi:10.2134/jeq2002.1090 PMID:11837415

Leblanc, M., Petit, D., Deram, A., Robinson, B., & Brooks, R. R. (1999). The phytomining and environmental significance of hyperaccumulation of thallium by Iberisintermedia from southern France. *Economic Geology and the Bulletin of the Society of Economic Geologists*, *94*(1), 109–113. doi:10.2113/gsecongeo.94.1.109

Li, Z., & Xing, D. (2010). Mitochondrial pathway leading to programmed cell death induced by aluminum phyto-toxicity in *Arabidopsis*. *Plant Signaling & Behavior*, *5*(12), 1660–1662. doi:10.4161/psb.5.12.14014 PMID:21512328

Liao, H., Wan, H., Shaff, J., Wang, X., Yan, X., & Kochian, L. V. (2006). Phosphorus and aluminium interactions in soybean in relation to alumunium tolerance. Exudation of specific organic acids from different regions of the intact root system. *Plant Physiology*, *141*(2), 674–684. doi:10.1104/pp.105.076497 PMID:16648222

Lombi, E., Terall, K. L., Howarth, J. R., Zhao, F. J., Hawkesford, M. J., & McGrath, S. P. (2002). Influence of iron status on cadmium and zinc uptake by different ecotypes of the hyperaccumulator *Thlaspi caerulescens*. *Plant Physiology*, *128*(4), 1359–1367. doi:10.1104/pp.010731 PMID:11950984

Lori, V., Pietrini, F., Massacci, A., & Zacchini, M. (2015). *Morphophysiological responses, Heavy metal accumulation and phytoremoval ability in four Willow clones exposed to cadmium under hydroponics*. Phytoremediation Springer International Publishing.

Lu, L., Tian, S. K., Yang, X. E., Li, T. Q., & He, Z. I. (2009). Cadmiumuptake and xylem loading are active processes in the hyperaccumulator *Sedum alfredii*. *Journal of Plant Physiology*, *166*(6), 579–587. doi:10.1016/j.jplph.2008.09.001 PMID:18937997

Ma, Y., Rajkumar, M., Luo, Y., & Freitas, H. (2011). Inoculation of endophytic bacteria on host and non-host plants-effects on plant growth and Ni uptake. *Journal of Hazardous Materials*, *195*, 230–237. doi:10.1016/j.jhazmat.2011.08.034 PMID:21872991

Maestri, E., Marmiroli, M., Visioli, G., & Marmiroli, N. (2010). Metal tolerance and hyperaccumulation: Costs and trade-offs between traits and environment. *Environmental and Experimental Botany*, *68*(1), 1–13. doi:10.1016/j.envexpbot.2009.10.011

Mahmood, T. (2010). Phytoextraction of heavy metals the process and scope for remediation of contaminated soils. *Soil & Environment*, *29*(2), 91–109.

Mariano, E. D., & Keltjens, W. G. (2005). Long term effects of aluminum exposure on nutrient uptake by maize genotypes differing in aluminum resistance. *Journal of Plant Nutrition*, *28*(2), 323–333. doi:10.1081/PLN-200047625

Meagher, R. B. (2000). Phytoremediation of toxic elemental and organic pollutants. *Current Opinion in Plant Biology*, *3*(2), 153–162. doi:10.1016/S1369-5266(99)00054-0 PMID:10712958

Meagher, R. B., Rugh, C. L., Kandasamy, M. K., Gragson, G., & Wang, N. J. (2000). Engineered phytoremediation of mercury pollution in soil and water using bacterial genes. In N. Terry & G. Banuelos (Eds.), *Phytoremediation of contaminated soil and water* (pp. 203–221). Lewis.

Meers, E., Ruttens, A., Hopgood, M., Lesage, E., & Tack, F. M. (2005). Potential of *Brassic rapa, Cannabis sativa, Helianthus annuus and Zea mays* for phytoextraction of heavy metals from calcareous dredged sediment derived soils. *Chemosphere*, *61*(4), 561–572. doi:10.1016/j.chemosphere.2005.02.026 PMID:16202810

Meers, E., Vandecasteele, B., Ruttens, A., Vangronsveld, J., & Tack, F. M. G. (2007). Potential of five willow species (*Salix spp.*) for phytoextraction of heavy metals. *Environmental and Experimental Botany*, *60*(1), 57–68. doi:10.1016/j.envexpbot.2006.06.008

Mehdawi, E. (2011). Selenium hyperaccumulators facilitate selenium-tolerant neighbors via phytoenrichment and reduced herbivory curt. *Current Biology*, *21*(17), 1440–1449. doi:10.1016/j.cub.2011.07.033 PMID:21856156

Memon, A. R., & Schroder, P. (2009). Implications of metal accumulations mechanisms to phytoremediation. *Environmental Science and Pollution Research International*, *16*(2), 162–175. doi:10.1007/s11356-008-0079-z PMID:19067014

Mokhtar, H., Morad, N., & Fizri, F. F. A. (2011). Phytoaccumulation of copper from aqueous solutions using *Eichhornia Crassipes* and *Centella asiatica. International Journal of Environmental Sciences and Development*, *3*(1), 89–95.

Moustakas, M., Ouzounidou, G., & Lannoye, R. (1995). Aluminum effects on photosynthesis and elemental uptake in an aluminium-tolerant and non-tolerant wheat cultivar. *Journal of Plant Nutrition*, *18*(4), 669–683. doi:10.1080/01904169509364930

Mugica-Alvarez, V., Cortes-Jimenez, V., Vaca-Mier, M., & Dommguez-Soria, V. (2015). Phytoremediation of mine tailings using *Lolium multiflorum. International Journal of Environmental Sciences and Development*, *6*(4), 246–251. doi:10.7763/IJESD.2015.V6.599

Mukhopadhyay, S., & Maiti, S. K. (2010). Phytoremediation of metal enrhiched mine wate: A review. *Global Environmental Research*, *4*, 135–150.

Nematian, M. A., & Kazemeini, F. (2013). Accumulation of Pb, Zn, Cu and Fe in plants and hyperaccumulator choice in Galali iron mine area. Iran. *International Journal of Agriculture and Crop Sciences*, *5*(4), 426–432.

Nichol, B. E., Oliveira, L. A., Glass, A. D. M., & Saiddiqi, M. Y. (1993). The effect of aluminum on the influx of calcium, potassium, ammonium nitrate and phosphate in an aluminum sensitive cultivar of barley (*Hordeum vulgare L.*). *Plant Physiology*, *101*(4), 1263–1266. doi:10.1104/pp.101.4.1263 PMID:12231781

Padmavathiamma, P. K., & Li, L. Y. (2007). Phytoremediation technology: Hyperaccumulation metals in plants. *Water, Air, and Soil Pollution*, *184*(1-4), 105–126. doi:10.1007/s11270-007-9401-5

Phukan, P., Phukan, R., & Phukan, S. N. (2015). Heavy metal uptake capacity of *Hydrilla verticillata*: A commonly available aquatic plant. *Int. Res. J. Environ. Sci.*, *4*(3), 35–40.

Pollard, A. J., Stewart, H. L., & Roberson, C. B. (2009). Manganese hyperaccumulation in *Phytolacca americana L.* from the southeastern United States. *Northeastern Naturalist*, *16*(sp5), 155–162. doi:10.1656/045.016.0513

Potysz, A., Lens, P. N. L., van de Vossenberg, J., Rene, E. R., Grybos, M., Guibaud, G., ... van Hullebusch, E. D. (2016). Comparison of Cu, Zn and Fe bioleaching from Cu-metallurgical slags in the presence of *Pseudomonas fluorescens* and *Acidithiobacillus thiooxidans*. *Applied Geochemistry*, *68*, 39–52. doi:10.1016/j.apgeochem.2016.03.006

Prasad, M. N. V., & Freitas, H. (2003). Metal hyperaccumulation in plants: Biodiversity prospecting for phytoremediation technology. *Electronic Journal of Biotechnology*, *6*(3), 275–321. doi:10.2225/vol6-issue3-fulltext-6

Purcino, A. A. C., Alves, V. M. C., Parentoni, S. N., Belele, C. L., & Loguercio, L. L. (2003). Aluminum effects on nitrogen uptake and nitrogen assimilating enzymes in maize genotypes with contrasting tolerance to aluminum toxicity. *Journal of Plant Nutrition*, *26*(1), 31–61. doi:10.1081/PLN-120016496

Qiao, X. Q., Zheng, Z. Z., Zhang, L. F., Wang, J. H., Shi, G. X., & Xu, X. Y. (2015). Lead tolerance mechanism in sterilized seedlings of *Potamogeton crispus L.* Subcellular distribution, poluamine and proline. *Chemosphere*, *120*, 179–187. doi:10.1016/j.chemosphere.2014.06.055 PMID:25033323

Raaijmakers, J. M., Paulitz, T. C., Steinberg, C., Alabouvette, C., & Moënne-Loccoz, Y. (2009). The rhizosphere: A playground and battlefield for soilborne pathogens and beneficial microorganisms. *Plant and Soil*, *321*(1-2), 341–361. doi:10.1007/s11104-008-9568-6

Ramamoorthy, V., Viswanathan, R., Raguchander, T., Prakasam, V., & Samiyappan, R. (2001). Induction of systemic resistance by plant growth promoting rhizobacteria in crop plants against pests and diseases. *Crop Protection (Guildford, Surrey)*, *20*(1), 1–11. doi:10.1016/S0261-2194(00)00056-9

Rengel, Z., & Elliott, D. C. (1992). Mechanism of aluminium inhibition of net ^{45}Ca2p Uptake by Amaranths Protoplasts. *Plant Physiology*, *98*(2), 632–638. doi:10.1104/pp.98.2.632 PMID:16668688

Ribeiro, M. A. Q., de Almeida, A. A. F., Mielke, M. S., Gomes, F. P., Pires, M. V., & Baligar, V. C. (2013). Aluminum effects on growth, photosynthesis, and mineral nutrition of cacao genotypes. *Journal of Plant Nutrition*, *36*(8), 1161–1179. doi:10.1080/01904167.2013.766889

Robinson, B., Fernandez, J. E., Madejon, P., Maranon, T., Murillo, J., Green, S., & Clothier, B. (2003). Phytoextraction: An assessment of biogeochemical and economic viability. *Plant and Soil*, *249*(1), 117–125. doi:10.1023/A:1022586524971

Rodriguez, E., Azevedo, R., Fernandes, P., & Santos, C. (2011). Cr (VI) induces DNA damage, cell cycle arrest and polyploidization: A flow cytometric and comet assay study in *Pisum sativum. Chemical Research in Toxicology*, *24*(7), 1040–1047. doi:10.1021/tx2001465 PMID:21667992

Roy, A. K., Sharma, A., & Talukder, G. (1988). Some aspects of aluminum toxicity in plants. *Botanical Review*, *54*(2), 145–178. doi:10.1007/BF02858527

Saba, G., Parizanganeh, A. H., Zamani, A., & Saba, J. (2015). Phytoremediation of heavy metals contaminated environments: Screening for native accumulator plants in Zanjan-Iran. *International Journal of Environmental of Research*, *9*(1), 309–316.

Salla, V., Hardaway, C. J., & Sneddon, J. (2011). Preliminary investigation of *Spartina alterniflora* for phytoextraction of selected heavy metals in soil from Southwest Louisiana. *Microchemical Journal*, *97*(2), 207–212. doi:10.1016/j.microc.2010.09.005

Salt, D. E., Smith, R. D., & Raskin, I. (1998). Phytoremediation. *Annual Review of Plant Physiology and Plant Molecular Biology*, *49*(1), 643–668. doi:10.1146/annurev.arplant.49.1.643 PMID:15012249

Sharma, D. M., Sharma, C. P., & Tripathi, R. D. (2003). Phytotoxic lesions of chromium in maize. *Chemosphere*, *51*(1), 63–68. doi:10.1016/S0045-6535(01)00325-3 PMID:12586157

Singh, O. V., Labana, S., Pandey, G., Budhiraja, R., & Jain, R. K. (2003). Phytoremediation: An overview of metallic ion decontamination from soil. *Applied Microbiology and Biotechnology*, *61*(5-6), 405–412. doi:10.1007/s00253-003-1244-4 PMID:12764555

Tong, X., Yuan, L., Luo, L., Yin, X., 2014. Characterization of a Selenium-Tolerant Rhizosphere Strain from a Novel Se-Hyperaccumulating Plant *Cardamine hupingshanesis*. *Scitific World J*, 1-8.

Ueno, D., Koyama, E., Yamaji, N., & Ma, J. F. (2011). Physiological, genetic, and molecular characterization of a high-Cd-accumulating rice cultivar, Jarjan. *Journal of Experimental Botany*, *62*(7), 2265–2272. doi:10.1093/jxb/erq383 PMID:21127026

Ullah, A., Heng, S., Munis, M. F. H., Fahad, S., & Yang, X. (2015). Phytoremediation of heavy metals assisted by plant growth promoting (PGP) bacteria: A review. *Environmental and Experimental Botany*, *117*, 28–40. doi:10.1016/j.envexpbot.2015.05.001

Wang, Q. Q., Wang, L. H., Han, R. B., Yang, L. Y., Zhou, Q., & Huang, X. H. (2015). Effects of bisphenol A on antioxidant system in soybean seedling roots. *Environmental Toxicology and Chemistry, 34*(5), 1127–1133. doi:10.1002/etc.2904 PMID:25651304

Wang, Y. S., & Yang, Z. M. (2005). Nitric oxide reduces aluminum toxicity by preventing oxidative stress in the roots of *Cassia tora L. Plant & Cell Physiology, 46*(12), 1915–1923. doi:10.1093/pcp/pci202 PMID:16179356

Wheeler, D. M., & Follett, J. M. (1991). Effect of aluminium on onions, *Asparagus* and *Squash. Journal of Plant Nutrition, 14*(9), 897–912. doi:10.1080/01904169109364251

Xia, B., Shen, S., & Xue, F. (2013). Phytoextraction of heavy metals from highly contaminated soils using *Sauropus androgynus*. Soil Sed. *Contamin., 22*(6), 631–640.

Yao, Z., Li, J., Xie, H., & Yu, C. (2012). Review on remediation technologies soil contaminated by heavy metals. *Procedia Environmental Sciences, 16*, 722–729. doi:10.1016/j.proenv.2012.10.099

Zhang, Y. F., He, L. Y., Chen, Z. J., Wang, Q. Y., Qian, M., & Sheng, X. F. (2011). Characterization of ACC deaminase-producing endophytic bacteria isolated from coppertolerant plants and their potential in promoting the growth and copper accumulation of *Brassica napus. Chemosphere, 83*(1), 57–62. doi:10.1016/j.chemosphere.2011.01.041 PMID:21315404

Zsoldos, F., Vashegyi, A., Bona, L., Pecsvaradi, A., & Szegletes, Z. (2000). Growth of and potassium transport in winter wheat and durum wheat as affects by various aluminium exposure times. *Journal of Plant Nutrition, 23*(7), 913–926. doi:10.1080/01904160009382070

Chapter 7
Technological Interventions in Management of Hg Contaminated Water

Vidushi Abrol
CSIR-Indian Institute of Integrative Medicine, India

Sharada Mallubhotla
Shri Mata Vaishno Devi University, India

Sundeep Jaglan
CSIR-Indian Institute of Integrative Medicine, India

ABSTRACT

Rising cases of environmental mercury hazards has led to a need for cost-effective mercury treatment techniques. Extensive use of mercury from ancient times has resulted in water contamination that may require remediation. Mercury contamination is tedious to treat and may pose a risk to human health and the environment. To deal with this threat of mercury contamination, industrial wastes and wastewaters containing mercury requires treatment for its removal and immobilization. This chapter provides a synopsis of the availability, performance, and technologies for management of mercury in water. It covers the innovative methods to treat the mercury contamination like biosorption. In this chapter, the technological aspects available for the mercury treatment technologies are reviewed. It describes the theory, design, and operation of the technologies; provides information on commercial availability and use; and includes data on performance, where available.

DOI: 10.4018/978-1-5225-4162-2.ch007

Copyright © 2018, IGI Global. Copying or distributing in print or electronic forms without written permission of IGI Global is prohibited.

INTRODUCTION

Heavy metal toxicity is the major environmental issue known in the emerging trends of life. Majority of the heavy metal ions are highly unsafe to living organisms and these metal ions are non-degradable but are also available in the nature in large quantity. Therefore, the removal of heavy metal ions from the contaminated wastewater is important to protect the environment and to reduce the public risk. Generally, industrial effluents/wastes are vital cause of heavy metal concentration, the activities like corrosion of water pipes, electroplating, electrolysis, electro-osmosis, mining, dumping of waste, surface finishing, energy and fuel production, metal surface treatments and atomic energy installations etc. Therefore, the elimination and recovery of the heavy metals from effluent streams are essential for the protection of the environment (Wang & Chen, 2009). Various methods comprise physical, chemical and biological technologies are involved in removing metal ions from the aqueous solutions. Conventional methods like chemical precipitation, lime coagulation, solvent extraction, membrane filtration, reverse osmosis, ion exchange and adsorption, are generally being used for the reduction of heavy metal ions from the aqueous wastes. However, each process has its own advantages and limitations in application, so these traditional metal removal methods have major disadvantage like incomplete metal removal, high reagent and energy requirements, generation of toxic sludge or other waste products. Moreover these methods are often ineffective or uneconomical at large scale process (Dzionek, Wojcieszyńska & Guzik, 2016). In recent years, microbial biomass has unfold as an option for developing economic and eco-friendly wastewater treatment process in controlling and removing metal pollution in the field of biotechnology and gradually preferred to control metal pollution due to its potential efficacy. The process is known as biosorption, which utilizes various certain natural materials of biological origin, including bacteria, fungi, yeast, algae, etc. (Wang & Chen, 2009).

Biosorption is a vast and hasty phenomenon of passive metal sequestration by the non-growing biomass/adsorbents. It has advantages compared with conventional techniques, in which microorganisms are used to eliminate and recover heavy metals from aqueous solutions, have been known for few decades (Kratochvil & Volesky, 1998; Mahbub et al., 2017). Biotechnological emerging trends and recent techniques can succeed in these areas and are designed to cover such niches. Microorganisms like fungi have demonstrated its capability to take up heavy metals like mercury from aqueous solutions, especially in different ranges of metal concentrations in the effluent (Wang & Chen, 2006). Besides versatility, selectivity to remove only the desired metals and the cost-effectiveness are some added merits of biological metal cleanup techniques. These factors have upgraded extensive research methodology on the biological interventions of heavy metal elimination. This chapter reviews the use of fungi as biosorbents to remove the heavy metals from the wastewaters; these biological biosorbents are highly effective as well as authentic in the removal of heavy metal ions from the wastewater. Hence, this work attempts to summarize recent literature studies in the removal of mercury (heavy metals) using biological biosorbents.

SOURCES OF MERCURY

Mercury exists naturally in the environment and can be found in elemental (metallic), inorganic, and organic forms, mainly in the form of Methylmercury. Contemporary uses for mercury include chemical manufacturing, thermometers, and lighting (mercury vapor and fluorescent lamps) (Figure 1). All the chemical and allied products in the industrial group are responsible for the largest quantity of mercury used and produced (Beckers & Rinklebe, 2017).

The scintillating liquid metal mercury is one of the chief natural elements that make up the Earth's composition. While a relatively negligible element comprising less than one-millionth of the Earth's crust, mercury may be found at much higher concentrations in certain regions. Mercury can neither be created nor destroyed. Mercury is known as a "heavy metal" because it is very dense (Gaffney & Marley, 2014; Organization, 1976). The procedure of methylation of inorganic mercury to organic mercury is important to the destiny of mercury in the environment. Living organism does assemble more organic mercury, becoming concentrated up the food chain. In a typical example of biomagnifications, the bulk of mercury present in fish is organic mercury (methyl-mercury), which crunch tightly to the proteins of all fish tissue, including muscle. Methylmercury boost in fish tissue to levels that are numerous times greater compared to the levels in the surrounding water. Mercury commute into the environment can alter between organic and inorganic forms (Organization, 1996).

Figure 1. Uses and applications of mercury (Obenauf and Skavroneck, 1997)

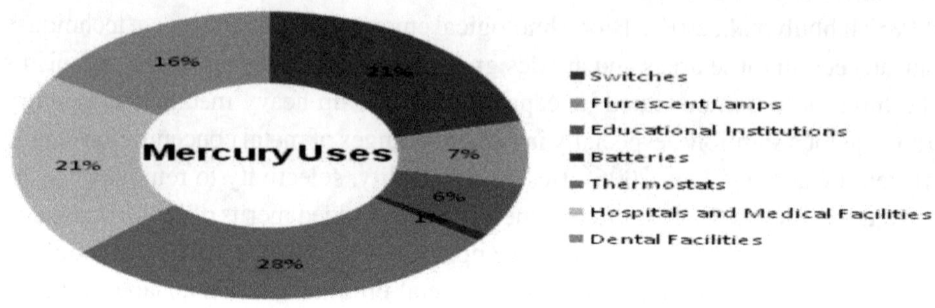

Naturally, Mercury exist in the Earth's crust is emitted via the weathering of rocks and from geological movements like volcanoes and geothermal sources, and known as primary natural source. These sources tend to increase the universal pool of mercury in the upper layer environment. Mercury re-emits secondary natural sources within ecosystems due to previously emitted mercury on vegetation, land, or water surfaces. Such re-scattering ends up with land use changes, biomass burning, meteorological conditions, and exchange mechanisms of gaseous mercury (Driscoll, Mason, Chan, Jacob, & Pirrone, 2013). Mercury mobilizes among various repository sources like atmosphere, lithosphere, biosphere, and hydrosphere results both from anthropogenic and natural emission reservoirs. Diversified global mercury assessments have analyzed the atmosphere as the dominant pathway for the trading of mercury on a universal level, although some of the atmospheric emitted mercury is locally precipitate on terrestrial and aquatic surfaces. Mercury travels from one part of the World to another, depending on the direction and speed of the wind, and also depends on the chemical behavior of the mercury (Cohen et al., 2016).

ENVIRONMENTAL AND HEALTH RISKS

Mercury, especially the organic form i.e. methyl mercury is a vigorous neurotoxin capable of impairing neurological development in fetuses and young children and of damaging the central nervous system of adults. Mercury has been identifies as one of the most hazardous heavy metals and is the most serious outbreaks of metal poisoning among the general population, e.g. Minamata disease. Elemental and methylmercury are threatening to the central and peripheral nervous systems and its inhalation of vapors can produce injurious effects on the nervous, digestive and immune systems, lungs and kidneys, and may be fatal. The inorganic form of mercury is skin, eyes and gastrointestinal tract damaging, and if ingested it can also induce kidney toxicity. After 1956's Minamata disaster was another similar methyl mercury calamity happened in Niigata prefecture of Japan in May 1965. Medical survey of the Niigata Minamata reports 64-908 parts per billion (ppb) mercury in blood, 56.8-570 parts per million (ppm) mercury in hair and 92-915 μg per day mercury in urine. A clear comparison was observed between the amount of river fish ingested and mercury content in hair. Another case of this disease happened in Ontario (Canada) during 1970. Illegal depositing and dumping of industrial chemical waste containing mercury into the river by Dryden pulp and paper industry was the reason behind the episode. It was calculated that there were at least seven tonnes of mercury released into the river system. People suffered were found to have more than 200 ppb of mercury accumulated in their blood in 1970s.(Harada,

1995) Survey conducted that "four decades later, the effects of that mercury are still present". Symptoms are hearing impairment, difficulty in walking in a straight line, headaches, insomnia, exhaustion, visual disturbances, fatigue, etc. were observed in nearly one-third of the people surveyed in 2010 (Bernhoft, 2011) (Table 1).

Table 1. Incidences of Mercury contamination in India

S. No.	Mercury Pollution Incidence	Source of information
1	India sanctioned the Minamata Convention for mercury pollution prevention in 2014. Dental amalgams used by dentist contain 50% mercury. Untold studies have showed that a single amalgam filling release 10 µg of mercury per day. WHO also proves that mercury emitted even in minor quantity may seed serious health problems, and is a threat to the development of the child in utero and early in life. Mercury has high risk on the nervous, digestive and immune systems, and on lungs, kidneys, skin and eyes.	SundayGuardianLive April, 2017
2	Paryavaran Suraksha Samiti in an exhaustive investigation In Ankhleshwar, (Gujarat) industrial estate, tested 120 ground water sources, affecting 100,000 people, were highly contaminated. Later studies conducted by DTE found high quantities of mercury, lead and zinc. As an effect of this pollution people are suffering with skin ailments, respiratory problems, headaches and many others which have not been noticed.	www.indiatogether.com September, 2001
3	The Tamil Nadu Pollution Control Borad (TNPCB) has confirmed that mercury contaminated glass pieces were dumped at various locations in Kodaikanal by a thermometer factory of Unilever, a multinational company. The waste relate as scrapyard contains more than 50 mg/kg which means mercury is approximately 1% of the total waste which works out to be 10,000 mg/kg. This clearly an offence under the Hazardous waste Management act.	Down to Earth 15th July, 2001
4	Study conducted in the Thane Bay area revealed that due to mercury peroxide, an effluent from a nearby paint factory caused serious health problem especially to children. Neurological disabilities developed in children. The mercury peroxide settled in the muscles of these children and rendered them permanently disabled.	Indian Express 10th Februaray,2000
5	The DTE/IIT test conducted on a water sample from a hand-pump in Mehak district of Andhra Pradesh showed that the level of mercury was 115 times the permissible limit. Arsenic and Nickel concentration level were also too high. Symptoms were plethora diseases such as epilepsy, skin and throat problem, respiratory diseases, cancer and paraplegia, while pregnant women are giving birth to still-born children.	Down to Earth 31st August,1999
6	After the Gwalior Rayons Factory was set up in Mavoor in Kozhikode district, the river turned dark, dead fish floated on the surface and the air smelled foul. Symptoms were high incidence of cancer and respiratory and skin diseases in the villages adjacent to the factory. After conducting test, major cause was contamination due to mercury.	The Week 14th December,1997

WATER TREATMENT TECHNOLOGIES

Among the mercury treatment, precipitation/co precipitation is the most widely used process to cure mercury polluted water. These management techniques are less likely to be affected by characteristics of the contaminants compared with other water treatment technologies effectiveness (Dhankhar & Hooda, 2011).

Surrogate mercury management technologies include adsorption and membrane filtration; however, these technologies are used less frequently because both are more likely to be affected by characteristics of the media and non-mercury contaminants than is precipitation/co precipitation. Biosorption technique to be used more often when mercury is the only contaminant to be treated, for relatively smaller systems, and as a polishing technology for effluent from larger systems whereas, Membrane filtration is used less frequently because it tends to produce a larger volume of residuals than do other mercury treatment technologies (Table 2).

BIOSORPTION

The search for novel methodology involving the removal of toxic metals from wastewaters indicated attention to biosorption, based on metal binding capacities of various biological materials. Biosorption can be defined as the property of

Table 2. Methods & technologies to treat mercury contaminated water

S.No.	Technologies	Process for treating water	References
1	Precipitation/ Coprecipitation	To transform soluble contaminants into an insoluble solid, or develop insoluble solids onto which dissolved contaminants are adsorbed generally involve pH adjustment and chemical precipitation. The insoluble solids are then separated from the liquid phase by clarification or filtration.	(Shafeeq et al., 2012)
2	Adsorption	Adsorbates are sustained on the surface by many types of chemical forces such as hydrogen bonds, dipole-dipole interactions, and vander Waals forces. Concentrates solutes at the surface of a sorbent, thereby decreases their concentration in the large liquid phase. The adsorption media is usually packed into a column. Contaminants are adsorbed as contaminated water is passed through the column.	(Henry, 2000)
3	Membrane Filtration	Water is passes through a semi-permeable, microporous membrane to concentrate contaminants into a smaller volume of water while it blocks others.	(Barakat, 2011)
4	Biological Treatment	This process uses a fixed-bed reactor with a combination of microbes, including proprietary microbial cultures and microorganisms isolated from the contaminated wastewater that act directly on contaminant species conversion of soluble mercury into a insoluble elemental form mercuric sulfide that cause the contaminant precipitate/coprecipitate from water.	(Wagner-Döbler, Von Canstein, Li, Timmis, & Deckwer, 2000)

biological materials to uptake heavy metals from wastewater through metabolically mediated or physico-chemical pathways of uptake. Algae, bacteria, fungi and yeasts have proved to be potential metal biosorbents (Vijayaraghavan & Balasubramanian, 2015; Volesky & Holan, 1995).

The potential advantage of biosorption process includes naturally abundant renewable biomaterials that can be cheaply produced. Its ability to treat large volumes of wastewater due to rapid kinetics, capacity to manage multiple heavy metals and mixed wastes, less need for additional expensive reagents which typically cause disposal and space problems. Relatively low capital investment and low operational cost, comparatively cheap method and various materials are available like bacteria, fungi, dead plants, reusable biological products, makes it more preferable. This technique is eco friendliness and high in availability. All the substances can be collected locally and the use of biologically inactive substances saves the production of virgin materials and avoids the production costs. The functionality is flexible as it does not require fixed condition and monitoring and processed in various ways like in columns or on carriers in water tanks. According to the literature survey, biosorption is easier to perform and are more versatile compared to bioaccumulation (Das, Das, & Guha, 2007; Veglio & Beolchini, 1997).

This biosorption process includes sorbent or biosorbent; biological material as solid phase material and solvent, normally water as a liquid phase containing a dissolved species to be sorbate, metal ions i.e. contaminated water. Different mechanisms are applied because of higher binding affinity of the sorbent for the sorbate species is attracted and bound with each other. The process goes on until equilibrium is maintained between the amount of solid-bound sorbate species and its portion remaining in the solution (Pagnanelli, Toro, & Vegliò, 2002).

BIOSORPTION OF MERCURY BY FUNGI

Fungi are one of the industrial fermentation waste biomass biological material which are really excellent metal sorbets. Fungi are good efficient and economical for sequestering heavy toxic metals from dilute aqueous solutions by biosorption because: It is having a high range of cell wall materials; depicts excellent metal binding properties; available in large quantities from food like industries and also provides an eco-friendly environment. Fungi are found naturally in the environment and important in industrial processes. The structures of fungi have morphologies from unicellular yeasts to polymorphic and filamentous fungi and also have complex macroscopic fruiting bodies. In the cell wall of fungi different types of ionizable sites affect the metal uptake capacity: phosphate groups, carboxyl groups on uranic acids and proteins, and nitrogen- containing ligands on protein as well as on chitin or chitosan (Bahobil, Bayoumi, Atta, & El-Sehrawey, 2017; Huang & Huang, 1996). Biosorption has been used to treat mercury in two pilot-scale studies which are included in this chapter. In one study, mercury ions were converted to elemental mercury by filamentous fungi and the adsorbed ions are transported across the membrane by which metabolically important ions such as potassium, magnesium, and sodium are conveyed. Filamentous fungi are included as a biosorption agent (Table 3). Diverse fungal species have been reported such as *Trichoderma atroviride, T. harzianum, T. virens* and *Aspergillus niger* etc., that are used in the process of cleaning polluted areas. In the other study, Factors influencing biosorption are important in the evaluation of the full biosorption potential of any biological treatment, was used to remove mercury from wastewater.

In the recent case study of Evi Kurniati 2014, he also proves successfully that the use of filamentous fungi is the most capable for biosorption of mercury contamination in a substrate. Six different fungal strains were preferred based on their capability to grow in 25mg/L Hg (2^{+}) on contaminated potato dextrose agar medium. KRP1 fungal strain exhibit the maximum ratio of growth diameter, 0.831, thus selected for additional observation and analysis. Then after selection, identification based on their phenotypic characteristics and cell morphology was executed by 18S rRNA analysis gave a 98% match to *Aspergillus flavus* strain KRP1. The fungal strain

Table 3. Fungi used for biosorption of mercury

S.No.	Fungi	References
1	*Aspergillus flavus I–V, Aspergillus fumigatus I-II, Helminthosporium sp., Cladosporium sp., Mucor rouxii mutant, M. rouxii IM-80, Mucor sp 1 and 2,* and *Candida albicans,*	(Martínez-Juárez, Cárdenas-González, Torre-Bouscoulet, & Acosta-Rodríguez, 2012)
2	*Candida* sp.	(Anaemene, 2012)
3	*Saccharomyces cerevisiae*	(Mapolelo & Torto, 2004)
4	*Penicillium purpurogenum*	(Say, Yılmaz, & Denizli, 2003)
5	*Aspergillus niger*	(Khambhaty, Mody, Basha, & Jha, 2009)
6	*Pleurotus eryngii*	(Amin et al., 2016)
7	*Aspergillus niger, Penicillum* sp., *Alternaria* sp., *Rhizopus* sp., *Monilia* sp. and *Trichoderma* sp.	(Zafar, Aqil, & Ahmad, 2007)
8	*Penicillum oxalicum, Tolypocladium* sp.	(Svecova, Spanelova, Kubal, & Guibal, 2006)
9	*Aspergillus niger, Trichoderma viride,* and *Humicola insolens*	(Javed, Ikram-ul-Haq, & Shahbaz, 2007)
10	*Acremonium strictum, Acremonium* sp. and *Penicillium* sp.	(Prigione, Grosso, Tigini, Anastasi, & Varese, 2012)

characteristics in mercury(II) contamination such as based on their factors like range of optimum pH, optimum temperature and tolerance level were 5.5-7 and 25-35°C and 100 mg/L respectively. Here, it is observed that the concentration of mercury in the media affected fungal growth during lag phases. Therefore, the potential of the fungal strain to remove the mercury(II) contaminant was analysed in 100 mL sterile 10mg/L Hg(2^+) contaminated potato dextrose broth medium inoculated with 108spore/mL fungal spore suspension at 30°C for 7 days incubation. The mercury(II) uptake capability was observed in shaking flasks at 130rpm in orbital shaker and also at static position with comparison to the control and observed that the strain was able to remove 97.50% and 98.73% mercury from shaken and static systems respectively. Hence, this study shows that *A. flavus* strain KRP1 have high potential use in bioremediation of aqueous substrates containing mercury (II) through a biosorption mechanism (Kurniati et al., 2014).

FACTORS EFFECTING BIOSORPTION OF HEAVY METALS

These factors are related to the biomass and metal and the others are related to environmental conditions. The major factors that affect the biosorption process are:

1. **Temperature:** Biosorption ability remains unaffected within the range 20-35 ° C, however high temperatures, like 50°C, may increase biosorption, but these high temperatures may cause permanent damage to microbial living cells and then decreasing metal uptake. Adsorption reactions are generally exothermic and the extent of adsorption increases with decreasing temperature. The maximum biosorption capacity for Hg by *M. rouxii* IM-80 fungus appears to be temperature-dependent over the temperature range tested (30–40°C) found to decrease as the temperature was increased to 40 oC (Martínez-Juárez et al., 2012).
2. **Biomass Concentrations:** Concentration of biomass in solution affects the specific uptake of heavy metals. The biomass adsorbs more metal ions at low cell densities than at high densities, electrostatic interaction between the cells plays an important role in metal uptake. At decreased biomass concentration, the specific uptake of metals is increased because an increase in biosorbent concentration leads to interference between the bindings sites. High biomass concentration restricts the access of metal ions to the binding sites.(Modak & Natarajan, 1995)
3. **pH:** The acidity (pH) is an important parameter for biosorption of metal ions from aqueous solution because it affects the solubility of the metal ions, concentration of the metal ions on the functional groups of the adsoption and

the degree of ionization of the adsorbate during process. The variation of pH affects the effectiveness as hydrogen ion itself is a tough competing adsorbate. The surface charge is predominantly negative over the pH range of 3.0-10.0. Below a pH of 3.0, the overall surface charge on microorganism's cells becomes positive.(Puranik, Modak, & Paknikar, 1999)

4. **Metal Affinity to the Biosorbent:** Permeability and surface charges of the biomass affects the physiochemical pretreatment and also makes metal binding groups accessible for binding. Studies also proved that, as biosorbent amount increases, the metal uptake decreases. This signifies an increase in the surface area of the adsorbent which results increase in the number of sites available for binding (Esposito, Pagnanelli, Lodi, Solisio, & Veglio, 2001). Due to the decrease in surface area per unit of adsorbent shows that the equilibrium concentration of metal ions increases with respect to the adsorbent concentration. Because of increase of pre-treated powder waste biosorbent concentration, the first order rate and second order rate constants also enhances, as a result of the increased binding sites of the biosorbent available. It also modify by pretreating the biomass with alkalis, acids detergents and heat, which may lead to the increase in the amount of metal uptake.(Ahalya, Ramachandra, & Kanamadi, 2003)

ONGOING TECHNIQUES

1. **Nanotechnology:** Nanotechnology for mercury treatment in an aqueous waste stream is the Thiol Self-Assembled Monolayers on Mesoporous Silica (Thiol-SAMMS). This new adsorbent was found by Pacific Northwest National Laboratory (PNNL) members. It comprises of a nanoporous ceramic substrate with a large surface area by a monolayer of thiol groups. It is synthesized by self-assembly process that uses sol-gel precursors and surfactant molecules. Gradually calcination to 500°C removes the surfactant templates and nanoporous ceramic substrate with a high surface area left behind, which can also be used for self-assembled monolayers of adsorptive functional groups. These Thiol functional groups have a high binding capacity for various heavy metals, including mercury. These are present in the ceramic substrate on one end and bind with mercury on the other end.These Thiol-SAMMS is proficient in selective affinity with mercury and can achieve a mercury loading capacity as high as 635 mg/g and mercury adsorption of SAMMS is not significantly affected by other cations and complexing anions present in waste solutions. (Domagalski, Alpers, Slotton, Suchanek, & Ayers, 2004)

2. **Air Stripping:** This technology is being identified to remove mercury from contaminated water. Generally, it is not been used to remove inorganic compounds such as mercury. However, study was conducted at the Savannah River Site in South Carolina to detect whether chemical reduction followed by collection of elemental mercury from the headspace air can remove low levels of mercury from groundwater and then groundwater was obtained from the feed as sample and effluent of an existing groundwater treatment system installed and developed to remediate a chlorinated solvents plume. This technique analyzed using stannous (Sn^{+2}) chloride to reduce Hg^{+2} to Hg^{0}, which is volatile, proceeded with collection of elemental mercury from the headspace air to remove the elemental mercury from water. Mercury concentrations in the groundwater ranged between 120 and 150 ng/L, and more than 95% of the mercury was Hg^{+2}. In groundwater samples Sn^{+2} chlorides were added in dosages from 0-766 mg/L ranging. After this, samples were sparged with air at an air-water ratio of 30 to 1 approximately. Stannous chloride doses greater than 0.011 mg/L resulted in more than 94% mercury removal, with the residual gross total mercury minimized to levels below 10 ng/L and vice versa (Looney, 2001).

This shows that chemical reduction along with air stripping is effective for treating mercury as this does not release any liquid or solid secondary wastes, and off-gas treatment may not be required for the expected air concentrations and mass release. Successful development of a chemical-reduction-based treatment system requires additional data on the key scientific questions and engineering evaluation of whether reliable long-term operation.(Looney, 2001)

CONCLUSION

Among the recent studies biosorption by filamentous fungi is the most frequently used technology to treat waste water contaminated with mercury. The literature shows that this technology has been used to meet regulatory cleanup levels, is commercially available to treat waste water, and generates a residual that typically does not require further treatment before disposal. There is always a scope of improvement and addition of new technologies or process improvement, the intent of this chapter is to call for action with the development of sound efficient and economical methods to reduce the contamination of mercury.

REFERENCES

Ahalya, N., Ramachandra, T., & Kanamadi, R. (2003). Biosorption of heavy metals. *Research Journal of Chemistry and Environment*, *7*(4), 71–79.

Amin, F., Talpur, F. N., Balouch, A., Chandio, Z. A., Surhio, M. A., & Afridi, H. I. (2016). Biosorption of mercury (II) from aqueous solution by fungal biomass Pleurotus eryngii: Isotherm, kinetic, and thermodynamic studies. *Environmental Progress & Sustainable Energy*.

Anaemene, I. (2012). The use of Candida sp. in the biosorption of heavy metals from industrial effluent. *European Journal of Experimental Biology*, *2*(3), 484–488.

Bahobil, A., Bayoumi, R., Atta, H., & El-Sehrawey, M. (2017). Fungal Biosorption for Cadmium and Mercury Heavy Metal Ions Isolated from Some Polluted Localities in KSA. *International Journal of Current Microbiology and Applied Sciences*, *6*(6), 2138–2154. doi:10.20546/ijcmas.2017.606.253

Barakat, M. (2011). New trends in removing heavy metals from industrial wastewater. *Arabian Journal of Chemistry*, *4*(4), 361–377. doi:10.1016/j.arabjc.2010.07.019

Beckers, F., & Rinklebe, J. (2017). Cycling of Mercury in the Environment: Sources, Fate, and Human Health Implications-A Review. *Critical Reviews in Environmental Science and Technology*. (accepted)

Bernhoft, R. A. (2011). Mercury toxicity and treatment: A review of the literature. *Journal of Environmental and Public Health*. PMID:22235210

Cohen, M. D., Draxler, R. R., Artz, R. S., Blanchard, P., Gustin, M. S., Han, Y.-J., ... Lei, H. (2016). Modeling the global atmospheric transport and deposition of mercury to the Great Lakes. *Elementa*, 4.

Das, S. K., Das, A. R., & Guha, A. K. (2007). A study on the adsorption mechanism of mercury on Aspergillus versicolor biomass. *Environmental Science & Technology*, *41*(24), 8281–8287. doi:10.1021/es070814g PMID:18200852

Dhankhar, R., & Hooda, A. (2011). Fungal biosorption–an alternative to meet the challenges of heavy metal pollution in aqueous solutions. *Environmental Technology*, *32*(5), 467–491. doi:10.1080/09593330.2011.572922 PMID:21877528

Domagalski, J. L., Alpers, C. N., Slotton, D. G., Suchanek, T. H., & Ayers, S. M. (2004). Mercury and methylmercury concentrations and loads in the Cache Creek watershed, California. *The Science of the Total Environment*, *327*(1), 215–237. doi:10.1016/j.scitotenv.2004.01.013 PMID:15172583

Driscoll, C. T., Mason, R. P., Chan, H. M., Jacob, D. J., & Pirrone, N. (2013). Mercury as a global pollutant: Sources, pathways, and effects. *Environmental Science & Technology*, *47*(10), 4967–4983. doi:10.1021/es305071v PMID:23590191

Dzionek, A., Wojcieszyńska, D., & Guzik, U. (2016). Natural carriers in bioremediation: A review. *Electronic Journal of Biotechnology*, *23*, 28–36. doi:10.1016/j.ejbt.2016.07.003

Esposito, A., Pagnanelli, F., Lodi, A., Solisio, C., & Veglio, F. (2001). Biosorption of heavy metals by Sphaerotilus natans: An equilibrium study at different pH and biomass concentrations. *Hydrometallurgy*, *60*(2), 129–141. doi:10.1016/S0304-386X(00)00195-X

Gaffney, J. S., & Marley, N. (2014). In-depth review of atmospheric mercury: Sources, transformations, and potential sinks. *Energy and Emission Control Technologies*, 2, 1–21. doi:10.2147/EECT.S37038

Harada, M. (1995). Minamata disease: Methylmercury poisoning in Japan caused by environmental pollution. *Critical Reviews in Toxicology*, *25*(1), 1–24. doi:10.3109/10408449509089885 PMID:7734058

Henry, J. R. (2000). An overview of the phytoremediation of lead and mercury. US Environmental Protection Agency, Office of Solid Waste and Emergency Response, Technology Innovation Office.

Huang, C., & Huang, C. (1996). Application of Aspergillus oryze and Rhizopus oryzae for Cu (II) removal. *Water Research*, *30*(9), 1985–1990. doi:10.1016/0043-1354(96)00020-6

Javed, M. M., Ikram-ul-Haq, , & Shahbaz, F. (2007). Biosorption of mercury from industrial effluent by fungal consortia. *Bioremediation Journal*, *11*(3), 149–153. doi:10.1080/10889860701548705

Khambhaty, Y., Mody, K., Basha, S., & Jha, B. (2009). Biosorption of inorganic mercury onto dead biomass of marine Aspergillus niger: Kinetic, equilibrium, and thermodynamic studies. *Environmental Engineering Science*, *26*(3), 531–539. doi:10.1089/ees.2008.0056

Kratochvil, D., & Volesky, B. (1998). Advances in the biosorption of heavy metals. *Trends in Biotechnology*, *16*(7), 291–300. doi:10.1016/S0167-7799(98)01218-9

Kurniati, E., Arfarita, N., Imai, T., Higuchi, T., Kanno, A., Yamamoto, K., & Sekine, M. (2014). Potential bioremediation of mercury-contaminated substrate using filamentous fungi isolated from forest soil. *Journal of Environmental Sciences (China)*, *26*(6), 1223–1231. doi:10.1016/S1001-0742(13)60592-6 PMID:25079829

Looney, B. (2001). Ultralow concentration mercury treatment using chemical reduction and air stripping (0013-936X). Academic Press.

Mahbub, K. R., Bahar, M. M., Labbate, M., Krishnan, K., Andrews, S., Naidu, R., & Megharaj, M. (2017). Bioremediation of mercury: Not properly exploited in contaminated soils! *Applied Microbiology and Biotechnology*, *101*(3), 963–976.

Mapolelo, M., & Torto, N. (2004). Trace enrichment of metal ions in aquatic environments by Saccharomyces cerevisiae. *Talanta*, *64*(1), 39–47. doi:10.1016/j.talanta.2003.10.058 PMID:18969566

Martínez-Juárez, V. M., Cárdenas-González, J. F., Torre-Bouscoulet, M. E., & Acosta-Rodríguez, I. (2012). Biosorption of mercury (II) from aqueous solutions onto fungal biomass. *Bioinorganic Chemistry and Applications*. PMID:23028382

Modak, J., & Natarajan, K. (1995). Biosorption of metals using nonliving biomass-A review. *Minerals & Metallurgical Processing*, *12*(4), 189–196.

Obenauf, P., & Skavroneck, S. (1997). *Mercury source sector asssessment for the Greater Milwaukee Area*. Retrieved from Organization, W. H. (1976). Environmental health criteria. 1. Mercury. *Environmental Health Criteria*, 1.

Organization, W. H. (1996). *Trace elements in human nutrition and health*. World Health Organization.

Pagnanelli, F., Toro, L., & Vegliò, F. (2002). Olive mill solid residues as heavy metal sorbent material: A preliminary study. *Waste Management (New York, N.Y.)*, *22*(8), 901–907. doi:10.1016/S0956-053X(02)00086-7 PMID:12423052

Prigione, V., Grosso, I., Tigini, V., Anastasi, A., & Varese, G. C. (2012). Fungal waste-biomasses as potential low-cost biosorbents for decolorization of textile wastewaters. *Water (Basel)*, *4*(4), 770–784. doi:10.3390/w4040770

Puranik, P., Modak, J., & Paknikar, K. (1999). A comparative study of the mass transfer kinetics of metal biosorption by microbial biomass. *Hydrometallurgy*, *52*(2), 189–197. doi:10.1016/S0304-386X(99)00017-1

Say, R., Yılmaz, N., & Denizli, A. (2003). Biosorption of cadmium, lead, mercury, and arsenic ions by the fungus Penicillium purpurogenum. *Separation Science and Technology*, *38*(9), 2039–2053. doi:10.1081/SS-120020133

Shafeeq, A., Muhammad, A., Sarfraz, W., Toqeer, A., Rashid, S., & Rafiq, M. (2012). Mercury removal techniques for industrial waste water. *World Academy of Science, Engineering and Technology*, *6*, 12–26.

Svecova, L., Spanelova, M., Kubal, M., & Guibal, E. (2006). Cadmium, lead and mercury biosorption on waste fungal biomass issued from fermentation industry. I. Equilibrium studies. *Separation and Purification Technology*, *52*(1), 142–153. doi:10.1016/j.seppur.2006.03.024

Veglio, F., & Beolchini, F. (1997). Removal of metals by biosorption: A review. *Hydrometallurgy*, *44*(3), 301–316. doi:10.1016/S0304-386X(96)00059-X

Vijayaraghavan, K., & Balasubramanian, R. (2015). Is biosorption suitable for decontamination of metal-bearing wastewaters? A critical review on the state-of-the-art of biosorption processes and future directions. *Journal of Environmental Management*, *160*, 283–296. doi:10.1016/j.jenvman.2015.06.030 PMID:26143501

Volesky, B., & Holan, Z. (1995). Biosorption of heavy metals. *Biotechnology Progress*, *11*(3), 235–250. doi:10.1021/bp00033a001 PMID:7619394

Wagner-Döbler, I., Von Canstein, H., Li, Y., Timmis, K. N., & Deckwer, W.-D. (2000). Removal of mercury from chemical wastewater by microoganisms in technical scale. *Environmental Science & Technology*, *34*(21), 4628–4634. doi:10.1021/es0000652

Wang, J., & Chen, C. (2006). Biosorption of heavy metals by Saccharomyces cerevisiae: A review. *Biotechnology Advances*, *24*(5), 427–451. doi:10.1016/j.biotechadv.2006.03.001 PMID:16737792

Wang, J., & Chen, C. (2009). Biosorbents for heavy metals removal and their future. *Biotechnology Advances*, *27*(2), 195–226. doi:10.1016/j.biotechadv.2008.11.002 PMID:19103274

Zafar, S., Aqil, F., & Ahmad, I. (2007). Metal tolerance and biosorption potential of filamentous fungi isolated from metal contaminated agricultural soil. *Bioresource Technology*, *98*(13), 2557–2561. doi:10.1016/j.biortech.2006.09.051 PMID:17113284

Chapter 8
Microalgae:
A Promising Tool for Remediation of Heavy Metals

Vinod Kumar
Uttaranchal University, India

Manisha Nanda
Dolphin (PG) Institute of Biomedical and Natural Sciences, India

ABSTRACT

The primary functions of ecosystems, wherever located, or whatever their constituents on earth, are to sustain life. They provide vital needs, commonly described as "ecological services," for all the living things residing within that particular system, or for those on its fringes. Such services include shelter, food, maintaining soils and hydrology, and providing livelihoods for its dwellers. Water is considered an important resource for mankind and other living organisms. Therefore, maintaining the water to a high quality is crucial. Bioremediation of heavy metals and micro pollutants by microorganisms (algae, bacteria, fungi, and yeast) has been used for the last two decades. All micro-organisms and microalgae are gaining increasing attention because microalgae can easily grow and is cheap to process and able to accumulate high metal content.

INTRODUCTION

All around the world natural water bodies are contaminated with different type of heavy metals and directly or indirectly discharge of wastewater by industries. These micro pollutants reached to our drinking water supplies, aquatic ecosystems and other

DOI: 10.4018/978-1-5225-4162-2.ch008

Copyright © 2018, IGI Global. Copying or distributing in print or electronic forms without written permission of IGI Global is prohibited.

organisms. Heavy metal toxicity has proven to be a major threat to living organisms. Bioremediation of waste water by microalgae is a technology that transfers harmful toxic chemicals to less harmful forms. Bioremediation is cost effective and efficient method for removal of heavy metals from waste water now-a- days.

The most abundant pollutants in the wastewater and in sewage are heavy metals (Hong *et al.*, 1996).Industrialization has proved to be one of the major reasons for the spread of heavy metal pollutants. Heavy metals are toxic, persistent and non-degradable that has a negative impact on the biosphere. These heavy metals leach into the drinking water reservoirs and freshwater habitats, altering the macro, as well as, microbiological communities (Teitzel & Parsek, 2003). Finally, they are accumulated through the food chain, leading to serious ecological and health problems. Phytoremediation appears to be a promising method that uses algae to clean up polluted water and soil. Algae have the natural ability to take up, accumulate and degrade the constituents that are present in their growth environment. Photosynthesis can be effectively exploited to generate oxygen from waste water remediation by algae. Algae are preferred to be used in wastewater treatment systems because of their efficiency to grow in and to take up nutrients from wastewater. Fresh-water microalgae are one of the potential organisms which can be safely used without any hazard to the environment. Instead of that the algal flora will help to reduce the carbon component in the waste water, as well as to remove the nitrate and phosphate component, whereas the oxygen cotent will increase in the water. There are various reports about the heavy metal uptake capacity of microalgae by different means, by adsorption or by intercellular uptake which are ultimately detoxified by metallothioneins or phytochelatins or are sequestered into the vacuoles or polyphosphate bodies. There are various Marine Algal flora which can really remove various heavy metals from aquatic environment, but we are more concerned about the heavy pollutants in rivers and waste water bodies where the fresh water microalgae can be used successfully. High Rate Algal Ponds or Algal Turf Systems are being used successfully for removal of pollutants in some places. But more studies should be done for using other combined technologies like immobilized algal cultures along with other to improve the different pollutants removal capacity by freshwater microalgae. Transgenic algae production must be attempted to increase their potentiality in these aspects by manipulating the detoxifying mechanisms.

ORIGIN OF HEAVY METALS IN AQUATIC ECOSYSTEMS

Most of the aquatic ecosystems are mainly used to industrial waste discharge. Pesticides and heavy metals are the major contaminants of both surface and groundwater. Metals enter into the aquatic ecosystems due to weathering of soil and rocks, volcanic eruptions and a number of human activities e.g., mining, processing and use of metals and/or substances containing metal contaminants (Jain 2004). Heavy metals may also be incorporated due to remobilization from natural soils as a consequence of variations in local redox conditions (Leung & Jiao 2006). González et al. (2007) studied the bioavailability and origin of heavy metals from the Nador lagoon sediments and reported that the most important traceelement anomalies (As, Cd, Co, Cu, Mn, Pb, Zn) were found, mainly around industry and old mining sites. Industrial activities has led to 100–1000 times higher concentrations of heavy metal in the environment (Carral et al., 1995). Okafor and Opuene, (2007) and Mohiuddin et al. (2010) observed the immobilisation of trace elements within the stream sediments. Thus, these could be involved in absorption, coprecipitation, and complex formation. Sometimes trace elements are co-adsorbed with other elements as oxides and hydroxides (Awofolu et al., 2005, Mwiganga et al., 2005). The levels of heavy metal levels can be controlled with proper treatment methods before they are released into the environment. This is important because heavy metal ions accumulate in living species and can pose a permanent toxic and/or carcinogenic effect (Manahan et al., 2000). The most commonly used treatment methods are chemical precipitation, oxidation/reduction, ion exchange, membrane technologies, reverse osmosis, and solvent extraction. Each method has its own advantages, disadvantages and ranges of applications which depends on the metal ion, initial concentration and raw water quality (Le Cloirec, 2005).

CONSEQUENCES OF HEAVY METAL CONTAMINATION

In recent years, ground soil and other materials polluted with heavy metals have become a serious environmental problem throughout the world due to their use in many manufacturing processes, which end up as waste in industrial effluent, through which heavy metals can enter water cycle, and are concentrated ultimately in the food chain reaching the toxic levels. Use of industrial wastewater for irrigation is a common practice in most of the third world countries which could alter the fertility of soil (Stillman & Presta, 2000). Moreover, accumulation of heavy metals in vegetation due to irrigation using waste water, a common practice in India, could affect human health.

Heavy metals from industrial processes are of special concern because they produce waste or effluent which causes chronic poisoning in aquatic animals (Ellis, 1989). While some heavy metals are purely toxic with no cellular role (Shi *et al.*, 2002), other metals are essential for life at low concentrations but become toxic at high concentrations (Badar *et al.*,2000). High concentrations of all heavy metals inhibit activity of sensitive enzymes. Heavy metals can damage the cell membrane, alter enzyme specificity, disrupt cellular functions and damage the structure of the DNA. Toxicity of these heavy metals occurs through the displacement of essential metals from their native binding sites or through ligand interactions. Toxicity can also occur due to alterations in the conformational structure of the nucleic acids and proteins and interference with oxidative phosphorylation and osmotic balance. Therefore, it is very important to take necessary steps to control it (Koropatnick & Leibbrandt, 1995).

Mercury is the most toxic heavy metal (Gerlach, 1981) and occupies the sixth position in the list of hazardous compounds (Nascimento & Chartone-Souza, 2003). Cadmium (Cd) is highly toxic even if it is present in low concentrations. It leads to DNA breaking (Bhatti & Noor, 2009). Conventional processes for removing metals from Industrial effluents include, oxidation or reduction, ion exchange, filtration, electrochemical treatment, reverse osmosis, membrane technologies and evaporation recovery.

Heavy metals whose density exceeds 5 g per cubic centimeter are generally considered as heavy metals. A large number of elements come under this category, but all are not toxic to human and animals Table 1. There are numerous heavy metals, some of which are highly toxic, like mercury, lead, arsenic, and cadmium. Others, like iron, manganese, chromium, and zinc are needed by the body in specific measurable amounts. Heavy metals cause health related problem like reduced growth and development, cancer, organ damage, nervous system damage, and in extreme cases, death.

BIOREMEDIATION

Conventionally, the removal of heavy metal ions from industrially polluted waters and soils as been achieved through a variety of ways centered on the chemical and physical nature of these pollutants. Such procedures include chemical precipitation, ion exchange, electrolytic technologies, chemical extraction and leaching, hydrolysis,

Table 1. Classification of metals based on toxicity

Heavy Metals	Level of Toxicity
Ca, Na, K	Non-Toxic
Fe, Mo, Mn	Low toxicity
Zn, Ni, Cu, V, Co, Cr	Average toxicity
As, Ag, Cd, Hg, Pb	High toxicity

polymer micro-encapsulation, reverse osmosis, excavation and land filling (Ahluwalia & Goyal, 2007; Kuan *et al.,* 2009).

The use of conventional chemicals for treating metal-bearing effluents and soils may also not be economically feasible, especially when dealing with low metal-ion concentrations. Most of the industries, however, cite high cost of technology transfer and the expenses associated with completely changing their manufacturing processes as reasons for not implementing clean-up technologies, or replacing the old system with cleaner, safer, environmental-friendly machinery. However, the possibility of employing, state-of-the-art biological treatments, or bioremediation techniques, as alternative methods for the treatment of contaminated soil and water is quite feasible (Khan *et al.,* 1997). These technological sound methods are both economically and socially attractive and have shown promising results in the removal of organics like petroleum hydrocarbons, as well as heavy metal ions from contaminated soil and water. As such these environmental-friendly proven techniques have a great potential and may be the better option for removing heavy metal-ions from contaminated sites.

The advancements in the field of biotechnology has made bioremediation one of the most rapidly developing fields of environmental restoration, utilizing microorganism to reduce, remove or detoxify various chemical pollutants, such as petroleum hydrocarbons, polycyclic aromatic hydrocarbons, polychlorinated biphenyls, esters, nitroaromatic compounds, industrial solvents, pesticides and heavy metals (Dua *et al.,* 2002).

Numerous bioremedial strategies have been developed for the treatment of contaminated sites and waste. However, the selection of the correct and most appropriate methodology is paramount for the given conditions and specific circumstances. Therefore, three basic guidelines have been designed to assist in this matter, which include: biochemical characterization of the pollutant, in other words the amenability of the pollutant to biological transformation to less toxic products; bioavailability, or accessibility of the contaminant to microorganisms; and bioactivity - the opportunity for optimization of the biological breakdown of the contaminant. Such criteria are made much easier with the added advancements in molecular genetics and genetic engineering for the bioaugmentation of the process of bioremediation (Dua *et al.,* 2002).

Biotechnological methods often involve using microorganisms that, initially, are not capable of tolerating polluted environments, but over a period of time are able to bio-transform, biodegrade and bioremediate them to non-toxic simpler substances or metabolites. The interactions of microorganisms in aqueous media and their ability to bioremediate heavy metals has been the focus of numerous studies in recent years and this technique is now employed by many industries and multinationals that either work with or produce, as a waste product, hazardous substances, which ultimately end up in the surrounding environment (White & Gadd, 1990; Fourest & Roux, 1992).

It is the understanding of the mechanisms of interaction that may control the microbial community structure, such as spatial isolation. This has important implications for the conservation of biodiversity, management of microbial communities for bioremediation, bio-control of plant diseases, and, of course, improved soil fertility, which is of paramount interest, especially for the agro-based economies, like that of Pakistan (Zhou *et al.,* 2002). It holds equally true for India.

Microorganisms can interact with metals via many mechanisms, some of which may be used as the basis of potential bioremediation strategies (Figure 1) (Ahmed, 2012).

Some metals such as Zinc (Zn), Copper (Cu), Nickel (Ni) and Chromium (Cr) are essential or beneficial micronutrients for plants, animals and microorganisms (Olson *et al.,* 2001) while others (e.g., Cadmium (Cd), Mercury (Hg), Arsenic (As) and Lead (Pb)) have no known biological and/or physiological functions (Gadd, 1992). Different microbes have been known to possess resistance to Cadmium (Cd) and Arsenic (As), two toxic heavy metal elements, widely distributed in nature.

Figure 1. Metal-microbe interactions affecting bioremediation
Source: Ahmed, 2012

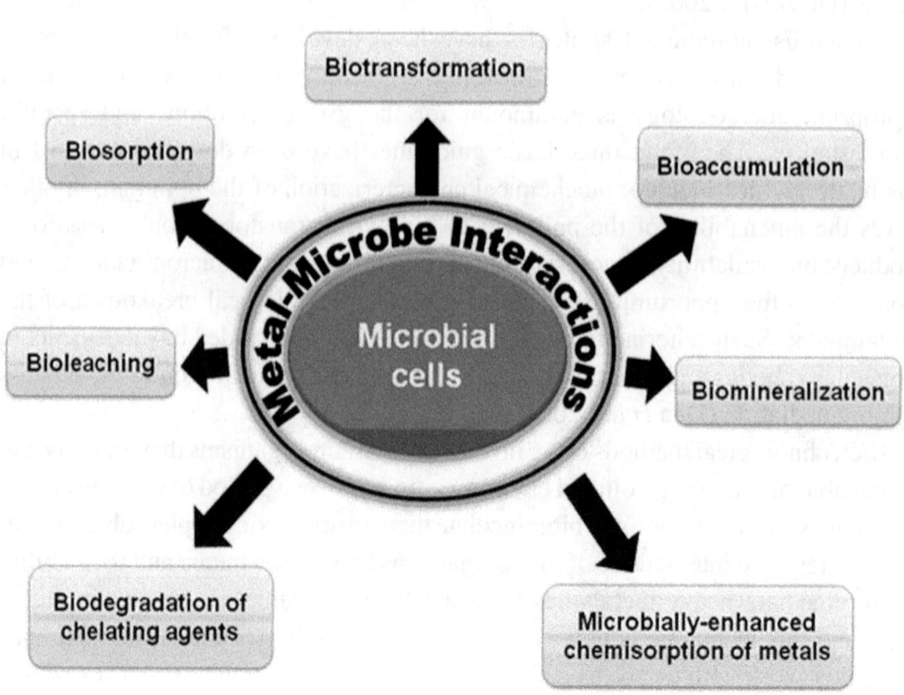

The microorganisms respond to these heavy metals by several processes; including transport across the cell membrane, biosorption to the cell wall, entrapment in extra cellular capsules, precipitation, complexation and oxidation-reduction reactions (Veglio *et al.*,1997).

METAL REMOVAL MECHANISMS ADAPTED BY ALGAE

On the basis of literature available how microalgae physiologically adapt to metal stress and removed heavy metals from water (Figure 2). This information is important because it can provide clues about how metals are toxic to cells.

Biosorption

Biosorption is an important for researcher to explore microbes from the ecological environment for the use in metal. Biosorption process is extensively studied *via* microbial biomass as a biosorbent for heavy metal removal. Biosorption of heavy metals by algae is two stage process (i) passive removal of metals ions by the algal cell (cell surface), this process occurs in both living and non-living cells. Passive removal includes physical adsorption, ion exchange, chemisorptions etc. (Monteiro

Figure 2. Mechanism of micro-algae adopting for heavy metals removal

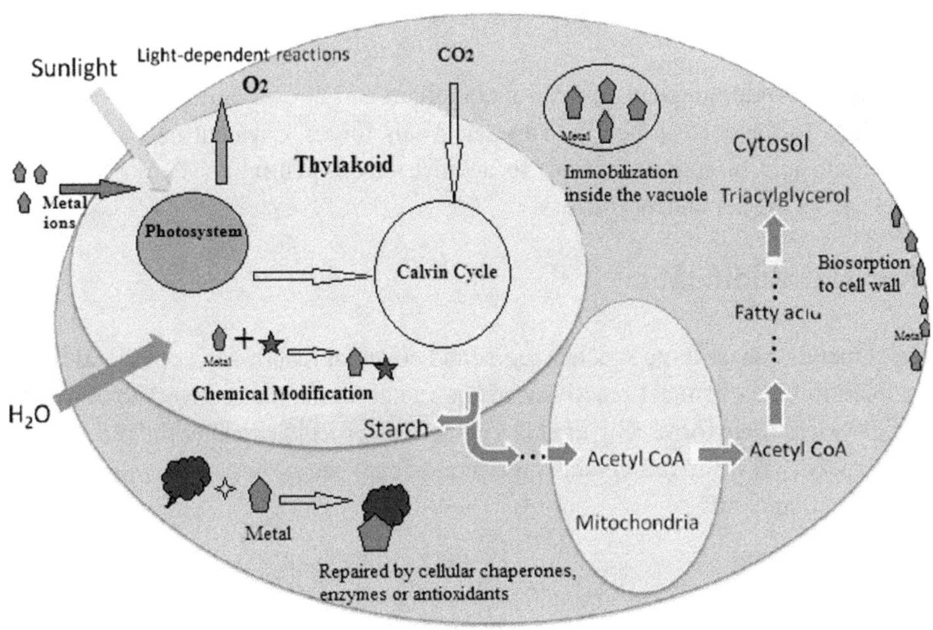

et al. 2012) (ii) Inside the algal cell, this process is occurring inside living cells and metabolism-dependent process (Kumar et al., 2015). Metabolism dependent removal includes precipitation, redox reactions, crystallization on the cell surface, or diffusion into the cell interior (Al-Qunaibit 2004).

Binding of Metals on the Cell Surface

Cell wall is a first defense line against heavy metals toxicity. Metal ions binding on the extracellular surface of the algae due to polymeric nature of its cell wall. The algal cell wall has negative charge due to many functional groups, such as, hydroxyl (OH), phosphoryl (PO_3O_2), amino (NH2), carboxyl (COOH), sulphydryl (SH), etc. which are responsible for the binding of heavy metals ions on the surface. It has been reported by many researchers that thicker cell wall better-modulated heavy metal toxicity. The proteins, carbohydrates and lipids present in cell wall of algae that could react with metallic species (Monteiro et al., 2012).

Ion Exchange

As mentioned earlier, the cell walls of microalgae consist mainly algal cell wall has negative charge due to many functional groups, such as, hydroxyl (OH), phosphoryl (PO3O2), amino (NH2), carboxyl (COOH), sulphydryl (SH), etc. which shows high affinity to metal cations of heavy metals (Mehta et al., 2002).

Physical Adsorption

Physical adsorption depends on cell metabolism and a reversible process. Metals ions binds to the cell surface by Van der Waals forces, covalent bonding, redox interaction, and biomineralization, to achieve electroneutrality (Perpetuo et al., 2011). It is a pH dependent process.

Chemical Modification

Heavy metals can undergo specific redox and covalent reactions in cells, and these reactions initiate chemical reactivity of the metal atoms that convert toxic metal species to less toxic forms (Silver et la 1996). This process can alter the redox state of the metal, create metal crystal precipitates or generate organometallic small-molecule compounds.

Efflux of Metal Ions

Cell membrane permeability of algae for metal ions leading to an efflux. According to Blaby-Haas and Merchant (2012), NRAMP (Natural Resistance- Associated Macrophage Proteins), ZIP (Zrt-, Irt-like Proteins), FTR (Fe TRansporter) and CTR (Cu TRansporter) families transporters are responsible for moving metal ions into the cytoplasm. These proteins are Group A transporters. Group B transporters includes CDF (Cation Diffu-sion Facilitator), P1B-type ATPases, FPN (Ferro Porti N) and Ccc1 (Ca (II)-sensitive Cross-Complementer1) /VIT1(Vacuolar Iron Trans- porter 1) families (Blaby-Haas & Merchant, 2012).

FACTORS AFFECTING HEAVY METAL REMEDIATION

Species

Heavy metal remediation by algae is varies to with each alga,i.e. it differs with genus and species. In contrast, even the algae belonging to the same group may have a different adsorption capacity. Algae of same genus but different in species respond differently to heavy metals like fresh water microalgae species C. *miniata*, C. *vulgaris*, and C. *reinhardtii* are reported to remove Hg, Cd, Pb, Ni, Cu and Zn. While Fe and Cr are removed by C. *vulgaris* and S. *platensis*; (González et al., 2011).

Tolerance Capacity

Though heavy metal toxicity to algae has been thoroughly reported by the number of researchers in the past couple of decades. Many algae growing in heavy metal contaminated environments display an ability to tolerate high concentrations of toxic metals. .Algal tolerance to heavy metals is highly dependent upon the defense response of algal species against possible oxidative damages and different mechanisms adopted by algae. These include the production of heavy metals binding factors and proteins (metallothionein, GSH, and phytochelatin conjugates), exclusion of toxic heavy metals from cells by ion-selective metal transporters (Arunakumara & Xuecheng, 2008; Hu et al., 2001).

CONCLUSION

Widely distributed heavy metals and other micro pollutants from industries, agriculture, mining are considered to be harmful to humans and the environment. Bioremediation of heavy metals by algae (micro and macro algae) is an economic and environmental friendly method. Mineral nutrients and heavy metals are removed by biosorption, binding of metals on the cell surface, ion exchange, physical adsorption, chemical modification, efflux of metal ions. Many researchers reported the algae tolerance towards heavy metals. Bioremediation of heavy metals by algae depends on different factors ranging from the anatomy of the algae species to the environmental conditions or growth medium of algae. For removal of heavy metals from wastewaters at industrial scale following points should be considered: (i) selection of strains with high metal sorption/removal capacity, (ii) Genetic modification of algae strains to surface area or efflux of metals, (iii) economic feasibility

REFERENCES

Ahluwalia, S. S., & Goyal, D. (2007). Microbial and plant derived biomass for removal of heavy metals from wastewater. *Bioresource Technology*, *98*(12), 2243–2257. doi:10.1016/j.biortech.2005.12.006 PMID:16427277

Ahmed, M. (2012). Implications of bacterial resistance against heavy metals in bioremediation: A review. *The IIOAB Journal*, *3*(3), 39–46.

Al-Qunaibit, M. H. (2004). A kinetic study of uptake of some cationic centities by the alga *Chlorella vulgaris*. *Chemindix*, *6*, 1–12.

Arunakumara, K. K. I. U., & Xuecheng, Z. (2008). Heavy metal bioaccumulation and toxicity with special reference to microalgae. *Journal of Ocean University of China*, *7*(1), 25–30. doi:10.1007/s11802-008-0060-y

Awofolu, O. R., Mbolekwa, Z., Mtshemla, V., & Fatoki, O. S. (2005). Levels of trace metals in water and sediment from Tyume River and its effects on an irrigated farmland. *Water S.A.*, *31*(1), 87. doi:10.4314/wsa.v31i1.5124

Badar, U., Ahmed, N., Beswick, A. J., Pattanapipitpaisal, P., & Macaskie, L. E. (2000). Reduction of Chromate by microorganisms isolated from metal contaminated sites of Karachi, Pakistan. *Biotechnology Letters*, *22*(10), 829–836. doi:10.1023/A:1005649113190

Bhatti, K. P., & Noor, A. R. (2009). Isolation and characterization of Chromium, Mercury and Cadmium Resistant bacteria. *Journal of the Chemical Society of Pakistan*, *31*(6), 928–936.

Blaby-Haas, C. E., & Merchant, S. S. (2012). The ins and outs of algal metal transport Biochimica et Biophysica Acta (BBA) -. *Molecular Cell Research.*, *1823*, 1531–1552.

Carral, E., Puente, X., Villares, R., & Carballeira, A. (1995). Background heavy metal levels in estuarine sediments and organisms in Galicia (northwest Spain) as determined by modal analysis. *The Science of the Total Environment*, *172*(2-3), 175–188. doi:10.1016/0048-9697(95)04788-3

Dua, M., Singh, A., Sethunathan, N., & Johri, A. (2002). Biotechnology and bioremediation: Success and limitations. *Applied Microbiology and Biotechnology*, *59*(2-3), 143–152. doi:10.1007/s00253-002-1024-6 PMID:12111139

Ellis, K.V. (1989). *Surface water pollution and its control.* Macmillan Press Ltd.

Fourest, E., & Roux, J. C. (1992). Heavy metal biosorption by fungal mycelia byproducts: Mechanism and influence of pH. *Applied Microbiology and Biotechnology*, *37*(3), 399–403. doi:10.1007/BF00211001

Gerlach, A. S. (1981). *Marine Pollution: Diagnosis and Therapy.* New York: Springer-Verlag Berlin Heidelberg. doi:10.1007/978-3-642-68182-0

González, F., Romera, E., Ballester, A., Blázquez, L., Muñoz, J. Á., & García-Balboa, C. (2011). Algal Biosorption and Biosorbents. In P. Kotrba, M. Mackova, & T. Macek (Eds.), *Microbial Biosorption of Metals* (pp. 159–178). Heidelberg, Germany: Springer Dordrech. doi:10.1007/978-94-007-0443-5_7

González, I., Águila, E., & Galán, E. (2007). Environ. Partitioning, bioavailability and origin of heavy metals from the Nador Lagoon sediments (Morocco) as a basis for their management. *Geol.*, *52*, 1581.

Hu, S., Lau, K. W. K., & Wu, M. (2001). Cadmium sequestration in *Chlamydomonas* re- inhardtii. *Plant Science*, *161*(5), 987–996. doi:10.1016/S0168-9452(01)00501-5

Khan, A. G., Bari, A., Chaudhry, T. M., & Qazilbash, A. A. (1997). Phytoremediation - a strategy to decontaminate heavy metal polluted soils and to conserve the biodiversity of Pakistan soils. In S. A. Mufti, C. A. Woods, & S. A. Hasan (Eds.), *Biodiversity of Pakistan* (pp. 133–144). Gainesville, FL: Pakistan Museum of Natural History, Islamabad & Florida Museum of Natural History.

Koropatnick, J., & Leibbrandt, M. E. I. (1995) Effects of Metal on Gene Expression. In Handbook of Experimental Pharmacology, Toxicology of metals, Biochemical Aspects. Springer Verlag. doi:10.1007/978-3-642-79162-8_5

Kuan, W. H., Hu, C. Y., & Chiang, M. C. (2009). Treatment of As (V) and As (III) by electrocoagulation using Al and Fe electrode. *Water Science and Technology*, *60*(5), 1341–1346. doi:10.2166/wst.2009.405 PMID:19717922

Kumar, K. S., Dahms, H. U., Won, E. J., Lee, J. S., & Shin, K. H. (2015). Microalgae – A promising tool for heavy metal remediation. *Ecotoxicology and Environmental Safety*, *113*, 329–352. doi:10.1016/j.ecoenv.2014.12.019 PMID:25528489

Leung, C. M., & Jiao, J. J. (2006). Heavy metal and trace element distributions in groundwater in natural slopes and highly urbanized spaces in Mid-Levels area, Hong Kong. *Water Research*, *40*(4), 753–767. doi:10.1016/j.watres.2005.12.016 PMID:16448684

Manahan, S. E. (1997). Environmental Science and Technology. Boca Raton, FL: Lewis Publishers.

Mehta, S. K., Singh, A., & Gaur, J. P. (2002). Kinetics of adsorption and up take of Cu2 by Chlorella vulgaris: Influence of pH, temperature, culture age and cations. J.En- viron. *Sci. Health Part A*, *37*(3), 399–414. doi:10.1081/ESE-120002837 PMID:11929076

Mohiuddin, K. M., Zakir, H. M., Otomo, K., Sharmin, S., & Shikazono, N. (2010). Geochemical distribution of trace metal pollutants in water and sediments of downstream of an urban river. *International Journal of Environmental Science and Technology*, *7*(1), 17–28. doi:10.1007/BF03326113

Monteiro, C. M., Castro, P. M. L., & Malcata, F. X. (2012). Metal up take by microalgae: Underlying mechanisms and practical applications. *Biotechnology Progress*, *28*(2), 299–311. doi:10.1002/btpr.1504 PMID:22228490

Mwiganga, M., & Kansiime, F. (2005). The impact of Mpererwe landfill in Kampala–Uganda, on the surrounding environment. *Physics and Chemistry of the Earth*, *30*(11-16), 744–750. doi:10.1016/j.pce.2005.08.016

Nascimento, A. M. A., & Chartone-Souza, E. (2003). Operon *mer*: Bacterial resistance to mercury and potential for bioremediation of contaminated environments. *Genetics and Molecular Research*, *2*, 92–101. PMID:12917805

Okafor, E. Ch., & Opuene, K. (2007). Preliminary assessment of trace metals and polycyclic aromatic hydrocarbons in the sediments. *International Journal of Environmental Science and Technology*, *4*(2), 233–240. doi:10.1007/BF03326279

Olson, J. W., Mehta, N. S., & Maier, R. J. (2001). Requirement of nickel metabolism protein HypA and HypB for full activity of both hydrogenase and urease in *Helicobacter pylori*. *Molecular Microbiology*, *39*(1), 176–182. doi:10.1046/j.1365-2958.2001.02244.x PMID:11123699

Perpetuo, E. A., Souza, C. B., & Nascimento, C. A. O. (2011). Engineering bacteria for bioremediation. In A. Carpi (Ed.), *Progress in Molecular and Environmental Bioengineering from Analysis and Modeling toTechnology Applications* (pp. 605–632). Rijeka, Croatia: In Tech Publishers.

Shi, W., Becker, J., Bischoff, M., Turco, R. F., & Konopka, A. E. (2002). Association of Microbial Community Composition and activity with Lead, Chromium and Hydrocarbon contamination. *Applied and Environmental Microbiology*, *68*(8), 3859–3866. doi:10.1128/AEM.68.8.3859-3866.2002 PMID:12147482

Silver, S., & Phung, L. T. (1996). Bacterial heavy metal resistance: New surprises. *Annual Review of Microbiology*, *50*(1), 753–789. doi:10.1146/annurev.micro.50.1.753 PMID:8905098

Stillman, M. J., & Presta, A. (2000). Characterizing Metal Ion Interaction with Biological Molecules—the Spectroscopy of Metallothionine. *Mol. Biol. Toxicol. Metals*, *1*, 276–290.

Teitzel, G. M., & Parsek, M. R. (2003). Heavy metal resistance of biofilm and planktonic *Pseudomonas aeruginosa*. *Applied and Environmental Microbiology*, *69*(4), 2313–2320. doi:10.1128/AEM.69.4.2313-2320.2003 PMID:12676715

Veglio, F., Beolchini, F., & Gasbarro, A. (1997). Biosorption of toxic heavy metals: An equilibrium study using free cells of *Arthrobacter* sp. *Process Biochemistry (Barking, London, England)*, *32*(2), 99–105. doi:10.1016/S0032-9592(96)00047-7

White, C., & Gadd, G. M. (1990). Biosorption of radionuclides by fungal biomass. *Journal of Chemical Technology and Biotechnology (Oxford, Oxfordshire)*, *46*, 331–343. PMID:1366965

Zhou, J., Xia, B., Treves, D. S., Wu, L. Y., Marsh, T. L., O'Neill, R. V., ... Tiedje, J. M. (2002). Spatial and resource factors influencing high microbial diversity in soil. *Applied and Environmental Microbiology*, *68*(1), 326–334. doi:10.1128/AEM.68.1.326-334.2002 PMID:11772642

Chapter 9
Phytoremediation of Heavy Metals in Vicinity of Industrially Polluted Sites Through Ferns:
An Overview

Alka Kumari
University of Lucknow, India

ABSTRACT

This chapter deals with screening of native Pteridophytes (ferns) in the vicinity of fly ash (FA) contaminated areas of National Thermal Power Corporation (NTPC) in Kanti and Kahalgaon, Bihar, India. These ferns were used for phytoremediation of heavy metals in the aforementioned areas. The metal tolerance ability and diversity of the native ferns were examined on several diversity indices (population density, concentration of dominance, heterogeneity, equitability) and the results were compared with reference site (nearby forests). The present data revealed that 11 species of ferns (10 terrestrial and 1 climbing fern) had rich frequency distribution and better performance on FA sites. The results also revealed that all ferns had accumulated significant metal content in their tissues. However, metal content was varied in different fern species. Effect of FA on photosynthetic pigments were also analyzed and compared with the reference site. The results of the above study suggest the suitability and usefulness of these ferns for reclamation of FA contaminated wastelands.

DOI: 10.4018/978-1-5225-4162-2.ch009

Copyright © 2018, IGI Global. Copying or distributing in print or electronic forms without written permission of IGI Global is prohibited.

INTRODUCTION

Coal is the chief source of energy and majority of power generating plants are coal based in India (Khan and Khan, 1996). As a result huge quantity of fly ash is generated as a by-product and besides, many toxic gases are emitted in the environment. In India, FA generation is expected to hike between 150 – 170 million tons per year by the end of 2012 (Pandey et al., 2011). It requires money and space for dumping. In addition, it has also been demonstrated to contain significant amount of noxious elements like Cr, Cd, Pb, Hg, As, Si, Ni, etc. has many deleterious effects are posing threat to the receiving environment. FA causes a number of human health related problems nearby inhabitants of coal-fired thermal power plants (Pandey et al., 2011). Besides, FA landfill decants into the nearby aquatic body and river through leaching increases turbidity, decreases primary productivity and adversely affect fishes and other aquatic biota (Pandey et al. 2011). Since, FA has been considered as the main source of pollution in nearby areas; therefore, its management is very important for the environmental perspective. Although FA is being used in various construction activities including land filling and query restoration (Pandey and Singh, 2010) as well as its safe utilization for agriculture (Pandey et al., 2009, 2010; Singh et al., 2011) and cultivation of non-agricultural crops would be helpful in its utilization but even then a huge quantity of FA remains for eco-friendly and cost-effective management. Therefore, cost-effective and green technology suggested for the management of FA is revegetation of the landfills by FA-tolerant plants, which serves the purpose of stabilization and provides a pleasant landscape (Rai et al., 2004; Pandey et al., 2009; Pandey and Singh, 2012). In this context, several plant species belonging to the family leguminosae has been found successful to revegetate fly ash landfills, (Gupta *et. al.* 2006; and Rai et. al., 2004) and various inorganic and organic blending of fly ash and inoculation of N_2 – fixing microbes (Blue Green Algae) has been found useful in enhancing growth of the plants. Besides, fern species *Pteris vittata* has been reported as hyper-accumulator of multimetals growing on fly ash dykes (Kumari et. al. 2011) including Arsenic hyper accumulator (Ma. *et. al,* 2001a, 2001b, Mehrag, 2002). Further, role of ferns in bioamelioration of fly ash has been done by (Kumari, 2007). Due to presence of high pH, negligible quantity of Nitrogen, Phosphorus and Potassium and high concentration of toxic metals in fly ash, crop plants show toxicity symptoms; whereas some common fern species like *Adiantum capillus-veneris* L., *Ampelopteris prolifera* (Retz.) Copel., *Diplazium esculentum* (Retz.) Sw., *Cyclosorus dentatus* (Forssk.) Ching and *Pteris vittata* L. do not show any visible toxicity symptoms during field surveys of both NTPC sites and it was also observed that these are luxuriantly growing in the vicinity of FA having high density cover. In this context, the feasibility of growing these ferns on FA dykes and their metabolic adaptation has been studied during present study by elucidating its diversity indices, metal accumulation and tolerance abilities. However, such studies on the ferns experiencing FA stress are meager and no attempt has been made to elucidate management systems in the ferns growing on various level of fly ash deposition. Present study is an attempt in this direction to utilize potential fern species in revegetating FA landfills/dykes.

MATERIALS AND METHODS

Study Site Description

The present study was conducted in vicinity of FA dumping sites of Kanti and Kahalgaon- NTPC units situated in Muzaffarpur and Bhagalpur districts of Bihar, India. The latitude and longitude of the study site is 25^{0}27’-26^{0}13’ N and 85^{0}27’-86^{0}1’ E and 25.8^{0}82’52” N and 87.2^{0}24’82” E respectively. This has installed capacity of 2340-3000 MW. FA generation is 1.82 million tons year^{-1} and discharged as slurry form in the FA pond. Temperature varies from 5°C to 46°C throughout the year. Total rainfall was 1100-1250 mm annually. During the sampling the humidity was 65-70%. Soil type of the study site is sandy alluvial soil with mixture of highly fertile calcareous soil of course of the Gandak and Ganga River respectively. The control reference sites used in this study is Vikramshila Gangetic Wildlife Sanctuary (VGWLS) situated in Bhagalpur district adjacent to Kahalgaon NTPC site having similar climatic condition and geographical position.

Sampling Materials and Techniques Used

FA samples and all native pteridophytes growing naturally on selected sites were collected and carried to laboratory for further studies. FA samples were collected randomly or in zigzag manner of sampling. Hand auger was used as a sampling tool to draw the sample. Composite samples were taken at a depth of 0-30 cm. to characterize the heavy metals and chemical parameters. The FA samples were air-dried for 7 days at room temperature and sieved and passed through a 2 mm sieve before laboratory analysis to remove stones and other undesired particles.

During field survey several diversity indices were also studied to compare the species density and basal cover of the ferns on contaminated sites with control reference sites. The several parameters of species diversity and community structure of ferns (frequency, density, abundance and relative values index, population density, Heterogeneity, equitability, Shannon Wiener’s index and concentration of dominance) were analyzed following the Misra (1968). Importance value index (IVI) was computed by summing up of relative frequency, relative density and relative dominance values of particular species. Species diversity index (H^-) was determined separately from the Shannon Wiener's information function (Shannon & Wiener, 1963). Concentration of dominance (Cd) was analyzed following Simpson (1949): (Cd) $=\Sigma$ (ni/N^2). Equitability or evenness (e) was calculated according to Pielou (1966) which refers to the degree of relative dominance of each species in that area.

Physico-Chemical Characterization of FA

To characterize the physical and chemical properties of FA, samples were analyzed for its physico-chemical properties like pH, electrical conductivity, bulk density, porosity and water holding capacity using distilled water with fly ash. The pH and electrical conductivity were determined through pH meter and conductivity meter by adding distilled water in fly ash in 1:5 ratio (Piper 1966), total organic carbon (Walkey & Black 1934); total nitrogen (Micro-kjeldahl digestion: Nelson & Sommers 1972); available phosphorus (Oleson & Sommers 1982) and cation exchange capacity (CEC) extracted with ammonium acetate (Allen *et al.* 1974). The samples were also analyzed for heavy metals (Fe, Mn, Cu, Zn, Cr, Cd, Pb and Ni) content.

Metal Uptake in FA and in Different Fern Species

For metal accumulation studies, FA and leaf samples were collected in plastic bags, washed on site with tap waters and dried at room temperature to carry inside laboratory and then dried in an oven at 60^0c for 24 hrs. The oven dried leaf tissues were digested in mixture of concentrated nitric acid (HNO_3) and perchloric acid ($HClO_4$) (V/V 3:1) at 140°C. The measurements of different heavy metals were performed with Atomic Absorption Spectrophotometer (Shimadzu AA-6300). The standered reference material of metals (E-Merck, Germany) was used for calibration. For blank either 0.1% or 1% nitric acid or deionised water was used. The blanks were run in duplicate to check the precision of the samples.

Estimation of Photosynthetic Pigments

For estimation of photosynthetic pigments, plant material (0.3 g) was ground in 80% ice-cold acetone in dark and centrifuged (4^0 C) (Eltek-RC 4100 F) at 10,000 rpm for 10 minutes. The absorbance of supernatants was measured on UV- visible spectrophotometer (Genesis-20, Thermospectronic). Chlorophyll and Carotenoid contents were calculated using the formula given by Arnon (1949) and Duxbury and Yentsch (1956) respectively.

Selection of Native Plants for Phytoremediation and Eco-Restoration of FA Landfills

The plant species selected for bioremediation and eco-restoration of FA landfills were based on their species richness and dominance as well as high biomass of the species on FA contaminated sites, which indicates its better adaptations towards climatologically condition of the area like resistance to high pH and salinity, heavy metal toxicity and low N and P content of FA.

RESULTS AND DISCUSSION

A total of 11 species of Pteridophytes were collected from both thermal power stations. *Ceratopteris thalictroides* and *Marsilea minuta* were embedded into mud and *Azolla pinnata* was free floating. *Adiantum capillus-veneris* L., *Adiantum incisum* L *Ampelopteris prolifera* (Retz.) Copel, *Christella parasitica* (L.) Lev., *Diplazium esculentum* (Retz.) Sw., *Pteris vittata* L., and *Thelypteris dentata* (Forssk.) Ching were luxuriantly growing in both the area. Single climbing fern *Lygodium flexuosum* L. was collected from Kahalgaon thermal power campus only. It was growing as epiphyte on *Accasia, Albizia labock* and *Shorea robusta* trees.

The habitat was examined with their terrestrial and other ecological parameters. Community structure of fern species on reference site and both the fly ash site were studied (Table-2). Fern species present on both sites were distributed frequently. The population was ranged from 156 plants per hectare (*Pteris vittata* L) to 198 plants per hectare (*Ampelopteris prolifera* Retz.) in fly ash site. *Ampelopteris prolifera* was the most dominated species in both sites with higher abundance (20.143 and 67.143) in reference as well as in fly ash site consequently. *Adiantum capillus-veneris* L was observed in low community structure in both fly ash sites. The importance values indexes were ranged from 41.68 (*Pteris vittata* L.) to 54.22 (*Ampelopteris prolifera*) whereas in fly ash site it ranged between 29.22 (*Thelypteris dentatus*) to 72.86 (*Ampelopteris prolifera*). Among the species present in both sites, such as *Pteris vittata, Ampelopteris prolifer*a, *Thelypteris dentatus, Diplazium esculentum,* and *Marsilea minuta* L. were the most dominating species. *Marsilea minuta* L. and *Ceratopteris thallicteroides* L. were also growing luxuriantly in muddy edge of fly ash pond. The fern *Christella parasitica* and *Lygodium flexuosum* were found in rare condition in all sites. Population size and plant diversity of fern species in both site were studied (Table 3). The overall population of reference site were measured 1035 plants per hectare with basal coverage of 110.5 m^2ha^{-1} and fly ash site the population were 1398 plant per hectare with basal coverage 149.9 m^2ha^{-1}. The population size and basal coverage both are higher in fly ash site than reference site. The low value of concentration of dominance were indicated that the dominance of species were equal in measured area. The low value of diversity index that is almost similar in both sites (1.78310and 1.78525) indicated that there is less variation in species in the studied area in both reference and fly ash site. The diversity index was comparatively low with rehabilitated tropical forest in North India ranged from 3.25 to 3.99 (Tripathi, 2001) and almost equal to natural terai forest (Kumari and Tripathi, 2007). Heterogeneity varied from 0.2189 (reference site) to 0.1957 (fly ash site) and show inverse trend with basal area. Measurement of biodiversity of specific area (local scale) on the basis of species richness does not provide a complete understanding about the individuals of the species as it suffers from the lack of evenness or equitability, which refers to the degree of relative dominance of each species in that area. The species were more evenly distributed in reference site than fly ash site. The Shannon Wiener's index of diversity was almost similar in both sites. It is the most polar index to measure the general diversity. Environmental conservation, sustainability and resilience of the ecosystems have been directly correlated with species diversity in several instances (Mooney *et al.*, 1995). Therefore, management of degraded lands should be looked with an ecosystem perspective to accommodate a high species richness and heterogeneity.

Physico-chemical properties of fly ash and soil of reference forest site (VGWLS) are depicted in Table-1. Fly ash of both sites used in the experiment was alkaline in nature (pH-8.8, pH-8.3) and having low total nitrogen and phosphorus content. The electrical conductivity of the fly ash was very high (7.7, 7.0) indicating high ionic concentration. Organic carbon was low in fly ash site as compared with reference site. Total nitrogen (0.01%) and Phosphorus (0.14%) contents were very low in FA. Although ferns can't directly contribute in enhancement of nitrogen but it enhances phosphorus status of fly ash landfills, which induces metal immobilization in fly ash (Cao, *et al.* 2003). Levels of almost all the elements were very high; however, Al, Si, Fe and Ni need special attention. Beside these fly ash contains rich micronutrients also. As fly ash contains several noxious metals and metalloids but ferns growing on fly ash show no any phytotoxic symptoms. All ferns growing in the vicinity of fly ash have accumulated significant metal content in comparison to control reference sites (Table-4). Since FA Contains high amount of essential elements as well as heavy metal contents, so significant uptake of these metals was observed in ferns. However, different species accumulated varied amount of these metals and maximum amount of toxic metals was found in leaves and least in roots. Roots are the first organ to contact with toxic metals but they translocate these elements into leaves through zinc and phosphorus mediated phytochelatins.

The effect of FA on photosynthetic pigments was studied (Table-5) and total chlorophyll content was found much affected in aquatic ferns but least affected in *Pteris vittata* plants grown in FA in comparison to reference sites. Similarly, *Ampelopteris prolifera, Diplazium esculentum* and *Thelypteris dentate* showed slight reduction in total chlorophyll content. It may be due to activity of δ- aminolevulinic acid dehydratase (ALAD) a metalloenzymes that maintain chlorophyll damage (Schoolingin & Cheeing, 1999). However, carotenoids were found increased in most of the ferns growing naturally in FA vicinity in comparison to ferns growing in control reference sites. Perhaps, it provides membrane bound stability/ internal metabolic adaptation to ferns (Singh et. al. 2005). Thus studies on photosynthetic pigments also indicate the suitability of these ferns for revegetating FA dykes.

CONCLUSION

The results of this study conclude that the ferns grow luxuriantly on fly ash without visual phytotoxicity having high population density; basal cover and concentration of dominance (Cd) on FA site in comparison to control site even in presence of high metal content and less total N and P. Besides, significant accumulation of metal contents like Fe, Si, Al, Ni, Cr, Cd, and As; and consequently slight decrease in photosynthetic pigments also indicates its metal tolerant ability. Thus the result of the above all studies suggest that these ferns are suitable candidate for revegetating FA landfills and also may be useful in management of other metal contaminated wastelands like FA dykes. Additionally, these ferns are also applicable for phytoremediation of heavy metals and thus minimize the pollution rate.

ACKNOWLEDGMENT

The author is thankful to the Directors, CSIR-IHBT and CSIR- NBRI providing laboratory facility and encouragement. Thanks are also due to DST (New Delhi) for financial support under Women Scientist Scheme (WOS-A- SR/FT/LS-117/2008).

REFERENCES

Arnon, D. I. (1949). Copper enzymes in isolated chloroplast. Polyphenol/oxidase in *Beta vulgaris. Plant Physiology, 24*(1), 1–15. doi:10.1104/pp.24.1.1 PMID:16654194

Cao Rocky, X., Lena, M. A., Chen Ming, Q., Singh Satya, P., & Harris Willie, G. (2003). Phosphate Induced metal immobilization in a contaminated site. *Environ. Pollut., 122*(1), 19–28. doi:10.1016/S0269-7491(02)00283-X PMID:12535592

Dexbury, A. C., & Yentch, C. S. (1956). Plankton pigment monograph. *Journal of Marine Research, 15*, 93–101.

Gupta, A. K., Dwivedi, S., Sinha, S., Tripathi, R. D., Rai, U. N., & Singh, S. N. (2007). Metal accumulation and growth performance of *Phaseolus vulgaris* grown in fly ash amended soil. *Bioresource Technology, 98*(17), 3404–3407. doi:10.1016/j.biortech.2006.08.016 PMID:17451948

Gupta, D. K., Tripathi, R. D., Rai, U. N., Dwivedi, S., Mishra, S., Srivastava, S., & Inouhe, M. (2006). Changes in amino acid profile and and metal content in seeds of *Cicer arietinum* L. (Chick pea) grown under various fly ash amendments. *Chemosphere, 65*(6), 939–945. doi:10.1016/j.chemosphere.2006.03.047 PMID:16682068

Khan, M. R., & Khan, M. W. (1996). The effect of Fly-ash on plant growth and yield of Tomato. *Environmental Pollution, 92*, 105–111. doi:10.1016/0269-7491(95)00098-4 PMID:15091390

Kumari, , & Lal, , Pakade, & Chand. (2011). Bioaccumulation of heavy metals by *Pteris vittata* in the vicinity of fly ash. *International Journal of Phytoremediation, 13*, 169–175. doi:10.1080/15226514.2010.525561

Kumari, A. (2007). Bioamelioration of fly ash by some fern species. *Ind. Fern J., 24*, 164–171.

Kumari, A., & Tripathi, K. P. (2008). Phytosociological studies on pteridophytic flora in tarai forest of north India. *Indian Journal of Forestry, 30*(4), 2008.

Ma, L. Q., Komar, K. M., Tu, C., Zhang, W., Cai, Y., & Kennelley, E. D. (2001a). A fern that hyperaccumulates arsenic. *Nature, 409*(6820), 579. doi:10.1038/35054664 PMID:11214308

Ma, L. Q., Komar, K. M., Tu, C., Zhang, W., Cai, Y., & Kennelley, E. D. (2001b). A fern that hyperaccumulates arsenic-addendum. *Nature, 411*(6836), 438. doi:10.1038/35078151 PMID:11373666

Mehra, A., Farago, M. E., & Banerjee, D. K. (1998). Impact of fly ash from coal fired Station in Delhi, with particular reference to metal contamination. *Environmental Monitoring and Assessment, 50*(1), 15–35. doi:10.1023/A:1005860015123

Mehrag, A. A. (2002). Variation in arsenic accumulation –hyperaccumulation in ferns and their allies. *The New Phytologist, 157*(1), 25–31. doi:10.1046/j.1469-8137.2003.00541.x

Mishra, R. (1968). *Ecology Work Book*. Oxford and IBH Publishing Company Culcutta.

Mooney, H. A., Lubchenco, J., Dirzo, R., & Sala, Q. E. (1995). *Biodiversity and Ecosystem functioning: Ecosystem analysis.* In V. H. Heywood & R. T. Watson (Eds.), *Global Biodiversity assessment* (pp. 326–345). UNEP.

Nelson & Sommers. (1972). Total carbon, organic carbon, and organic matter. In Methods of soil analysis, part 2 Chemical and microbiological properties. ASA.

Olson, S. R., & Sommers, L. E. (1982). Phosphorus. In Methods of soil analysis, Part 2 Chemical and Microbiological properties. ASA.

Pandey, V. C., Abhilash, P. C., Upadhyay, R. N., & Tewari, D. D. (2009). Application of fly ash on the growth performance and translocation of toxic heavy metals within *Cajanus cajan* L.: Implication for safe utilization of fly ash for agricultural production. *Journal of Hazardous Materials*, *166*(1), 255–259. doi:10.1016/j.jhazmat.2008.11.016 PMID:19111395

Pandey, V. C., & Singh, B. (2012). Rehabilitation of coal fly ash basins: Current need to use ecological engineering. *Ecological Engineering*, *49*, 190–192. doi:10.1016/j.ecoleng.2012.08.037

Pandey, V. C., Singh, J. S., Kumar, A., & Tewari, D. D. (2010). Accumulation of heavy metal by chickpea grown in fly ash treated soil: Effect on antioxidants. CLEAN – Soil, Air. *Water (Basel)*, *38*, 1116–1123.

Pandey, V. C., Singh, J. S., Singh, R. P., Singh, N., & Yunus, M. (2011). Arsenic hazards in coal FA and its fate in Indian scenario. *Resources, Conservation and Recycling*, *55*(9-10), 819–835. doi:10.1016/j.resconrec.2011.04.005

Pielou, E. C. (1966). The measurement of diversity in different types of biological collections. *Journal of Theoretical Biology*, *13*, 131–144. doi:10.1016/0022-5193(66)90013-0

Piper, C. S. (1966). *Soil and Plant Analysis*. New York: Inter Science.

Rai, U. N., Pandey, K., Sinha, S., Sinh, A., Saxena, R., & Gupta, D. K. (2004). Revegetating fly-ash landfills with Prosopis *julifora* L.: Impact of different amendments and *Rhizobium* inoculation. *Environment International*, *30*(3), 293–300. doi:10.1016/S0160-4120(03)00179-X PMID:14987858

Schoolinglin-Jordan, P.M., & Cheeing, K.M. (1999). Biosynthesis of heme in comprehensive natural products vol. 4 Aminoacid, Porphyrins and alkaloid chemistry. Elsevier Science Ltd.

Shannon, C. E., & Weaver, W. (1963). *The Mathematical theory of Communication.* Urbana, IL: University of Illinois Press.

Simpson, E. M. (1949). Measurement of diversity. *Nature*, *163*(4148), 688. doi:10.1038/163688a0

Singh, J. S., Pandey, V. C., & Singh, D. P. (2011). Coal fly ash and farmyard manure amendments in dry-land paddy agriculture field: Effect on N-dynamics and paddy productivity. *Applied Soil Ecology*, *47*(2), 133–140. doi:10.1016/j.apsoil.2010.11.011

Singh, N., Ma, L. Q., Srivastava, M., & Rathinasabapathi, B. (2006). Metabolic adaptations to arsenic-induced oxidative stress in *P. vittata* L. and *Pteris ensiformis* L. *Plant Science*, *170*(2), 274–282. doi:10.1016/j.plantsci.2005.08.013

Tripathi, K. P. (2001). *Ecology of a rehabilitated forest on sodic wasteland* (Ph.D Thesis). Lucknow University, Lucknow, India.

Walkely, Y. A., & Black, I. A. (1934). An examination of the Degtjareff method for determining soil organic matter and a proposed modification of the chromic acid titration method. *Soil Science*, *37*(1), 29–38. doi:10.1097/00010694-193401000-00003

APPENDIX

Table 1. Physico-chemical properties of fly ash sites of Kanti and Kahalgaon with reference site at Vikramshila Gangetic Wildlife Sanctuary of Bihar

Parameter	Fly-ash site (Kanti)	Fly ash site (Kahalgaon)	Reference site
PH	8.7±0.39	8.3±0.35	7.3±0.22
EC	7.7±0.35	7.01±0.18	1.7±0.05
CEC	1.29±0.33	1.32±0.21	1.42±0.31
Total Nitrogen (%)	0.02±0.01	0.01±0.002	0.10±0.01
Total Phosphorous (%)	0.04±0.03	0.07±0.005	0.97±0.03
Organic Carbon (%)	1.74±0.05	2.12±0.051	1.83±0.06
WHC (%)	36.53±0.27	43.36±0.35	55.6±0.75
Metals (μg/g)			
Zn	85±2.3	78±2.1	32.5±1.2
Fe	4310±211	3997±197	2910±139
Si	5760±175	4980±187	ND
Ni	225±11.5	181±9.7	24.7±1.2
Mn	79±5.3	67±3.9	46.5±2.1
Cu	59±3.5	53.6±3.83	33.5±1.1
Cd	39±1.1	35.3±2.01	ND
Pb	43±2.5	29.7±3.5	ND
B	31±2.5	22.5±2.06	ND
Al	4555±221	3775±247	ND

Mean ± SD (n=3, student's t test significant ($p<0.05$) as compared to control

Table 2. Community structure of fern species on reference and fly ash site at Kanti and Kahalgaon thermal power station, Bihar

Species	% F	D (No/ha)	Ab	IVI	H
Reference Site:					
Pteris vittata L.	100	156.7933	16.521	45.65	0.271274
Ampelopteris prolifera (Retz.)	100	189.2363	20.033	53.31	0.305567
Diplazium esculentum (Retz.) Sw.	100	194.051	21.317	55.56	0.375843
Adiantum capillus-veneris L.	100	169.035	17.732	46.47	0.27245
Adiantum incisum L.	100	137.0541	13.021	42.34	0.21022
Christella parasitica (L.) Lev.	100	141.1245	12.783	41.55	0.21342
Ceratopteris thalictroides L.	100	149.541	15.013	43.95	0.210305
Thelypteris dentata (Forssk.) St.John	100	160.544	16.857	48.09	0.290916
Lygodium flexuosum L.	100	147.067	14.913	41.325	0.310123
Marsilea minuta L.	100	187.7551	19.714	53.42	0.311064
Azolla pinnata R. Br.	100	168.7075	17.714	49.69	0.297503
Fly Ash Site:					
Pteris vittata L. KT	100	475.5215	46.756	58.2	0.393676
Pteris vittata L. KG	100	435.2305	47.547	57.5	0.354239
Ampelopteris prolifera Copel. KT	100	635.5657	66.233	73.54	0.456105
Ampelopteris prolifera Copel. KG	100	647.3451	70.012	75.25	0.0432505
Diplazium esculentum (Retz.) Sw. KT	100	527.6783	55.536	63.46	0.392366
Diplazium esculentum (Retz.) Sw. KG	100	547.347	55.352	65.453	0.392351
Adiantum capillus-veneris KT	100	127.245	25.234	25.255	0.13546
Adiantum capillus-veneris KG	100	135.249	25.795	27.125	0.13424
Adiantum incisum KT	100	128.231	24.413	26.578	0.12915
Adiantum incisum KG	100	135. 342	25.012	27.125	0..131053
Christella parasitica KT	100	123.543	23.567	25.975	0.145661
Christella parasitica KG	100	119.451	22.451	24.657	0.134578
Ceratopteris thalictroides KT	100	131.217	23.547	24.531	0.14793
Ceratopteris thalictroides KG	100	133.254	25.451	25.257	0.13575
Thelypteris dentatus KT	100	149.7973	15.59	29.48	0.183517
Thelypteris dentatus KG	100	139.213	14.51	28.59	0.16947
Lygodium flexuosum L. KG	100	123.2351	14.25	25.75	0.13456
Marsilea minuta L. KT	100	255.5401	26.176	38.69	0.179379
Marsilea minuta L. KG	100	267.2305	25.251	35.97	0.13541
Azolla pinnata R. Br. KT	100	311.6437	31.537	43.54	0.275078
Azolla pinnata R. Br. KG	100	291.5415	28.421	39.7	0.234565

KT= Kanti thermal power, KG= Kahalgaon thermal power.

Table 3. Population dynamics of fern species on reference and fly ash site at Kanti and Kahalgaon thermal power station, Bihar

Parameters	Reference Site	Fly Ash Site (Kanti)	Fly Ash Site (Kahalgaon)
Population density (No.ha^{-1})	1035	1277	1398
Basal cover (m^2 ha^{-1})	110.55	138.97	149.95
Heterogeneity	0.2189	0.1778	0.1957
Equitability (e)	36.70055	39.7667	41.1276
Shannon Wiener's index (H)	1.78310	1.74510	1.78525
Concentration of dominance (Cd)	0.165125	0.168303	0.171516

Table 4. Effect of FA on photosynthetic pigments of different fern species growing on FA dykes

Species	Treatment	Chl. a	Chl. b	Total Chl.	Carotenoids
Adiantum capillus-veneris	Control	1.64±0.02	0.97±0.01	2.61±0.02	2.13±0.02
	KT	1.34±0.03	0.78±0.02	2.12±0.03	2.27±0.03
	KG	1.43±0.01	0.81±0.01	2.24±0.01	2.33±0.01
Adiantum incisum	Control	1.46±0.01	0.85±0.02	2.31±0.02	1.89±0.01
	KT	1.27±0.02	0.63±0.01	1.90±0.01	1.93±0.02
	KG	1.32±0.01	0.71±0.03	2.03±0.02	1.99±0.01
Ampelopteris prolifera	Control	1.78±0.02	1.13±0.01	2.91±0.01	2.61±0.01
	KT	1.57±0.03	0.89±0.01	2.46±0.03	2.78±0.05
	KG	1.49±0.01	0.81±0.02	2.30±0.02	2.93±0.04
Azolla pinnata	Control	1.54±0.02	1.07±0.03	2.61±0.02	2.29±0.03
	KT	1.44±0.03	0.84±0.01	2.28±0.01	2.78±0.02
	KG	1.47±0.01	0.81±0.01	2.28±0.01	2.69±0.02
Christella parasitica	Control	1.33±0.02	0.87±0.01	2.20±0.02	2.38±0.04
	KT	1.14±0.05	0.62±0.02	1.76±0.04	2.63±0.01
	KG	1.17±0.02	0.66±0.01	1.83±0.03	2.57±0.02
Ceratopteris thalictroides	Control	1.24±0.02	0.77±0.01	2.01±0.02	1.89±0.02
	KT	1.11±0.01	0.55±0.02	1.66±0.02	2.13±0.01
	KG	1.23±0.01	0.49±0.01	1.72±0.03	2.09±0.01
Diplazium esculentum	Control	1.89±0.02	104±0.03	2.93±0.02	2.71±0.04
	KT	1.64±0.03	0.98±0.02	2.52±0.03	2.98±0.05
	KG	1.73±0.04	0.91±0.02	2.64±0.04	2.86±0.04
Lygodium flexuosum	Control	1.14±0.01	0.87±0.01	2.01±0.01	1.69±0.02
	KT	1.03±0.02	0.61±0.02	1.64±0.02	1.78±0.01
	KG	1.11±0.01	0.62±0.01	1.73±0.01	1.73±0.03
Marsilea minuta	Control	1.36±0.02	0.77±0.01	2.13±0.03	1.58±0.03
	KT	1.24±0.01	0.59±0.02	1.83±0.02	1.88±0.01
	KG	1.13±0.01	0.61±0.01	1.74±0.01	1.91±0.02
Pteris vittata	Control	2.24±0.03	1.37±0.01	3.61±0.04	2.77±0.03
	KT	1.94±0.02	1.03±0.02	2.97±0.01	2.89±0.04
	KG	1.83±0.01	0.81±0.01	2.64±0.01	2.97±0.01
Thelypteris dentata	Control	1.92±0.03	1.16±0.02	3.08±0.02	2.45±0.01
	KT	1.67±0.01	0.98±0.01	2.65±0.01	2.57±0.02
	KG	1.73±0.02	0.93±0.02	2.66±0.02	2.64±0.03

Mean ± SD (n=3, student's t test significant ($p<0.05$) as compared to control

Table 5. Metal accumulation (in µg/g) by different fern species growing on both FA dykes and control sites

Species	Sites	Fe	Si	Al	Ni	Cr	Cd	As
Adiantum capillus-veneris	Control	515±113	15±1.5	07±1.1	27±2.1	BDL	BDL	BDL
	KT	921±215	218±11.5	146.±5.1	64±5.2	47±7.5	21±1.5	08±0.6
	KG	41±3.6	246±79	225±11.5	79±5.3	59±3.5	29±1.1	13±1.1
Adiantum incisum	Control	436±89	11±1.2	13±2.1	17±2.3	BDL	BDL	BDL
	KT	478±85	179±2.5	87±2.1	57±2.5	33±1.3	18±1.1	5±0.3
	KG	489±73	168±2.6	107±2.1	71±5.4	46±2.2	15±0.3	14±1.3
Ampelopteris prolifera	Control	568±93	28±1.4	09±1.0	31±2.5	BDL	BDL	BDL
	KT	616±23	249±21	109±11	89±1.7	89±3.1	32±1.1	16±0.1
	KG	596±21	253±24	138±13	117±6.1	83±2.5	35±1.2	25±1.1
Azolla pinnata	Control	495±76	12±1.1	11±1.1	17±1.1	BDL	BDL	BDL
	KT	567±31	113±1.2	35±0.3	68±2.1	92±11	25±0.1	9±1.1
	KG	607±53	125±2.1	48±1.1	75±1.5	116±22	28±1.1	16±1.2
Christella parasitica	Control	305±53	13±1.2	04±0.1	25±1.3	BDL	BDL	BDL
	KT	378±45	92±11	24±1.1	48±1.2	36±1.3	21±1.1	4±0.1
	KG	403±31	106±13	32±1.2	53±1.5	39±1.1	18±12	11±0.3
Ceratopteris thalictroides	Control	411±25	11±1.1	09±1.1	18±1.1	BDL	BDL	BDL
	KT	465±31	56±1.3	38±1.2	34±0.1	55±1.3	26±1.2	6±0.1
	KG	476±33	64±2.1	43±1.3	39±0.1	61±1.5	29±1.1	9±0.1
Diplazium esculentum	Control	587±123	28±1.3	11±0.1	23±1.3	BDL	BDL	BDL
	KT	691±89	339±41.5	137±3.5	136±2.3	39±0.3	33±0.2	06±0.1
	KG	697±83	347±43.6	147±3.6	147±2.2	46±0.4	41±0.1	14±1.0
Lygodium flexuosum	Control	467±96	21±0.5	07±1.1	27±2.1	BDL	BDL	BDL
	KT	485±93	225±21.2	135±3.6	127±2.1	27±0.3	21±0.1	13±1.0
	KG	493±84	237±33.6	135±3.6	127±2.1	33±0.2	24±0.1	16±1.1
Marsilea minuta	Control	418±94	16±1.5	05±0.1	19±1.1	BDL	BDL	BDL
	KT	385±79	215±21.3	85±3.3	89±2.3	19±0.1	13±0.1	09±0.1
	KG	405±91	233±21.5	94±2.7	97±2.5	23±0.1	16±0.2	13±0.2
Pteris vittata	Control	625±112	35±1.3	19±1.2	39±2.3	BDL	BDL	BDL
	KT	973±111	349±34.5	159±6.4	183±3.5	54±2.8	47±0.4	31±0.6
	KG	987±97	378±29.5	175±3.5	195±7.3	69±1.2	39±0.3	28±1.3
Thelypteris dentata	Control	507±103	29±1.2	09±0.1	21±1.1	BDL	BDL	BDL
	KT	815±113	331±33.6	145±5.3	133±3.2	43±2.5	36±0.3	17±0.1
	KG	785±93	325±31.5	135±3.6	127±2.1	39±1.3	27±0.1	23±1.2

Mean of triplicate±SD, BDL= Beyond detection limit.

Chapter 10
Application of Plant Tissue Culture in Management of Hg Contaminated Water

Srijan Goswami
Indian School of Complementary Therapy and Allied Sciences, India

Debraj Modak
Institute of Genetic Engineering, India

Munmun Modak
Institute of Genetic Engineering, India

ABSTRACT

In the chapter, the authors describe how a plant can be selected as a natural indicator of heavy metal (mercury) contamination, how one can select a plant species to monitor a particular type of heavy metal and use those plants to remove the contaminants from the area under consideration. The chapter also contains a brief idea of environmental contamination by heavy metals and how the situation can be managed by the techniques of modern plant biotechnology. The authors add some reports and data of their in-vitro studies of mercury toxicity on Ananas sp., generally known as pineapple, for better understanding of the text.

INTRODUCTION

Since the genesis of the Earth and its atmosphere, the nature has its own mechanisms to deal with the naturally occurring pollutants or contaminants. Early human beings used natural resources like air, water, shelter and food mainly to satisfy their basic needs. The residues generated by the use of these resources were generally compatible

DOI: 10.4018/978-1-5225-4162-2.ch010

Copyright © 2018, IGI Global. Copying or distributing in print or electronic forms without written permission of IGI Global is prohibited.

with the environment, or were readily assimilated by it. Primitive humans usually ate raw plant and animals without disturbing the atmosphere with the smoke from fire. When the use of fire became common, the small amount of smoke generated by these activities were rapidly dispersed and easily assimilated by the atmosphere. Early civilizations drank from the same rivers and water bodies in which they bathed and deposited their wastes. However, the impact of such use of natural water resources was relatively low since natural cleansing mechanisms or self-purification processes easily restored the water's natural quality. The early humans used natural shelters like caves or built their homes from wood, mud or animal skins. They left behind items that were readily decomposed by microorganisms and absorbed by the nature. With the rise of Industrial Revolution, humans were better able than ever before to satisfy their natural needs for air, water, food or shelter. By the early twentieth century automobiles, appliances, processed food and beverage became popular as necessities. Meeting these acquired needs became a major thirst of modern industrial society. Unlike natural needs the acquired needs are usually met by the products that must be processed or manufactured or refined. The production, distribution and the use of such products usually result in more complex residues, many of which are not compatible with the environment, or readily assimilated by it. These residues are less compatible with the environment and less likely to be readily assimilated into the biosphere. As modern societies ascend the socioeconomic ladder, the list of acquired needs or luxuries increases and so do the complexities of the production chain and the amounts and varieties of pollutants generated. As a result, the impact of modern human population upon the environment is of major concern to the environmental engineers and scientists.

If one considers the present state of the environment, the exponential rise in industrialization and urbanization over past years has resulted in contamination of the air one breathes, the water one drinks, and even the foods and the soil and water in which the food grows. These are the consequences of use of synthetic organic chemicals such as solvents, pesticides, refrigerants and chemical intermediates etc. The contamination of environment has also resulted due to various industrial operations like fugitive emissions, accidental spills and leaks, dumping of hazardous substances etc. The excessive use of these synthetic chemicals over the years which are not degradable by natural means turns out to be hazardous or toxic. These issues can be managed to an extent with the proper use of Genetic Engineering Technologies and their application in Plant Tissue Culture and Plant Biotechnology with respect to Environmental Engineering. In other words application of plants in bioremediation processes. The process of performing bioremediation with the help of plants are known as phytoremediation. Phytoremediation is the term, coined in 1991, given to a group of technologies that employs plants for cleaning up of the contaminated sites.

EFFECTS OF MERCURY: THE HEAVY METAL PRODUCED AS INDUSTRIAL WASTE

In today's date Mercury is one of the most harmful pollutants faced by the human and wildlife globally. Toxic mercury gets emitted from various industries and gets accumulated in rivers, lakes and forests. Mercury is a highly potent neurotoxin that impacts the function and development of the central nervous system in both people and wildlife. Exposure to mercury is particularly dangerous for pregnant and breastfeeding women, as well as children, since mercury is most harmful in the early stages of development. As reported by National Wildlife Federation, Merrifield, researchers have found very high levels of mercury accumulated in a wide range of wildlife species, resulting in dangerous reproductive and neurological problems. Fishes have difficulty schooling and had decreased spawning success. Birds laying fewer eggs and having trouble caring for their chicks. Mammals are having impaired motor skills that in turn effecting their ability to hunt and find food thus threatening their survival. Along with these findings, some evidence indicated that elevated mercury levels adversely affect a species' immune systems. All these effects together possess a severe threat to the survival of wildlife and human population. Mercury is one of the naturally occurring, toxic heavy metal. But, human activity has significantly increased mercury levels in the environment over the past several centuries. Once emitted to the air, mercury falls to the earth and gets accumulated in water and soil where it gets transformed into a compound known as methyl mercury, which is a highly toxic form that gets accumulated in the tissues of wildlife and people.

According to Minnesota Pollution Control Agency and The U.S. Environmental Protection Agency (EPA) there are three widely available forms of mercury namely elemental, inorganic and methyl mercury. When present in high concentrations these forms of mercury can cause serious damage to both environment and health. Inhaling elemental mercury, is highly dangerous for health as it damages the nervous system in permanent manner. The pathogenesis of mercury is extremely precise. The exposure to mercury causes trembling and tingling of hands, limbs and extremities. Higher the concentration of exposure more prominent will be the effect. If exposed to higher concentration, results in death.

PLANT BIOTECHNOLOGY AND BIOREMEDIATION

Plants are the silent workers of the planet. It absorbs natural constituents from its surrounding environment. Plants are also the natural indicator of the condition of the environment around them. So if the changes in the plant's structure and function can be identified and studied carefully, then several of the current environmental issues could be managed and regulated. Due to the rise of the discipline of plant tissue culture studying the plant behavior with respect to a particular environmental condition has become much easier. The term “plant tissue culture”, broadly refers to the in vitro cultivation of all plant parts, whether a single cell, a tissue, organs, under completely sterile condition. The underlying principles involved in plant tissue culture are very simple. Firstly, a plant part from the intact plant is isolated. Secondly, it is necessary to provide the plant part with appropriate environment in which it can express its intrinsic or induced potential. Finally, the above procedures must be carried out aseptically. Insert History. Application of biotechnology for better environment has also been made using plants for identification and filtration of industrial wastes that are introduced in to the environment. Bioremediation programs using plants involves the use of various species of plants for cleaning the contaminated air, land, lakes and waterways. Recombinant DNA Technology helped in improving the efficacy of these processes. Recombinant DNA Technology are also being used for transforming plants into bio indicators, in which the organism under consideration are genetically modified so that it gives a special type of response in presence of several chemical pollutants and hazardous chemicals. The mercury resistance gene (*mer*) or toluene degradation gene (*tol*) are used for genetically modifying organisms.

Phytoremediation is the implementation of plants for in situ removal, degradation or containment of contaminants present in soil, sludges, surface and ground water etc. Phytoremediation can be stated as the efficient and well planned use of plants for removal, detoxification or immobilization of environmental contaminants, with the help of natural biological, chemical or physical activities and processes of the plants. Plants are unique organisms equipped with remarkable metabolic and absorption capabilities, as well as transport systems that can take up nutrients or contaminants selectively from the environment, be it soil or water. The plants can be subsequently harvested, processed and disposed. Phytoremediation can be employed for cleaning up metals, pesticides, solvents, explosives, crude oil, polyaromatic hydrocarbons, and other contaminants. According to United Nation Environment Programme (UNEP), the phytoremediation process is classified into following basic types: Phytodegradation, Phytostimulation or Rhizodegradation, Phytovolatilization, Phytoextraction, Rhizofiltration, Phytostabilization depending on their mechanism of action.

Plants as Bio indicators

Bio indicators are living organisms like microorganisms, plants or animals, in their natural or genetically modified form, implemented for screening the health of natural ecosystem in the environment. Based on the reports published by Khatri and Tyagi in 2015, naturally occurring bio indicators are being implemented for assessment of environmental condition and are also useful in monitoring changes in the environment and their subsequent impact of wildlife and human population. There are several advantages of using plant as bio indicators and some of them are as follows:

1. Plants are easily available and can survive in variety of ecosystems.
2. Biological impacts can easily be determined through plants.
3. Monitoring synergistic and antagonistic impacts of various pollutants on living organisms.
4. Easy monitoring and detection of harmful effects of pollutants on plants as well as animals at an early stage.
5. Cost effective procedure.

In 1992 Batiuk et al. and in 2006 Joanna et al. reported that plants are being used as very sensitive tool for environmental status recognition and prediction. According to Walsh 1978 and Peterson in 1986, the state and health of an environment can be determined by the presence and absence of certain plants because they react to ecological changes, climatic conditions and quality of air.

PLANT TISSUE CULTURE: LABORATORY ORGANIZATION

The overall design in any laboratory organization must focus on maintaining complete sterile environment. The specific design can vary but an effective laboratory organization for plant tissue culture and related works must include certain elements irrespective of whether used by a scientist or a group of workers. The size and proportion of different parts of the facility will be dependent upon the function, purpose and size of operation. The basic rule essential in the planning of an efficient and functional laboratory is the order and continuous maintenance of cleanliness because all operations requires aseptic conditions.

A general guideline for setting up a facility is to focus on a design that different units and activities need to be arranged to make operational steps possible. Thus a tissue culture facility should have following fundamental features:

Washing Facility

An area with chemical resistant sinks and proper drainage systems is crucial. The general method for washing laboratory glassware involves treatment and soaking with chromic and sulphuric acid followed by thorough washing with tap water and then washing with distilled water.

Because acids are corrosive in nature, it is recommended that for routine procedures glasswares should be soaked in a 2% detergent cleaner for 16 hours followed by washing with 60°C to 70°C hot tap water, and finally with distilled water. The washed glassware should then be dried at 150°C to 200°C in a hot air oven for around 1 to 4 hours. The highly contaminated glassware should be first treated in chromic acid for about 16 hours followed by regular washing methods.

Sterile Transfer Area

The laminar airflow cabinet is the most essential equipment used for aseptic manipulation. The cabinet should be designed with horizontal air flow system for general tissue culture purposes. But in cases where the tissue culture process involves the use of pathogenic microorganisms, vertical air flow instead of horizontal one is highly recommended.

Culturing Facilities and Incubation Room

Plant tissue cultures should be incubated under conditions of well controlled temperature, illumination, photoperiod and humidity and air circulation. Incubation culture rooms should be contamination free.

Green Houses

Green houses are required to grow regenerated plants for further propagation and for growing plants to maturity. This facility is required as a transitional step of taking plant materials from culture containers present in the controlled room to the field. Thus in the greenhouse, plants are acclimatized and hardened before being transferred to the field conditions. The plants are grown in the greenhouse to develop adequate root system and leaf structures to withstand the field environment. The greenhouse should be equipped with cooling and heating systems to control temperatures.

NUTRIENT MEDIUM FOR TISSUE CULTURE

The greatest progress towards the development of nutrition media for plant cells grown in culture took place in 1960s and 1970s. The fundamental nutritional requirements of cultured plant cells are very similar to those utilized by plants in nature. The nutritional composition varies according to the cells, tissues, organs and protoplasts and also with respect to particular plant species.

Table 1. Preparation of stock solution of MS medium

Constituent	Concentration in Stock Solution
Macronutrients NH4NO3 KNO3 MgSO4.7H2O CaCl2.2H2O KH2PO4	16500 mg/L 19000 mg/L 3700 mg/L 4400 mg/L 1700 mg/L
Micronutients H3BO3 MnSO4.4H2O ZnSO4.7H2O Kl Na2MoO4.2H2O CuSO4.5H2O CoCl2.6H2O	620 mg/L 2230 mg/L 860 mg/L 83 mg/L 25 mg/L 2.5 mg/L 2.5 mg/L
Vitamins Nicotinic acid Thiamine HCl Pyridoxine HCl **Others** Glycine	50 mg/100mL 50 mg/100mL 50 mg/100mL 50 mg/100mL

Different types of culture have unique requirements of one or more nutritional components. A nutrient media is defined by its mineral salt composition, carbon source, vitamins, phytohormones and other organic supplements. Any number and concentration of amino acids, vitamins, phytohormones or organic supplements can be added in an infinite variety of compositions to a given salt composition to achieve the desired results.

In the year 1962, Scientists named Murashige and Skoog developed the composition of a plant tissue culture media known as MS media. This media is the most widely used plant tissue culture media till date.

Preparation of Stock Medium

Above Listed Constituents

- **Macronutrients and Micronutrients:** All the chemical components for the preparation of macronutrients and micronutrients, as listed in Table 1, were weighed one by one and dissolved in double distilled water accordingly and ultimately the volume was made up to 1 lit by addition of double distilled water.

- **Vitamins:** The 3 vitamins listed in Table 1, namely Nicotinic Acid 50mg, Thiamine Hydrochloride 50mg and Pyridoxine Hydrochloride 50mg were weighed and dissolved in 100 ml double distilled water separately.
- **Other Constituent:** 50 mg of Glycine was weighed and dissolved in 100 ml double distilled water and mixed uniformly.

Growth Hormones

For preparation of stock medium 10 mg powder of N6 Benzyl Adenine, a type of cytokinin was dissolved in 1 ml of 1(N) HCl. Volume adjusted up to 10 ml by addition of 9 ml double distilled water.

10 mg powder of Indole Acetic Acid, a type of Auxin dissolved in 1 ml of absolute alcohol. The volume adjusted up to 10 ml by addition of 9 ml double distilled water.

All the containers containing stock solutions are stored in a refrigerator at 4°C.

Preparation of Working Medium (1 L)

All stock solutions are brought to room temperature from the refrigerator approximately one hour before preparation of working media. First approximately 400 ml of double distilled water is taken in a conical flask with capacity of holding one liter liquid. Then the following components are weighed and added to the conical flask one by one as indicated below:

- 30 grams of Sucrose (added fresh)
- 100 ml macronutrient (from stock solution)
- 10 ml micronutrient (from stock solution)
- 40 mg Fe.EDTA-Na salt (added fresh)
- 2 ml Glycine (from stock solution)
- 500 µl Nicotinic Acid (from stock solution)
- 500 µl Pyridoxine (from stock solution)
- 100 µl Thiamine Hydrochloride (from stock solution)
- 100 mg Inositol (added fresh)
- 100 mg Adenine Sulphate (added fresh)
- 1 ml N6 Benzyl Adenine (1 mg/L from stock solution)
- 0.1 ml Indole Acetic Acid (0.1 mg/L from stock solution)

After dissolving mercuric chloride with alcohol in different concentrations was added to the culture media.

Then the volume was made up to 1 liter by adding double distilled water.

The pH of the media was adjusted to 5.7 strictly by using 1(N) KOH or 1(N) HCl

Then the 20 ml of the prepared culture media was dispensed into a plant tissue culture test tubes (20 ml/tube).

A filter paper was folded and inserted inside the tube such that it gets immersed in the media for making a paper bridge that provide support to the floating plant that are being cultured.

The test tubes were sealed with cotton plugs, followed by wrapping with brown paper and then autoclaving at 121°C temperature with 15 psi pressure for 15 to 20 minutes. This step is essential for complete sterilization of the culture media.

STERILIZATION TECHNIQUES

Sterilization is a technique used for removal of microorganisms. The proper maintenance of aseptic condition are critical for successful tissue culture. The need for aseptic conditions requires that all culture vessels, media, and instruments used in handling tissues including explant itself be sterilized. The emphasis is being laid on keeping the air, working surface and floor free from excess dust. All the actions are to be performed under laminar airflow chamber maintaining all aseptic conditions. The laminar chamber used for performing plant tissue cultures should not be shared with any other disciplines like that of microbiology, animal cell culture or pathology etc. In this section the authors explains the specific procedures or guidelines in brief for the sterilization of nutrient media, containers, explant and small instruments and also for maintaining a sterile environment. The most accepted way of sterilizing culture media and required apparatus is by autoclaving the material using steam or dry sterilization.

- **Steam Sterilization:** Most culture media are sterilized by using an autoclave machine. The standardized criteria for autoclaving culture media are 121°C with a pressure of 15 psi for 20 minutes. An autoclave has a temperature range of 115°C to 135°C. Proper time, pressure, temperature, and volume of the object to be sterilized are critical for optimum sterilization. The test tubes or flasks containing 20 to 50 ml of media should be treated at 121°C with 15 psi for about 20 minutes. For the flasks containing 50 to 500 ml nutrient media should be treated at 121°C with 15 psi for 25 minutes. For the flasks containing 500 to 5000 ml nutrient media should be treated at 121°C with 15 psi for 35 minutes. It must be remembered that the heat penetration is very important in an autoclave and more the volume more time should be given for proper sterilization. After autoclaving, the sterilized materials are kept in the laminar airflow and tops of bottles are tightened when they are cool. Speed, simplicity and destruction of viruses are the major advantages of an autoclave. While the disadvantages are changes in pH, components might separate out and chemical reactions can occur resulting in a loss of activity of media constituents. During and after autoclaving, the following issues should be considered that, autoclaving at too high temperature can caramelize sugars, which may be toxic. Volatile substances can be destroyed by the use of an autoclave. Autoclaving for too long periods can precipitate salts and depolymerize the agar. Care should be taken to use the correct duration and temperature.

- **Dry Sterilization:** The sterilization of components can be carried out in dry heat for about three hours at 160°C to 180°C. An exposure of 160°C dry heat for two hours is considered equivalent to moist heat sterilization at 121°C for 15 minutes with respect to effectivity. Hot air ovens are generally employed for performing this kind of sterilization.
- **Filter Sterilization:** Certain components such as the growth factors, amino acids, vitamins, and toxins are highly heat sensitive and might get degraded while autoclaving. These type of chemical substances should therefore be sterilized with simple filtration processes.
- **Ultraviolet Sterilization:** The nutrient media and other components used for tissue culture procedures are sterilized using UV radiation for a limited period of time.
- **Alcohol Sterilization:** It is crucial that the hand of the laboratory personnel be relatively aseptic during manipulation work. Washing hand with antibacterial detergent followed by spraying of 70% ethanol on hands are recommended. The laminar air flow cabinet should be sprayed with 70% ethanol before performing tissue culture related work.
- **Flame Sterilization:** This technique is used for sterilization of instruments that are continuously used during manipulation work. Instruments are soaked in 70% to 80% alcohol followed by flaming with burner inside a laminar airflow chamber, and should be carried out repeatedly while aseptic manipulation work is in progress.
- **Sterilization of Plant Material:** Plant material can be surface sterilized with a variety of chemicals. It is performed for complete eradication of microorganisms. The surface sterilization of plant material is explained below in detail.

METHODOLOGY: SURFACE STERILIZATION AND INOCULATION OF THE PLANT MATERIAL

Surface Sterilization

The technique of surface sterilization is extremely crucial and should be performed with extra care. All the trace of contaminants should be removed from the surface of the plant material to be used as tissue culture component, known as explant by using certain chemicals known as sterilant. The concentration of the chemicals used for sterilizing and the duration of treatment using these chemicals should be controlled so that it should be appropriate to remove all the contaminants but does not damage the plant material under consideration.

Surface Sterilization and Inoculation of Explant

1. Healthy plants, known as explant are collected from the garden.
2. Thoroughly washed under tap water for removing soil and other loose surface contaminants.
3. 100 ml distilled water taken in a sterilized container and four drops of Tween 20, a type of liquid detergent was added to it and the container was shaken for 6 minutes thoroughly under aseptic condition.
4. The content of the container was decanted and washed well by distilled water for about three times.
5. 99 ml of distilled water and 1 ml of 0.5% mercuric chloride and mixed well thoroughly, this mixture to be used as surface sterilant for the plant material. For preparing 0.5% of mercuric chloride, 0.5 gm. of HgC12 was dissolved in 1 ml of absolute alcohol. Then the container was shaken for 6 minutes thoroughly under aseptic condition.
6. After sterilization of the plant material, also known as explant, the solution is decanted and washed thoroughly by distilled water for minimum three times.
7. The explant is excised by sterile forceps and scalpel and inoculated in the culture tubes containing the liquid media with different concentration of heavy metal, in this case mercuric chloride.
8. All the procedures from step 3 to 7 was performed under complete aseptic condition inside Laminar Airflow Chamber to avoid any type of contamination.
9. The culture tubes are kept inside the incubation room for plant tissue culture for optimal growth.
10. The different physical conditions were strictly maintained during these incubation period and these includes:

- **Photoperiod:** The photoperiod maintained by adjusting 17 hours day mode and 7 hours night mode in the plant tissue culture incubation room.
- **Temperature:** The temperature is maintained around 28°C, more or less 2°C.
- **Humidity:** Around 80% approx.

OBSERVATION AND RESULTS

The authors performed a simple, very common yet effective procedure of determining the minimum lethal dose of mercury, and used easily available Ananas plant, commonly known as Pineapple, that can be used as an indicator for presence of heavy metals, in this case Mercury. The authors studied the growth and stress response of Ananas plant when exposed to different concentration of mercury compounds in vitro in two sets. The two sets are labeled SET I and SET II and each set had the cultures in the gradual concentration of 150 mg/L, 130 mg/L, 100 mg/L, 70 mg/L, 50 mg/L and 0 mg/L of Mercuric Chloride in each culture tube.

INTERPRETATION OF OBSERVATIONS

Performing the basic screening tests in the laboratory makes the decision making strategy much easier. The authors cultured the Ananas plants in liquid media containing different concentration of mercury compound in each culture tubes and observed the outcomes that are interpreted as follows:

Table 2. Observation Table

Concentration of HgCl2	Status of the Plant (SET I)	Status of the Plant (SET II)
Control	Normal Growth	Normal Growth
50 mg/L	Alive (Less healthy w.r.t Control)	Alive (Less healthy w.r.t Control)
70 mg/L	Alive (Less healthy w.r.t previous concentration)	Alive (Less healthy w.r.t previous concentration)
100 mg/L	Barely Alive	Barely Alive
130 mg/L	Lethal	Lethal
150 mg/L	Lethal	Lethal

1. In both SET I and SET II, the culture tube labeled as Control, which did not contain any mercury compound, the plant material received a healthy growth.
2. In both SET I and SET II, the culture tube labeled 50mg/L, the plant materials were alive but was less healthy and the leaves were less lustrous as compared to the plants grown as Controls.
3. In both SET I and SET II, the culture tube labeled 70mg/L, the plant materials were alive but was much weaker and a mild discoloration of the leaf documented as compared to the plants grown in the tubes containing 50mg/L of mercury.
4. In both SET I and SET II, the culture tube labeled 100mg/L, the plant materials were barely alive and discoloration of the leaves were more prominent as compared to previous concentration.
5. In both SET I and SET II, of the culture tube labeled 130 mg/L and 170 mg/L, both the plant materials did not survive and there were complete discoloration of the entire plant material. No green pigmentation left in these samples.

CONCLUSION

The authors used the pineapple plant as a model for performing tissue culture procedures for detection and management of environmental health because the plant is easily available and can grow in diverse range of environment and requires no special maintenance. The *Ananas sp.* commonly known as pineapple, can be used as a natural indicator or bio indicator of mercury or any other heavy metal toxicity by planting the plants around the water bodies. As evident through the simple plant tissue culture experiment that presence of mercuric compounds interferes with normal physiological activities of the plant resulting in decrease in growth, loss of vitality, depigmentation and finally death of the plant species. And if the study of the plant's genetic material were performed, the degradation of the plant DNA could be observed, indicating that mercury also damages the genetic material of an organism permanently. So if the plantations around the water bodies starts exhibiting these features even in the slightest form, is indicative of mercury toxicity. As evident through the experiment that if the concentration mercury compound exceeds 50 mg/L with respect to the water body under consideration, the plants will be showing the effects indicative of the mercury contamination. After that the conventional methods can be applied to measure the exact amount of mercury concentration. This procedure is useful because planting easily available plants as bio indicators are more reliable and cost effective as compared to installing a toxicity monitoring apparatus. Further toxicity studies of heavy metal under consideration could be studied in a large extent and in a faster rate. The rate of uptake of heavy metals by the plants from their surrounding environment can be studied *in vitro* which will help in deciding which plant species to be utilized for optimum removal of ground water contaminants. Finally it can be concluded that the technique of tissue culture is thus very essential for managing environmental pollution using plants. And there is a serious need for environmental engineers with expertise on Plant tissue culture and associated Recombinant DNA Technologies for creation of novel variety of plants that can be utilized for phytoremediation purposes in an environment friendly manner.

REFERENCES

Ball, E. (1946). Development in sterile culture of stem tips and adjacent regions of *Tropaeolum majus* L. and of *Lupinus albus* L. *American Journal of Botany*, *33*(5), 301–318. doi:10.2307/2437117

Barton, K. A., Whiteley, H. R., & Yang, N.-S. (1987). Bacillus thurigiensis δ-endotoxin expressed in transgenic Nicotiana tabacum provides resistance to Lepidoptera insects. *Plant Physiology*, *85*(4), 1103–1109. doi:10.1104/pp.85.4.1103 PMID:16665812

Bergmann, L. (1960). Growth and division of single cells of higher plants in vitro. *The Journal of General Physiology*, *43*(4), 841–851. doi:10.1085/jgp.43.4.841 PMID:13799400

Carlson, P. S. (1970). Induction and isolation of auxotrophic mutants in somatic cell culture of Nocatiana tabacum. *Science*, *168*(3930), 487–489. doi:10.1126/science.168.3930.487 PMID:17838129

Evans, D. A., Sharp, W. R., Ammirato, P. V., & Yamada, Y. (1983). *Handbook of Plant Cell Culture*. New York: Macmillan Publishing Company.

Fodor, S. P. A. (1997). Massively parallel genomics. *Science*, *270*(5324), 393–394. doi:10.1126/science.277.5324.393

Gautheret, T. J. (1939). Culture du tissue cambial. *C.R. Acad. Sci.*, *198*, 2195–2196.

Guha, S., & Maheshwari, S. C. (1964). In vitro production of embryos from anthers of Datura. *Nature*, *204*(4957), 497. doi:10.1038/204497a0

Haberlandt, G. (1902). Culturversuchemit isolierten Pflanzenzellen. *Sitzungsber. Math. Naturrwiss. Kl. Kias. Akad. Wiss. Wien*, *111*, 69–92.

Han, H. (1987). Application of pollen derived plants in crop improvement: Natesh, S. et al. Biotechnology in Agriculture. Oxford and IBH Publishing Co. Pvt. Ltd.

Horsch, R. B., Fraley, R. T., Rogers, S. G., Sanders, P., Lloyd, A., & Hoffman, N. (1984). Inheritance of functional foreign gene in plants. *Science, 223*(4635), 496–498. doi:10.1126/science.223.4635.496 PMID:17781445

Pierik, R.L.M. (1987). *In vitro Culture of Higher Plants*. Academic Press.

Reinert, J., & Bajaj, P. P. S. (1977). *Applied and Fundamental Aspect of Plant Cell, Tissue and Organ Culture*. Berlin: Springer-Verlag. doi:10.1007/978-3-662-02279-5

Street, H. S. (Ed.). (1973). *Plant Tissue and Cell Culture*. Oxford, UK: Blackwell Scientific Publications.

Thorpe, T. A. (1981). *Plant Tissue Culture: Methods and Applications in Agriculture*. New York: Academic Press Inc.

Chapter 11
Interaction of Heavy Metal Ions With Nanomaterials:
A Remediation Process

Suphiya Khan
Banasthali University, India

Sonu Kumari
Banasthali University, India

ABSTRACT

Increasing water pollution due to heavy metals is a major global concern, and favorable remediation techniques are required. Heavy metal contamination affects both flora and fauna as it enters the food web. The development of nanotechnology and novel nanomaterials production has attracted researchers worldwide. Both carbon- and metal-based nanomaterials proved great adsorbents for heavy metal remediation because of their unique properties such as thermal and chemical stability and high surface. Novel green route for nanomaterials synthesis make the nanomaterials production an environmentally friendly, low cost, and user-friendly approach. This chapter reviews the heavy metal pollution causes and utilization of different nanomaterials for the remediation process.

INTRODUCTION

Today approximately 663 million people lack access to safe drinking water all around the world (WHO, 2015). Heavy metal contamination is the major contributor to the world drinking water crisis, affecting millions of people worldwide. Major contributors for heavy metal pollution are lead (Pb), cadmium (Cd), arsenic (As),

DOI: 10.4018/978-1-5225-4162-2.ch011

Copyright © 2018, IGI Global. Copying or distributing in print or electronic forms without written permission of IGI Global is prohibited.

mercury (Hg) and chromium (Cr) as are most toxic and nonbiodegradable. Their presence above permissible limits in body can cause severe damages to vital organs of the body, such as kidney, liver and brain, reproductive and nervous system (Moreno-Castillaet al., 2004; Zhao et al., 2011). These heavy metals are widely utilized by many industries such as metal finishing, iron and steel, metallurgical, mining, battery producing industries etc. and are released in various water sources (Chen et al., 2007). Various technologies reported for remediation of metal ions from wastewater are solvent extraction, reverse osmosis, separation by membrane, ion exchange etc (Walsh et al., 1994; Xing et al., 2007; Ersahin et al., 2012). The available techniques have disadvantages such as high cost, high required investment and disposal of metal sludge; therefore, efforts have been made to synthesis low cost materials for removing heavy metals from aqueous solutions. Nowadays, nanomaterials (NMs) are recommended as effective, low cost and eco-friendly substitute to available treatment materials, in both resource conservation and environmental remediation (Zare et al., 2013; Gupta et al., 2015). The elimination of heavy metals from polluted sites using nanomaterials formed by plant, fungi and bacteria with the assistance of nanotechnology is called nanobioremediation. Green route for synthesis of nanoparticles has developed markedly to form novel materials that are eco-friendly, cost effective and stable. Although nanoparticles can be manufactured through a range of conventional processes but the biological route of synthesizing is more beneficial because of ease of fast production, low toxicity, control of size characteristics, low cost and eco-friendly approach (Mei et al., 2014). In the development of various remediation technologies for heavy metals, adsorption has already been proven to be an effective approach towards contamination removal (Tripathy et al., 2006; Onyangoa et al., 2006). The adsorption process mainly depends upon the applicability of the absorbent. The quality of the effluent generated is also better than the rest of the processes because the adsorbent has a high affinity towards the metal ions. Various types of adsorbents such as zeolites, activated carbons, chelating materials and nanoparticles have been developed for the remediation applications. The main selection parameters of adsorbents are the high adsorption capacity, fast kinetics and low cost (Wang et al., 2010; Cai et al., 2013; Gupta et al., 2015). With the development of technology, nanomaterials are utilized for heavy metal remediation due to their more surface area and low-cost synthesis and proved efficient adsorbents. Among all reported nanomaterials, remediation of heavy metals ions by metal oxide based nanocomposites and iron oxide based magnetic nanomaterials has been broadly utilized. Nano sized iron oxide particles exhibit super paramagnetism, an additional special property from common nanoparticles. Super paramagnetic nanoparticle have large surface area, are biocompatible, less toxic, chemically inert, offer small diffusion resistance and their surface can be modified with organic molecules, inorganic ions or some functional groups, which render them surfaces having good potential for removing heavy metals (Hua et al.,2012). Nanotubes synthesized are used for the removal of Cu (II), Zn (II), Hg (II), Ni (II), Cr (VI) and Pb (II). Manganese oxides (MnO) nano-particles, zinc oxide (ZnO) nano-particles, magnesium oxide (MgO) nano-particles, carbon nanotubes (CNTs) and graphene based nano-adsorbents are utilized for the remediation of different heavy metal (Mubarak et al., 2014). Heavy metal removal from contaminated water through the use of nano-adsorbents is influenced by several parameters such as temperature, pH, adsorbent dose and contact period. The enhancement in contact period elevates the adsorption of heavy metals from polluted water as it gives more time for the adsorption. At first, the adsorption of heavy metals is more due to the high concentration of metals, which declines as time period increases due to blockage of active sites. Adsorption is also influenced by nanomaterials properties, such as surface charge, BET surface area, hydrophobicity and addition of new functional groups (Matlochová et al., 2013). There is no doubt of efficiency of utilization nano-materials in wastewater treatment; however, this technology has some serious downsides that need to be negotiated. Since nanoparticles might release into the environment during preparation and treatment processes, where they can accumulate for long time and cause serious risks. In order to reduce the health risk there is need a future research to prepare such catalysts having least toxicity to the environment. More work is required to re-evaluate the ecotoxicity potential for each new modification in catalyst and for existing materials. In addition, life cycle assessment of nanomaterials is crucially required to address their overall benefits and risks. Nanotechnology is rarely adapted to mass processes. Given that up until now, most of the nanomaterials have not been cost-competitive when compared with conventional materials such as activated carbon, and thus future applications will focus on efficient processes where only small quantities of nanomaterials are required. Moreover, further work is required on developing a cost effective methods of synthesizing nanomaterials and testing the efficiency at large scale for successful field application.

POLLUTION BY HEAVY METALS

Heavy metals pollution in environment occurs through pedogenetic process of weathering of parent materials and also through anthropogenic sources. Recently, the anthropogenic activities, like combustion of fossil fuels, mining, ore refining, metal-working industries, insecticides and phosphate fertilizers, etc., have resulted to the release of heavy metals and the addition of these complexes in environment (Weast et al., 1984; Goldbold and Huttermann, 1985; Nies et al., 1999) causing major risk to the ecosystem. The different sources of heavy metals in environment are summarized in Figure 1. The highly toxic group of metals as Cu, Cr, Cd, and Pb is one of the key contaminant which includes metals with an atomic density greater than 6 g/cm^3. The heavy metals in the present world become main contaminants in the soil and water systems because of their increased generation by natural and man-made activities, transfer from mines to other sites where easily comes in contact with living beings, discharged as metal wastes in high amount from several industries, and greater bioavailability.

EFFECTS OF HEAVY METAL ON LIFE FORMS

Recently, researchers detected that the microbial biomass is seriously inhibited in soil system contaminated by Pb, Cu, Zn and other known heavy metals (Kandeler et al., 1997). The microbial biomass is low in area near to mines and increases with increasing the distance from mines. The effect of heavy metals on microbial biomass varies with heavy metal types and concentration in soil. Chander et al. (1995) identified the effect of different amount of heavy metals on soil microbial biomass, and results showed that if the amount of heavy metals in soil is three times more than environmental standard, established by the European Union, it can restrain

Figure 1. Different sources of heavy metals in environment

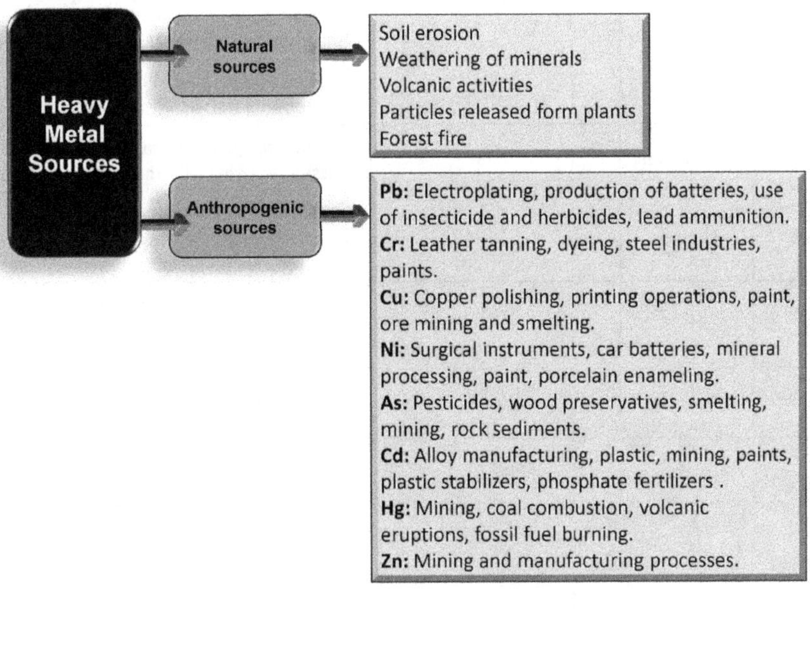

microbial biomass. Another study concluded that low concentration of heavy metals is responsible for stimulating microbial growth and enhanced microbial biomass, whereas microbial biomass decreases with increase in heavy metal concentrations (Fliepbach et al., 1994). Studies have revealed that the activities of several enzymes were also affected by heavy metal contamination in soil. The activities were reduced by 10 to 50 times of almost all enzymes with the enhancement in heavy metal concentrations in soil systems (Chander et al., 1995). Research have also been done on heavy metals effect on plant systems and human beings. The growth of bean and cabbage seedlings was inhibited when treated with 30 μ mol/ L Cd concentrations, the plant height and root length were also dropped due to Cd treatment (Qin et al., 1994). Studies also showed that photosynthesis and protein synthesis were also affected due to Cd contamination in soil (Acar and Alshawabkeh, 1993). Heavy metals can enter in water supplies through acid rain or through industrial wastes and enters in the food chain affecting plants, animals and humans of food chain. The concentrations of heavy metal above permissible limits can cause several harmful effects on humans as described in Table.1.

Table 1. Permissible limit for heavy metal and their toxic effects (Farooq et al., 2010; WHO, 2015)

S.No	Element	Permissible limit (ppm)	Effects
1.	Arsenic	0.01	Cardiovascular and nervous system trouble, bone marrow depression, hemolysis, hepatomegaly, gastrointestinal problem, melanosis, tumor formation in liver.
2.	Cadmium	0.003	Renal instability, lung function affected cancer, hypertension, Itai–Itai disease, weight reduction.
3.	Chromium	0.05	Cancer, teratogenic, epigastric pain, nausea, queasiness, serious diarrhea, tumors in lungs.
4.	Copper	2.0	Reproductive abnormality, neurotoxicity, and acute toxicity, wooziness, diarrhea, developmental problem.
5.	Lead	0.01	Anemia, brain damage, anorexia, malaise, low appetite, IQ decrement.
6.	Mercury	0.006	Neurological and renal instability, pulmonary function affected, corrosive to skin, eyes, muscles, dermatitis, and kidney damage.
7.	Nickel	0.07	Chronic bronchitis, low lung function, lung cancer.

NANOTECHNOLOGY FOR HEAVY METAL REMEDIATION

Nanomaterials are defined as particles of size in range of 1 nm to 100 nm. The first scientific report explaining the nature of nanoparticles was given by Michael faraday in his pioneering work, "Experimental relations of gold to light" in 1857. The removal of heavy metals from contaminated sites using nanomaterials formed by plant, fungi and bacteria with the help of nanotechnology is called nanobioremediation (Yadav et al., 2017). The metal nanoparticles can be synthesized through various techniques such as sonochemical process, microemulsion process, radiolysis, microwave, laser ablation method and green synthesis method (Latha and Gowri, 2014). Due to the limitations of other conventional methods green synthesis method is now widely used in which nanoparticles are synthesized by using different plant parts (Shameli et al., 2012). The advantages of green synthesis of nanoparticles are clean and eco-friendly approach, the active biological component itself act as reducing and capping agent, cost effective and large scale production. A very wide range of biological resources like microorganisms (bacteria, yeast, fungi, algae, and viruses) and plants can be used for nanoparticle synthesis (Mohanpuria et al., 2008). While microbe-based protocols have been developed from the cumulative research efforts of several authors, plant-mediated green route for synthesis of nanoparticles has gained importance in the present world (Gardea-Torresdey et al., 2002). The time required for reduction process of metal ions is reported less as compared to microbes.

Carbon-Based Nanomaterials

Carbon based materials having controllable pore size and high surface area proved more advantageous than available materials. The carbon nanomaterials application for heavy metal adsorption studies showed fast equilibrium rates with increased adsorption capacities. Carbon nanomaterials proved effective over a broad range of pH with favorable Langmuir, or Freundlich isotherms (Mauter et al., 2008). The carbon nanomaterials prepared through chemical vapour deposition process (CVD) were compared with normal activated carbon materials for adsorption process of Pb, Zn and Cd heavy metals by Ruparelia and his coworkers. Another study also conducted for comparision of adsorption capacity of activated carbon, nanoporous carbon and nanomaterial synthesized through cobalt catalyst. The nanoporous carbon demonstrated higher heavy metal adsorption capacity compared to other two. The alteration in surface morphology and adsorption capacity is reported due to activation of nanomaterials. The metallic impurities were dissolved and new functional groups were formed during the activation process. Carbon nanotubes were activated using H_2O_2, NaOH, HNO_3 and $KMnO_4$ in oxidizing environment (Ruparelia et al., 2008).

The adsorption capacity of carbon nanotubes enhances with the removal of amorphous carbon (Gotovac et al., 2007). Lu et al. (2006) conducted an experiment and reported that concentration of Zn^{+2} adsorbed on carbon nanotubes enhanced with increase in temperature. The carbon nanotubes adsorption capacity for Zn^{+2} reported higher as compared to powdered activated carbon, reflecting that the single walled carbon nanotubes (SWNTs) and multi walled carbon nanotubes (MWNTs) are effective sorbents. The structure of SWCNTs and MWCNTs is shown in Figure 2. The adsorption of Zn^{+2} heavy metal on CNTs proved to be spontaneous and endothermic in nature. The SWCNTs and MWCNTs were utilized up to 10 cycles through desorption method by using 0.1 mol/L nitric acid solution for the detachment of Zn^{+2} heavy metal. This shows that use of CNTs in water purification process is favorable due to its reusable properties and high adsorption capacity (Lu et al., 2006). MWCNTs were also reported for the remediation of Cr^{+6} from water system. Pseudo-second order reaction is favored in Cr^{+6} remediation and maximum adsorption is observed at low pH. During the Cr^{+6} adsorption, redox reaction occurs on the surface of MWCNTs and Cr^{+6} is converted in Cr^{+3} for the adsorption process to take place on MWCNTs (Hu et al., 2009). The remediation of Pb^{+2} was also reported; the removal percent achieved was maximum up to 85% through CNTs at pH 5 and 40 mg L^{-1} CNTs. In the process, Langmuir isotherm is favored which shows the adsorption capacity of 102 mg g^{-1} (Kabbashi et al., 2009). Another study discussed the adsorption of Cd^{+2} using CNTs and the pH plays a major role in adsorption. The equilibrium state is achieved through the oxidation of H_2O_2 for high adsorption

Figure 2. Structures of (a) SWCNTs and (b) MWCNTs

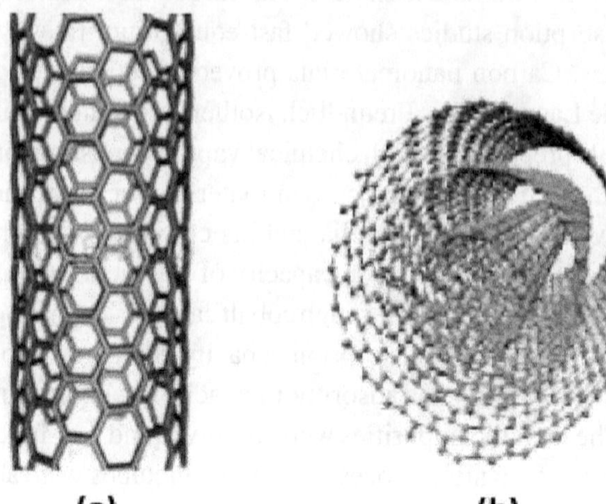

Table 2. Nanomaterial and parameters tested

Nanomaterial	Contaminant	Parameters Tested	References
Carbon-encapsulated iron nanoparticles	Cu^{+2}, Cd^{+2}, Co^{+2}	functional groups, pH	4
MWCNTs	Pb^{+2}	pH, temperature	28
As-produced CNTs	Ni^{+2}	Contact time, pH	22
Oxidized CNTs "with H2O2, KMnO4, and HNO3"	Cd^{+2}	pH, adsorbent morphology	26
SWCNTs MWCNTs PAC	Zn^{+2}	contact time, pH	30
As applied CNTs	Cu^{+2}	Adsorbent dose, pH, temperature	53
Normal AC, CNTs impregnated AC	Cr^{+6}	contact time, agitation speed, pH	3
Oxidized CNTs	Cu^{+2}, Pb^{+2}, Cd^{+2}, Zn^{+2}	pH, temperature	47
CNTs	Pb^{+2}	pH, adsorbent concentration, contact period	21
HNO_3-modified and NaOCl modified CNTs	Cu^{+2}	temperature, adsorbent amount, pH	53
ZnO	Pb^{+2}, Cu^{+2}	pH, contact time	33, 49
CeO_2	Cr^{+6}, Pb^{+2}	pH	5, 41

process at low pH. The conducted experiment concluded that CNTs are capable to remove Cd^{+2} with high adsorption capacity (Li et al., 2003). The adsorption of Pb^{+2} through Mn coated CNTs in water system was studied by Wang and his coworkers. The Pb^{+2} remediation ability of MnO_2 coated CNTs declined with the reduction of pH in aqueous solution. The Langmuir isotherm is favored with adsorption capacity of 78.74 mg/g, showed that MnO_2 coated CNTs are better adsorbent as compared to the normal CNTs. The Pseudo-second order kinetic model is favored for adsorption (Wang et al., 2007).

Metal and Metal Oxide Nanoparticles

Along with CNTs, metal and metal oxide nanoparticles are widely utilized for the heavy metal remediation process in water systems. The nanoparticles were favored due to their high surface area and affinity, the major metal oxide nanoparticles utilized broadly are Ag, Fe, Mn, Ti, Mg, Cu etc (Luo et al., 2010; Fabrega et al., 2011; Gupta et al., 2011; Feng et al., 2012). Studies demonstrated that Fe_3O_4 can be utilized to diffuse nano-FeO and thus maintain Cr mitigation by nano-FeO. The Fe_3O_4 nanoparticles were added in reaction volume during the synthesis of FeO nanoparticles for the impregnation of FeO nanoparticles on Fe_3O_4. The high remediation of Cr heavy metal is achieved after the impregnation of FeO onto Fe_3O_4 nanoparticles as it restricts the FeO aggregation. The obtained data represents that more concentrations of Fe_3O_4 in the nanocomposites can enhance the rate of Cr reduction, and the optimal ratio of Fe_3O_4:FeO for Cr reduction was detected to be 40:1. Results obtained also depicts that pH plays a significant role in the reduction rate of Cr, it concludes the presence of fast reaction rate in acidic environment as compared to basic. The data suggests that pH plays important role in that reduction rate of Cr, with reaction process occur as high speed in acidic as compared to basic environment (Wu et al., 2009). The titanate nanoflowers were prepared having high surface area and further utilized for heavy metal remediation process by Huang and his coworkers. The conducted experiment resulted that adsorption capacity of titanate nanoflower is obtained higher than nanotubes and nanowires. The titanate nanoflower also displays high selectivity for heavy metals ions (Huang et al., 2012). In another study conducted by Zhang et al. showed the synthesis of magnesium hydroxide nanotubes to form $Mg(OH)_2/Al_2O_3$ composites for the remediation of Ni ions from polluted water (Zhang et al., 2012).

Graphene Based Nanomaterials

Graphene is considered as one of the allotropy of carbon and have favorable properties for heavy metal remediation purpose. The two-dimensional nanostructure of graphene oxide is formed through the oxidation process of graphite layer by using most common Hummer's method (Lingamdinne et al., 2016). The hydroxyl and carboxyl functional groups presence in graphene oxide make them better heavy metal adsorbents (Li et al., 2009). Other favorable properties present in graphene oxide are chemical stability, high surface area, and flexibility (Gopalakrishnan et al., 2015; Taherian et al., 2013). Graphene oxide has two major features which make them better applicants for heavy metal removal. First, catalyst is not required in the chemical synthesis method along with the use of simple instrument. Secondly, maximum heavy metal adsorption occurs as single layers graphene oxide has two-dimensional basal planes (Santhosh et al., 2016). Graphene enabled sand filter has been utilized for heavy metal removal in column reactor by Ding and his coworkers (Ding et al., 2014). Titanium oxide has been used for the modification of graphene oxide for the adsorption of zinc, cadmium and lead ions. The adsorption capacity obtained upto 88.9, 65.6 and 72.8 mg g^{-1} for zinc, lead and cadmium respectively (Lee et al., 2012). The application of graphene based adsorbents is increasing with technology for heavy metal remediation process.

MECHANISM OF HEAVY METAL ADSORPTION

The adsorption process of heavy metals is reported by several researchers. The effect of various parameters such as pH, coexisting ions and ionic strength has been investigated (Li et al., 2002; Al-Degs et al., 2006). In the adsorption process using carbon nanotubes, the presence of different functional groups such as –OH, -COOH and –C=O provides active sites for heavy metal adsorption process and enhances the efficiency of carbon nanotubes (Shim et al., 2001; Long and Yang, 2001). The –COOH group present on CNTs forms the complexes during heavy metal remediation with metal ions and enhances adsorption as shown in Figure 3. The mechanism of heavy metal removal includes the interaction between heavy metal ions and functional groups available on CNTs. Figure 3 shows the protons exchange of -COOH and phenolic groups with the heavy metal ions in water remediation. This concludes that CNTs surface modification is favorable for achieving the high heavy metal ions adsorption (Pan and Xing, 2008; Cho et al., 2009).

ADVANTAGES AND DISADVANTAGES

Nanotechnology proved a major landmark in the heavy metal remediation from aqueous solutions. The low toxicity and high adsorption capacity of carbon based nanomaterials makes them favorable material for removal of heavy metals form water system. CNTs were widely utilized for heavy metal removal due to their high

Figure 3. Heavy metal adsorption mechanism by CNTs

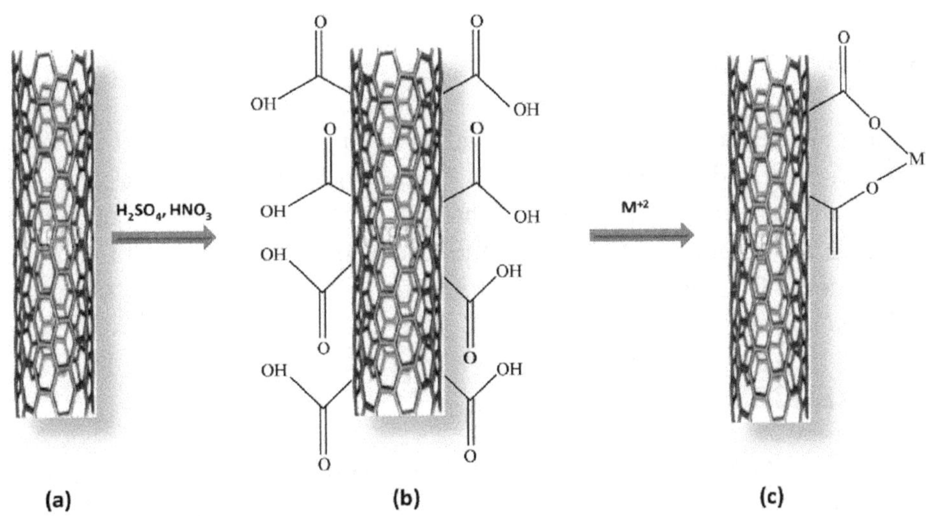

thermal and chemical stabilities along with high surface area presence. The green synthesis approach for nanoparticles make them potential environment friendly and cost effective materials for remediation process. Along with the advantageous nature, the nanomaterials also exhibit harmful effects which should be reduced and eliminated through major researches. The nanomaterials leaching during the adsorption process in water system gives them gate to enter in the food chain and food web. Through food chain nanomaterials can be accumulated in plants, animals and human beings. This is the major concerning issue that should be precisely described. It is a requirement of today's nanotechnology to develop reliable, eco-friendly, cheap and clean materials for heavy metal remediation.

CONCLUSION

Heavy metals are considered highly toxic to environment and thus, cost effective, environment friendly and effective remediation techniques must be develop. Still, the technology for remediation of heavy metal ions has not reached the optimum conditions. Emerging utilization of nanomaterials for remediation is proved beneficial due to their special properties. High adsorption properties were reported by several researchers through use of carbon and metal nanomaterials. This chapter illustrate that researchers in most cases utilize carbon nanotubes (MWCNTs and SWCNTs) as CNM sorbents. The green synthesis approach for synthesis of nanoparticles proved favorable and in it further developments are required. This concludes the application of nanotechnology in heavy metal remediation processes. Furthermore, development in technology is required due to some disadvantages.

ACKNOWLEDGMENT

Authors gratefully acknowledge Prof. Aditya Shastri for providing us research facilities and the Bioinformatics Centre, Banasthali University, Rajasthan (India) for extensive use of computational facilities and also thankful to MHRD Department of Higher Education, Government of India under the scheme of Establishment of Centre of Excellence for Training and Research in FAST for their generous financial support.

REFERENCES

Acar, Y. B., & Alshawabkeh, A. N. (1993). Principles of electrokinetic remediation. *Environmental Science & Technology*, *27*(13), 2638–2647. doi:10.1021/es00049a002

Al-Degs, Y. S., El-Barghouthi, M. I., Issa, A. A., Khraisheh, M. A., & Walker, G. M. (2006). Sorption of Zn (II), Pb (II), and Co (II) using natural sorbents: Equilibrium and kinetic studies. *Water Research*, *40*(14), 2645–2658. doi:10.1016/j.watres.2006.05.018 PMID:16839582

Atieh, M. A. (2011). Removal of chromium (VI) from polluted water using carbon nanotubes supported with activated carbon. *Procedia Environmental Sciences*, *4*, 281–293. doi:10.1016/j.proenv.2011.03.033

Bystrzejewski, M., Pyrzyńska, K., Huczko, A., & Lange, H. (2009). Carbon-encapsulated magnetic nanoparticles as separable and mobile sorbents of heavy metal ions from aqueous solutions. *Carbon*, *47*(4), 1201–1204. doi:10.1016/j.carbon.2009.01.007

Cai, X., Li, J., Zhang, Z., Yang, F., Dong, R., & Chen, L. (2013). Novel Pb^{2+} ion imprinted polymers based on ionic interaction via synergy of dual functional monomers for selective solid-phase extraction of Pb^{2+} in water samples. *ACS Applied Materials & Interfaces*, *6*(1), 305–313. doi:10.1021/am4042405 PMID:24344795

Cao, C. Y., Cui, Z. M., Chen, C. Q., Song, W. G., & Cai, W. (2010). Ceria hollow nanospheres produced by a template-free microwave-assisted hydrothermal method for heavy metal ion removal and catalysis. *The Journal of Physical Chemistry C*, *114*(21), 9865–9870. doi:10.1021/jp101553x

Chander, K., Brookes, P. C., & Harding, S. A. (1995). Microbial biomass dynamics following addition of metal-enriched sewage sludges to a sandy loam. *Soil Biology & Biochemistry*, *27*(11), 1409–1421. doi:10.1016/0038-0717(95)00074-O

Chen, C., & Wang, X. (2006). Adsorption of Ni (II) from aqueous solution using oxidized multiwall carbon nanotubes. *Industrial & Engineering Chemistry Research*, *45*(26), 9144–9149. doi:10.1021/ie060791z

Cho, H. H., Wepasnick, K., Smith, B. A., Bangash, F. K., Fairbrother, D. H., & Ball, W. P. (2009). Sorption of aqueous Zn [II] and Cd [II] by multiwall carbon nanotubes: The relative roles of oxygen-containing functional groups and graphenic carbon. *Langmuir*, *26*(2), 967–981. doi:10.1021/la902440u PMID:19894751

Ding, Z., Hu, X., Morales, V. L., & Gao, B. (2014). Filtration and transport of heavy metals in graphene oxide enabled sand columns. *Chemical Engineering Journal*, *257*, 248–252. doi:10.1016/j.cej.2014.07.034

Ersahin, M. E., Ozgun, H., Dereli, R. K., Ozturk, I., Roest, K., & van Lier, J. B. (2012). A review on dynamic membrane filtration: Materials, applications and future perspectives. *Bioresource Technology*, *122*, 196–206. doi:10.1016/j.biortech.2012.03.086 PMID:22534368

Fabrega, J., Luoma, S. N., Tyler, C. R., Galloway, T. S., & Lead, J. R. (2011). Silver nanoparticles: Behaviour and effects in the aquatic environment. *Environment International*, *37*(2), 517–531. doi:10.1016/j.envint.2010.10.012 PMID:21159383

Farooq, U., Kozinski, J. A., Khan, M. A., & Athar, M. (2010). Biosorption of heavy metal ions using wheat based biosorbents–a review of the recent literature. *Bioresource Technology*, *101*(14), 5043–5053. doi:10.1016/j.biortech.2010.02.030 PMID:20223652

Feng, L., Cao, M., Ma, X., Zhu, Y., & Hu, C. (2012). Superparamagnetic high-surface-area Fe_3O_4 nanoparticles as adsorbents for arsenic removal. *Journal of Hazardous Materials*, *217*, 439–446. doi:10.1016/j.jhazmat.2012.03.073 PMID:22494901

Fliessbach, A., Martens, R., & Reber, H. H. (1994). Soil microbial biomass and microbial activity in soils treated with heavy metal contaminated sewage sludge. *Soil Biology & Biochemistry*, *26*(9), 1201–1205. doi:10.1016/0038-0717(94)90144-9

Gardea-Torresdey, J. L., Parsons, J. G., Gomez, E., Peralta-Videa, J., Troiani, H. E., Santiago, P., & Yacaman, M. J. (2002). Formation and growth of Au nanoparticles inside live alfalfa plants. *Nano Letters*, *2*(4), 397–401. doi:10.1021/nl015673+

Godbold, D. L., & Hüttermann, A. (1985). Effect of zinc, cadmium and mercury on root elongation of Picea abies (Karst.) seedlings, and the significance of these metals to forest die-back. *Environmental Pollution. Series A. Ecological and Biological*, *38*(4), 375–381. doi:10.1016/0143-1471(85)90108-4

Gopalakrishnan, A., Krishnan, R., Thangavel, S., Venugopal, G., & Kim, S. J. (2015). Removal of heavy metal ions from pharma-effluents using graphene-oxide nanosorbents and study of their adsorption kinetics. *Journal of Industrial and Engineering Chemistry*, *30*, 14–19. doi:10.1016/j.jiec.2015.06.005

Gotovac, S., Yang, C. M., Hattori, Y., Takahashi, K., Kanoh, H., & Kaneko, K. (2007). Adsorption of polyaromatic hydrocarbons on single wall carbon nanotubes of different functionalities and diameters. *Journal of Colloid and Interface Science*, *314*(1), 18–24. doi:10.1016/j.jcis.2007.04.080 PMID:17604043

Gupta, K., Bhattacharya, S., Chattopadhyay, D., Mukhopadhyay, A., Biswas, H., Dutta, J., & Ghosh, U. C. (2011). Ceria associated manganese oxide nanoparticles: Synthesis, characterization and arsenic (V) sorption behavior. *Chemical Engineering Journal, 172*(1), 219–229. doi:10.1016/j.cej.2011.05.092

Gupta, V., Nayak, A., Bhushan, B., & Agarwal, S. (2015). A critical analysis on the efficiency of activated carbons from low-cost precursors for heavy metals remediation. *Critical Reviews in Environmental Science and Technology, 45*(6), 613–668. doi:10.1080/10643389.2013.876526

Gupta, V. K., Tyagi, I., Sadegh, H., Shahryari-Ghoshekandi, R., Makhlouf, A. S. H., & Maazinejad, B. (2015). Nanoparticles as adsorbent; a positive approach for removal of noxious metal ions: A review. *Sci. Technol. Dev, 34*(3), 195–214. doi:10.3923/std.2015.195.214

Hu, J., Chen, C., Zhu, X., & Wang, X. (2009). Removal of chromium from aqueous solution by using oxidized multiwalled carbon nanotubes. *Journal of Hazardous Materials, 162*(2), 1542–1550. doi:10.1016/j.jhazmat.2008.06.058 PMID:18650001

Hua, M., Zhang, S., Pan, B., Zhang, W., Lv, L., & Zhang, Q. (2012). Heavy metal removal from water/wastewater by nanosized metal oxides: A review. *Journal of Hazardous Materials, 211*, 317–331. doi:10.1016/j.jhazmat.2011.10.016 PMID:22018872

Huang, J., Cao, Y., Liu, Z., Deng, Z., Tang, F., & Wang, W. (2012). Efficient removal of heavy metal ions from water system by titanate nanoflowers. *Chemical Engineering Journal, 180*, 75–80. doi:10.1016/j.cej.2011.11.005

Kabbashi, N. A., Atieh, M. A., Al-Mamun, A., Mirghami, M. E., Alam, M. D. Z., & Yahya, N. (2009). Kinetic adsorption of application of carbon nanotubes for Pb (II) removal from aqueous solution. *Journal of Environmental Sciences (China), 21*(4), 539–544. doi:10.1016/S1001-0742(08)62305-0 PMID:19634432

Kandah, M. I., & Meunier, J. L. (2007). Removal of nickel ions from water by multi-walled carbon nanotubes. *Journal of Hazardous Materials, 146*(1), 283–288. doi:10.1016/j.jhazmat.2006.12.019 PMID:17196328

Kandeler, F., Kampichler, C., & Horak, O. (1996). Influence of heavy metals on the functional diversity of soil microbial communities. *Biology and Fertility of Soils, 23*(3), 299–306. doi:10.1007/BF00335958

Latha, N., & Gowri, M. (2014). Bio Synthesis and Characterisation of Fe_3O_4 Nanoparticles Using Caricaya Papaya Leaves Extract. *Synthesis, 12*, 17.

Lee, Y. C., & Yang, J. W. (2012). Self-assembled flower-like TiO_2 on exfoliated graphite oxide for heavy metal removal. *Journal of Industrial and Engineering Chemistry*, *18*(3), 1178–1185. doi:10.1016/j.jiec.2012.01.005

Li, J., Guo, S., Zhai, Y., & Wang, E. (2009). Nafion–graphene nanocomposite film as enhanced sensing platform for ultrasensitive determination of cadmium. *Electrochemistry Communications*, *11*(5), 1085–1088. doi:10.1016/j.elecom.2009.03.025

Li, Y. H., Wang, S., Luan, Z., Ding, J., Xu, C., & Wu, D. (2003). Adsorption of cadmium (II) from aqueous solution by surface oxidized carbon nanotubes. *Carbon*, *41*(5), 1057–1062. doi:10.1016/S0008-6223(02)00440-2

Li, Y. H., Wang, S., Wei, J., Zhang, X., Xu, C., Luan, Z., ... Wei, B. (2002). Lead adsorption on carbon nanotubes. *Chemical Physics Letters*, *357*(3), 263–266. doi:10.1016/S0009-2614(02)00502-X

Lingamdinne, L. P., Koduru, J. R., Roh, H., Choi, Y. L., Chang, Y. Y., & Yang, J. K. (2016). Adsorption removal of Co (II) from waste-water using graphene oxide. *Hydrometallurgy*, *165*, 90–96. doi:10.1016/j.hydromet.2015.10.021

Long, R. Q., & Yang, R. T. (2001). Carbon nanotubes as superior sorbent for dioxin removal. *Journal of the American Chemical Society*, *123*(9), 2058–2059. doi:10.1021/ja003830l PMID:11456830

Lu, C., Chiu, H., & Liu, C. (2006). Removal of zinc (II) from aqueous solution by purified carbon nanotubes: Kinetics and equilibrium studies. *Industrial & Engineering Chemistry Research*, *45*(8), 2850–2855. doi:10.1021/ie051206h

Luo, T., Cui, J., Hu, S., Huang, Y., & Jing, C. (2010). Arsenic removal and recovery from copper smelting wastewater using TiO_2. *Environmental Science & Technology*, *44*(23), 9094–9098. doi:10.1021/es1024355 PMID:21053910

Ma, X., Wang, Y., Gao, M., Xu, H., & Li, G. (2010). A novel strategy to prepare ZnO/PbS heterostructured functional nanocomposite utilizing the surface adsorption property of ZnO nanosheets. *Catalysis Today*, *158*(3), 459–463. doi:10.1016/j.cattod.2010.07.013

Matlochová, A., Plachá, D., & Rapantová, N. (2013). The application of nanoscale materials in groundwater remediation. *Polish Journal of Environmental Studies*, *22*(5), 1401–1410.

Mauter, M. S., & Elimelech, M. (2008). Environmental applications of carbon-based nanomaterials. *Environmental Science & Technology*, *42*(16), 5843–5859. doi:10.1021/es8006904 PMID:18767635

Mie, R., Samsudin, M. W., Din, L. B., Ahmad, A., Ibrahim, N., & Adnan, S. N. A. (2014). Synthesis of silver nanoparticles with antibacterial activity using the lichen Parmotrema praesorediosum. *International Journal of Nanomedicine*, *9*, 121. PMID:24379670

Mohanpuria, P., Rana, N. K., & Yadav, S. K. (2008). Biosynthesis of nanoparticles: Technological concepts and future applications. *Journal of Nanoparticle Research*, *10*(3), 507–517. doi:10.1007/s11051-007-9275-x

Moreno-Castilla, C., Alvarez-Merino, M. A., López-Ramón, M. V., & Rivera-Utrilla, J. (2004). Cadmium ion adsorption on different carbon adsorbents from aqueous solutions. Effect of surface chemistry, pore texture, ionic strength, and dissolved natural organic matter. *Langmuir*, *20*(19), 8142–8148. doi:10.1021/la049253m PMID:15350085

Mubarak, N. M., Sahu, J. N., Abdullah, E. C., & Jayakumar, N. S. (2014). Removal of heavy metals from wastewater using carbon nanotubes. *Separation and Purification Reviews*, *43*(4), 311–338. doi:10.1080/15422119.2013.821996

Nies, D. H. (1999). Microbial heavy-metal resistance. *Applied Microbiology and Biotechnology*, *51*(6), 730–750. doi:10.1007/s002530051457 PMID:10422221

Onyango, M. S., Kojima, Y., Kumar, A., Kuchar, D., Kubota, M., & Matsuda, H. (2006). Uptake of fluoride by Al^{3+} pretreated low-silica synthetic zeolites: Adsorption equilibrium and rate studies. *Separation Science and Technology*, *41*(4), 683–704. doi:10.1080/01496390500527019

Pan, B., & Xing, B. (2008). Adsorption mechanisms of organic chemicals on carbon nanotubes. *Environmental Science & Technology*, *42*(24), 9005–9013. doi:10.1021/es801777n PMID:19174865

Qin, T., Wu, Y., & Wang, H. (1993). Effect of cadmium, lead and their interactions on the physiological and biochemical characteristics of Brassica chinensis. *Acta Ecologica Sinica*, *14*(1), 46–50.

Recillas, S., Colón, J., Casals, E., González, E., Puntes, V., Sánchez, A., & Font, X. (2010). Chromium VI adsorption on cerium oxide nanoparticles and morphology changes during the process. *Journal of Hazardous Materials*, *184*(1), 425–431. doi:10.1016/j.jhazmat.2010.08.052 PMID:20832170

Ruparelia, J. P., Duttagupta, S. P., Chatterjee, A. K., & Mukherji, S. O. U. M. Y. A. (2008). Potential of carbon nanomaterials for removal of heavy metals from water. *Desalination, 232*(1), 145–156. doi:10.1016/j.desal.2007.08.023

Santhosh, C., Velmurugan, V., Jacob, G., Jeong, S. K., Grace, A. N., & Bhatnagar, A. (2016). Role of nanomaterials in water treatment applications: A review. *Chemical Engineering Journal, 306*, 1116–1137. doi:10.1016/j.cej.2016.08.053

Shameli, K., Ahmad, M. B., Zamanian, A., Sangpour, P., Shabanzadeh, P., Abdollahi, Y., & Zargar, M. (2012). Green biosynthesis of silver nanoparticles using Curcuma longa tuber powder. *International Journal of Nanomedicine, 7*, 5603–5610. doi:10.2147/IJN.S36786 PMID:23341739

Shim, J. W., Park, S. J., & Ryu, S. K. (2001). Effect of modification with HNO_3 and NaOH on metal adsorption by pitch-based activated carbon fibers. *Carbon, 39*(11), 1635–1642. doi:10.1016/S0008-6223(00)00290-6

Taherian, F., Marcon, V., van der Vegt, N. F., & Leroy, F. (2013). What is the contact angle of water on graphene? *Langmuir, 29*(5), 1457–1465. doi:10.1021/la304645w PMID:23320893

Tofighy, M. A., & Mohammadi, T. (2011). Adsorption of divalent heavy metal ions from water using carbon nanotube sheets. *Journal of Hazardous Materials, 185*(1), 140–147. doi:10.1016/j.jhazmat.2010.09.008 PMID:20926186

Tripathy, S. S., Bersillon, J. L., & Gopal, K. (2006). Removal of fluoride from drinking water by adsorption onto alum-impregnated activated alumina. *Separation and Purification Technology, 50*(3), 310–317. doi:10.1016/j.seppur.2005.11.036

Walsh, F. C., & Reade, G. W. (1994). Electrochemical techniques for the treatment of dilute metal-ion solutions. *Studies in Environmental Science, 59*, 3-44.

Wang, S., & Peng, Y. (2010). Natural zeolites as effective adsorbents in water and wastewater treatment. *Chemical Engineering Journal, 156*(1), 11–24. doi:10.1016/j.cej.2009.10.029

Wang, S. G., Gong, W. X., Liu, X. W., Yao, Y. W., Gao, B. Y., & Yue, Q. Y. (2007). Removal of lead (II) from aqueous solution by adsorption onto manganese oxide-coated carbon nanotubes. *Separation and Purification Technology, 58*(1), 17–23. doi:10.1016/j.seppur.2007.07.006

Wang, X., Cai, W., Lin, Y., Wang, G., & Liang, C. (2010). Mass production of micro/nanostructured porous ZnO plates and their strong structurally enhanced and selective adsorption performance for environmental remediation. *Journal of Materials Chemistry*, *20*(39), 8582–8590. doi:10.1039/c0jm01024c

Weast, R. C. (1984). *CRC handbook of chemistry and physics* (64th ed.). Boca Raton, FL: CRC Press, Inc.

WHO's drinking water standards. (1993). Retrieved from http://www.lenntech.com/applications/drinking/standards/who-s-drinking-water-standards.htm

World Health Organization (WHO). (2015). *Progress on sanitation and drinking-water–2015 update*. Geneva, Switzerland: World Health Organization and UNICEF.

Wu, C. H. (2007). Studies of the equilibrium and thermodynamics of the adsorption of Cu^{2+} onto as-produced and modified carbon nanotubes. *Journal of Colloid and Interface Science*, *311*(2), 338–346. doi:10.1016/j.jcis.2007.02.077 PMID:17442333

Xing, Y., Chen, X., & Wang, D. (2007). Electrically regenerated ion exchange for removal and recovery of Cr (VI) from wastewater. *Environmental Science & Technology*, *41*(4), 1439–1443. doi:10.1021/es0614991 PMID:17593754

Yadav, K. K., Singh, J. K., Gupta, N., & Kumar, V. (2017). *A Review of Nanobioremediation Technologies for Environmental Cleanup: A Novel Biological Approach*. Academic Press.

YanJun, W., JingHui, Z., YiFei, T., & XinHua, X. (2009). Chromium (VI) reduction in aqueous solutions by Fe_3O_4-stabilized Fe0 nanoparticles. *Journal of Hazardous Materials*, *172*(2/3), 1640–1645. PMID:19740609

Zare, K., Najafi, F., Sadegh, H., & Ghoshekandi, R. (2013). Studies of ab initio and Monte Carlo simulation on interaction of fluorouracil anticancer drug with carbon nanotube. *Journal of Nanostructure in Chemistry*, *3*(1), 71. doi:10.1186/2193-8865-3-71

Zhang, S., Cheng, F., Tao, Z., Gao, F., & Chen, J. (2006). Removal of nickel ions from wastewater by $Mg(OH)_2$/MgO nanostructures embedded in Al_2O_3 membranes. *Journal of Alloys and Compounds*, *426*(1), 281–285. doi:10.1016/j.jallcom.2006.01.095

Zhao, G., Li, J., Ren, X., Chen, C., & Wang, X. (2011). Few-layered graphene oxide nanosheets as superior sorbents for heavy metal ion pollution management. *Environmental Science & Technology*, *45*(24), 10454–10462. doi:10.1021/es203439v PMID:22070750

Chapter 12
Nano-Bioremediation:
Nanotechnology and Bioremediation

Sandeep Tripathi
NIMS University, India

R. Sanjeevi
NIMS University, India

J. Anuradha
NIMS University, India

Dushyant Singh Chauhan
NIMS University, India

Ashok K. Rathoure
Independent Researcher, India

ABSTRACT

The functional aspect of nanotechnology (NBT) is driven either to accelerate the performance of materials and/or to reduce the quantity of materials that are used for the purpose. Most significantly, its potential attribute to the environment includes the treatment and remediation, sensing and detection, and pollution prevention. In general nano-bio remediation (NBR) involves the use of nano-materials either in in-situ (in place), or ex-situ (off-place) treatment of contaminated materials. To accomplish this, the elemental or zero-valent metals and like materials in nano-form (1-100 nm) have been applied as an instinctive need to embrace sustainable environment. The use of nanomaterials initially reduces the biodegradable contaminants and then it promotes to achieve the standard levels. Thus, the role of nano-materials could be an efficient, effective approach to remediate the environmental contaminant sustainably. However, further research is required to record the detailed fate of the nano-materials that are used in environment remediation.

DOI: 10.4018/978-1-5225-4162-2.ch012

Copyright © 2018, IGI Global. Copying or distributing in print or electronic forms without written permission of IGI Global is prohibited.

INTRODUCTION

Globally, urbanization as well as industrial revolution ever since 19th Century brought great economical *cum* technological advancements. The technological innovations raised the course of humanity, through the process of excessive resource extraction, product dissemination and in addition greater disposal of waste materials into the environment with no proper diligence (Cecchin *et al.*, 2016). Generation and non-diligent disposal of materials led to several serious contamination issues challenging environmental, social and economy of world. In order to mitigate these concerns, efficient managerial strategies need to be adopted to achieve the environmental sustainability. In recent years sustainable remediation techniques gained greater importance, primarily aiming at reduction of contaminants concentration their by lowering the risk level as well as to eliminate the global environmental impacts induced by greenhouse gas emissions (Reddy and Adams, 2010). Bio-based approach has already substantially affected environmental protection. Thus bioremediation is considerably a clean, green and sustainable solution for treatment of contaminated materials and is in general considered as 'environmentally appropriate'. However, bio-based techniques are time consuming and at higher contaminant concentration toxic to the organisms involved for the purpose.

With emergence of bio-mimetic nanotechnology in the recent past has attracted the researches to intensively explore this concealed avenue. Arena of nano-biotechnology intersects the knowledge from all branches of science, technology and involves all forms of living organisms (Baker and Satish, 2012). The scope of nano-biotechnology has higher potential to support extensively at managing major global environmental challenges. Thus integration of nano-biotechnology along with bioremediation possibly achieves efficient, effective and sustainable solution for clean environment. This chapter reveals the recent trends, in the context of nano-biotechnology in pollution prevention and treatment either in place (*in-situ*) or off place (*ex-situ*) of contaminants. Hence, disclosure of nano-bioremediation technology might have far-reaching beneficial portent for the protection against large tracts of polluted sites that are overwhelmed with the waste materials.

NANO-BIOTECHNOLOGY AND UNIQUENESS OF NANOPARTICLES

Nanotechnology as we know in its modern and post-modern forms, is the young branch of knowledge but mankind has empirically used since thousands of years (Pradeep, 2007, Anuradha, 2013). It is an umbrella term covering wide range of technologies comprising structures and processes at nanometer scale (Abbasi *et al*., 2009). The potential of nanotechnology is often believed to revolve round the nanoparticles. The core versatility of the nanoparticles is the fact that they possess significant properties than that of bulk counterparts.

Nanoparticles or nanomaterials are in general termed engineered material possessing size between 1 nm and 100 nm, at least in one dimension. It includes the two groups either organic or and inorganic. The first category comprises carbon nanoparticles (fullerenes), while the second with magnetic, noble metal (*e.g.* palladium, gold and silver) and semiconductor nanoparticles (*e.g.* titanium dioxide and zinc oxide). Nanoparticles are fabricated either *top down* or *bottom up* approach. The traditional method of nanoparticles generation includes chemical or physical techniques. Of these, the former often uses hazardous reagents while the later involve energy-intensive and is highly expensive. In contrast, biotechnological approach emerged in the recent past involving biomolecules contained in microbes, algae or higher vascular and non-vascular plants to fabricate nanoparticles of desired shape and size in a similar way that mimics the nature (Figure 1). Thus biomimetic nanoparticles formations are much 'clean' and 'green'owing great relevance to the field of nano-biotechnology (Abbasi *et al*., 2012; Anuradha *et al*., 2015a; 2015b).

The synthesized nanoparticles with larger fraction of surface area, higher surface energy, spatial confinements and perfect dimensions were found to have potential benefits in the field based on catalytic, magnetic, electronic and optical properties. The realm of nanotechnology does extend with bio-based nanoparticles, finding a promising benefit on protecting the environment through pollution prevention, treatment, and cleanup of contaminates, among others. Thus environmental nano-biotechnology is one among the fast growing research interest in the technological development in the recent years.

BIOTECHNOLOGICAL APPROACH ON FABRICATION OF NANOPARTICLES

Bio-fabrication of Nanoparticles are in general achieved either through reduction/oxidation process. The biomolecules present in microbes or botanical species were found be responsible for *reduction-cum-stabilization* of metal ions into their respective nano structures (Figure 2). Generation of nanoparticles occurs in a very similar way to that which occurs in nature – i.e. by biomimetics – is much 'cleaner' and 'greener'. The living systems involved in nanoparticles production has been recapitulated on several occasions (Abbasi *et al*., 2011; Anuradha, 2013; Singh *et al*., 2016).

Of the living species, plants based nanofabrication of particles is reckoned to be facile, benign, quicker, and can be easily scaled up for large-scale production than microbe based nanoparticles synthesis. Since the latter rely on maintenance of microbial culture and might also carry toxic moieties, which can endanger the environment as well as the human health (Anuradha*et al.*, 2015a). The difficulties associated with this synthesis route further depreciate the value and favors the plant-based system for nanoparticles synthesis. Beginning with the pioneer work reporting on metal accumulating plants dates back to 1855, several other reports pertaining to abnormal high accumulation of metal ions in plant tissues appeared more than five decades (Reevs and Baker, 2000). Plant based nanoparticles are generated either intracellular or by extracellular method. The first report on intracellular nanoparticles production and characterization was reported by Gardae-Torresday and co-workers (2002). Whereas, extracellular production for the first time using plant broth was reported by Sastry and co-workers (Shanker *et al.*, 2003). Utilization of weeds and wastes for nanoparticles synthesis has been detailed with a special reference and is reported by Abbasi and co-workers (Abbasi *et al.*, 2011; 2012a; 2012b; Anuradha, 2013). Most of these processes on nanoparticles production are complicated and required rigorous controls such as maintenance of various factors comprising pH, agitation, illumination, bioagent *cum* metal ion concentration and like, for shape and size control (Abbasi *et al.*, 2015). Authors have innovated systems for production of particles with desired morphology by varying the mixing proportion as the process variable (Anuradha *et al.*, 2010; 2011; 2015a; 2015b; Anuradha, 2013; Abbasi *et al.*, 2014; 2016a; 2016b). General mechanism involved in the nanoparticles formation is illustrated in the figure 3.

APPLICATION OF NANOPARTICLES WITH SPECIAL REFERENCE TO POLLUTION CONTROL

The advancement in the field of nanoscience, *i.e.* controlled molecular assembly into nanoscale materials and its unique properties brings control of matter. Due to this aspect, nanoparticles are of commercial importance, and are being synthesized in a single-pot on bringing metal ion solutions, in contact with bioagent broth. In the manufacturing community, nanomaterials is at prime importance and finds wider applications bestowed by catalytic activity and its anti-microbial, optical, magnetic, electronic and medicinal properties.Biogenic nanoparticles like silver and gold were demonstrated to possess the antimicrobial activity and are applied as anti-dermatophyte, larvicide, anti-cancerous, wound-healing agents in pharmacology.

Studies revealed that antimicrobial activities are shown to have maximum zone of inhibition against the studied microbes (*i.e.* bacteria and fungi) (Nisha *et al.*, 2014). Apoptosis of these microbes might have occurred due to the generation of ROS (Reactive oxygen species) and reduced hydroxyl radicals induced by the capping/ reducing agents on the nanoparticles (Firdhouse and Lalitha, 2015). In another study, appreciable larvicidal activity was recorded against *Anopheles subpictus* and *Culex tritaeniorhynchus* (Rajakumar and Rahuman, 2011).The incubation of cancerous cell lines was induced to generate ROS resulting in apoptosis, by suppressing cell growth, arresting cell cycle phases and reduces DNA synthesis (Gajendran *et al.*, 2014). Investigations on the wound-healing activity of biocompatible nanoparticles were found with positive effect in excision animal model, owing to promising benefit in clinical practices and sterilization of the medical instruments (Arunachalam *et al.*, 2013).Nevertheless, recent research adds up the application of biogenic nanoparticles in medicine, agriculture, food technology. Gold particles is reported, to induce the insulin secretion for diabetes management (Mubayi *et al.*, 2017); in-situ stimulation of germination and growth rate in plants and can be potentially used in agriculture (Kumar *et al.*, 2013) and selenium nanoparticles with antimicrobial and anti-cancerous activity against food borne pathogens and human cancer cells (Ahmed *et al.*, 2015; Nguyen *et al.*, 2017).

The domain were these biosynthesized nano-materials are applied is not restricted to agriculture, medicine and food technology, but also extends to environmental pollution control (Figure 4). The functional aspect of nanotechnology relies on driving force, either to accelerate the performance of materials and/or to reduce the quantity of materials that are in contact with the nanostructures materials. The reactivity of bio-fabricated nanoparticles either destroys the contaminants, for instance, or sequesters them. Here, environmental clean-up strategy or sequestration of pollutants such as heavy metals, organic and inorganic pollutants that are achieved through nano-biotechnological approach is entitled with nano-bioremediation technology.

NANOBIOREMEDIATION

In the recent past, nano-biotechnology has also contributed significantly in remediating the soil, water and air pollutants into environment friendly compounds. Its attribute, depends on the fact that comprises three categories, *viz. (i) treatment and remediation; (ii) sensing and detection; and (iii) pollution prevention* (Carata *et al.*, 2017). Most significantly, the emergence of nano-bioremediation could nurture the environment through pollution removal and cleanup strategies. Aforesaid, integration of conventional bioremediation and nano-biotechnological approach or direct nano-remediation techniques could be the feasible option that could inundate the contaminants from environment. Nanoparticles utilized to degrade or sequester the heavy metals/ herbicides/ pesticides/ insecticides (*organic/ inorganic pollutants*) from the polluted environment.

In general, recent research reveals a promising scope for pollution control either direct or integrated application of nanoparticles. For an example the removal of organic pollutants like herbicides (*i.e.* atrazine, molinate) and pesticides (*i.e.* chlorpyrifos) are degraded directly using nZVI (*Zerovalentnano ions*). Whereas, confluence of nanoparticles degrades organic pollutants along with the phyto-remediation has also been reported. Here, long-chain hydrocarbons and halocarbons that remained persistent with bioremediation were broken down nanoparticles mediated reaction. There by influencing the induction of degradability of the living organisms in the pollution control. The possible mechanism that entails, improved reactivity of nano-materials for the degradation and detoxification of contaminants is achieved through the conversion of nZVI/ nano scale Fe^0 (Sakulchaicharoen *et al.*, 2010). Thus, application of this reaction mechanism occurring through detection, extraction and degradation of environmentally toxic contaminants like halo and hydrocarbons in general, could be a novel approach in achieving sustainable environment.

ADVANCES IN ENVIRONMENTAL APPLICATION OF NANOBIOTECHNOLOGY

The distinguishing feature of nano-bioremediation relies on the applicability of nano-biotechnology, that encompassing three significant attributes such as, (i) utilization of benign, clean and green nanomaterials; (ii) provides solution for the removal of hazardous materials from contaminated site; and (iii) used as the environmental sensor agent (Tratnyek and Johnson, 2006). In particular, efforts on achieving 'environmental improvement' primary depend on the nano-remediation phenomenon known as: adsorptive or reactive techniques that are applied at on-site (*in-situ*) or off site (*ex-situ*) treatment of pollutants. The farmer technique involve in removal of heavy metal contaminants by sequestration, whereas the later technology degrades the organic contaminants leaving behind no harmful by products like CO_2 and H_2O. Gradual growth and development of these versatile remediation technologies brings scope towards achieving the sustainable mitigation of pollutants from the environment.

Biogenic nanomaterials that have been explored for the remediation strategies, designed for the purpose includes, variants of metal oxides, noble metalsin particular the bimetallic form and titanium dioxides. Of these, nanoscale zero-valent iron (nZVI) and its derivatives were found to have wider application in the nano-remediation technologies. The use of these nanostructures such as iron, nickel, gold, silver and Pd, has shown encouraging results towards removal of toxic substances like chromium, arsenic and degradation of persistent organic compounds (Thome *et al.*, 2015).

In-situ treatment of contaminantsis the feasible and most preferred approach due to higher efficacy, cost effectiveness and large scale application. Here, nZVI were injected into the contaminated porous sites like soil, sediments and aquifers. Several studies reveal that higher reactivity enhances the degradability of pollutants, while agglomeration and loss of reactivity with aging limited the use of nZVI in remediation (Reddy, 2010; Yan *et al.*, 2013). The surface coatings of organic polymers like gaur gum, lactate etc., on nZVI were found to improve the mobility, stability and reactivity resulting in enhanced site decontamination (Reddy *et al.*, 2014). In this way, remediation is achieved either that nanoparticles gets relatively immobile; or that gets migrated to the contaminated sites with sufficient mobility.

In some studies, results were found to have negative effect on tested microbial communities like *Escherichia coli*, *Staphyloccoides sp.* and *Dehalococcoides sp.* (Li and Jiang, 2009; Auffan and Bottero, 2009; Gordon and Margel, 2011). While, other studies shown to have stimulating effect on methanogens, sulfate reducing and bacterial colonies in contaminated aquifers (Xiu and Alvarez, 2010; Kirschling and Tilton, 2010). In case of fungal growth, studies demonstrated to have neither structural modification, nor growth reduction even after the adhesion of nZVI on the individual (Gonzalez *et al.*, 2013). But Lefevre et al.(2016) reported that formation of ROS (reactive oxygen species) inside the cell damages the cell structure. Thus, controversial results and limitations on types of nanoparticles and type of tested microbes were reckoned in the microbial inhibition studies.

For instance, remediation at *in-situ* site does not occur with pumping of ground water or transportation of soil separately above the ground level or to other sites for treatment purpose. But injection of nano-materials brings treatment efficacy of contaminant concentration to near zero at all on-sites. Nanoparticles generated from bio-agent binding finds a potential role in soil and ground water remediation. It is possibly achieved due to the unique properties of nano-materials like high reactivity and stronger sorption. Due to the colloidal nature, they penetrate deep into the contaminated sites and treats water inevitably than the conventional techniques.

The recapitulated studies on utilization of nanoparticles include the adsorption of various pollutants like heavy metals and dyes. This adsorption technique combined with magnetic separation has far reaching benefits in water treatment and environmental clean-up. Mechanism of in-situ technique involving nZVI in treatment of polluted soil and aquifers include, *(i) injection of nanoscale ZVI into contaminated site; (ii) injection of nZVI with high mobility inducing nZVI plume; (iii) spiking of nanoparticles onto the topsoil to adsorb/degrade the organic and inorganic pollutants*.

Nano-remediation that is influenced with contaminant degradation – rather than adsorption – is of greater significance, particularly in treatment of organic contaminants. In *ex-situ* technology, biogenic nanoparticles such as TiO_2, ZnO, nZVI and noble metal nanoparticles in particular, has potential role in remediating the organic contaminants. The investigation on the bimetallic gold-palladium NPs serves as an active catalyst, by enhancing the degradability of major water pollutants like trichlorothane (TCE) and oxidation of mercury. Gold nanoparticles have also been used for designing gas sensors for detecting pollution levels of carbon monoxide (CO) and oxides of nitrogen (NOx). Metal oxide nanoparticles like ZnO, TiO_2 and CeO_2 have stimulated the growth of both root and shoot of plant species such as soybean, wheat, corn and alfalfa, suggesting the nano-biotechnology showing significant impact on phytoremediation. TiO_2 at nanoscale increased the reactivity of several enzymes and promoted nitrate adsorption, conversion of inorganic into organic nitrogen.

In another study, silver nanoparticles at nano form incubated with Congo red containing *Aspergillus niger* broth, showed more than 85.8% decolorization within 24 hours (Nithya *et al.*, 2010). The reduction and adsorption of lead (Pb) and hexachlorohexanes (HCHs) from contaminated water was found to be remediated with 99.9% and 95% of removal using nano-scale ZVI (Xi *et al.*, 2010; Daniel *et al.*, 2008).Summary of nano-bioremediation method employed for decontamination of pollutants is presented in table 1. The process of nano-bioremediation initiates with reduction of contaminants levels and then promotes the biodegradability of contaminants below the risk level.

In addition to soil, water/ wastewater treatment, air pollution monitoring and treatment is under progress with the advancement in nanotechnology. Carbon nanotubes are found to be efficient than activated carbon on absorbing dioxin to a multifold due to the increased surface area in nano form. Technological development in the field of nano-biotechnological approach in nano-remediation are still in theirinfant stage, since the size, shape and surface chemistry are not yet optimal. The use of these particles in nano-remediation will inevitably lead to release of nanomaterialsinto the environment. But mankind could be exposed to these nanoparticles through inhalation, ingestion, dermal contact, and/ or injection. In order to confirm the safety of employing fabricated nanostructures, in depth research is required to know nano toxicity on the biological systems.

CONCLUSION

The research innovation in nano-biotechnology opens up an avenue for designing the nano-remediation strategies. Special emphasis on this technology depends on the potentials benefits received on the applicability in the field of medicine, electronics, agriculture and environment for social wellbeing. A far reaching sustainable process development has been achieved for tailoring the nanoparticles on size and shape control through nano-biotechnology. Adaptation of nano-remediation has the efficacy to reduce the overall cost for environmental clean-up at large scale application.

The reduction in the contaminant level even could reach to near zero in in-situ site nano-remediation. Whereas, ex-situ nano-remediation are still under infant stage and are shown to have promising results in decontamination of microbes, organic and inorganic pollutants. Nano-remediation being in exciting stage of development is considered as potential tool for mitigating environmental pollution. These technological improvements are likely to be a favorable proxy for conventional treatments, but potential risks associated with environmental nano-toxicity are still unclear. Once these research gaps were disclosed, it is certain that nano-remediation will no doubt be a promising technology for achieving environmental sustainability.

REFERENCES

Abbasi, S.A., Abbasi, T., & Anuradha, J. (2012). *A process for synthesis of metal nanoparticles from aquatic weeds.* Off. J. Patent Office. Application No.458/CHE/2012A, p.6184.

Abbasi, S.A., Abbasi, T., Anuradha, J., Neghi, N., Prathiba, S., & Ganaie, S.U. (2011). *Gainful utilization of four otherwise worthless and problematic weeds for silver nanoparticle synthesis.* Off. J. Patent Office. Application No.2175/CHE/2011A, 11869.

Abbasi, T., Anuradha, J., Ganaie, S. U., & Abbasi, S. A. (2015). Biomimetic synthesis of nanoparticles using aqueous extracts of plants (botanical species). *Journal of Nano Research, 31*, 138–202. doi:10.4028/www.scientific.net/JNanoR.31.138

Abbasi, T., Anuradha, J., & Abbasi, S. A. (2009). Nanotechnology and its potential in revolutionizing the pollution control scenario. *J. Institution of Public Health Engineers, India, 10*, 5–12.

Abbasi, T., Anuradha, J., & Abbasi, S. A. (2014). Utilization of the terrestrial weed guduchi (*Tinosporacordifolia*) in clean-green synthesis of gold nanoparticles. *Nanoscience and Technology, 1*(3), 1–7.

Abbasi, T., Anuradha, J., & Abbasi, S. A. (2016a). Clean and rapid biomimetic synthesis of gold nanoparticles using all parts of the aquatic weed water hyacinth (*Eichhorniacrassipes*). *Journal of the Indian Chemical Society, 93*, 851–865.

Abbasi, T., Anuradha, J., & Abbasi, S. A. (2016b). Utilization of the pernicious aquatic weed salvinia (*Salviniamolesta* D. S. Mitchell) in generating gold nanoparticles. *Indian Journal of Biotechnology, 15*, 382–391.

Ahmad, M. S., Yasser, M. M., Sholkamy, E. N., Ali, A. M., & Mehanni, M. M. (2015). Anticancer activity of biostabilized selenium nanorods synthesized by Streptomyces bikiniensis strain Ess_amA-1. *International Journal of Nanomedicine*, *10*, 3389–3401. PMID:26005349

Anuradha, J. (2013). *Use of some aquatic and terrestrial weeds in the 'green' synthesis of gold nanoparticles* (PhD thesis). Pondicherry (Central) University.

Anuradha, J., Abbasi, T., & Abbasi, S. A. (2015a). An eco-friendly method of synthesizing gold nanoparticles using an otherwise worthless weed pistia (*Pistiastratiotes* L.). *Journal of Advanced Research*, *6*(5), 711–720. doi:10.1016/j.jare.2014.03.006 PMID:27563461

Anuradha, J., Abbasi, T., & Abbasi, S. A. (2015b). Rapid and Green synthesis of gold nanoparticles by the use of an otherwise worthless weed lantana (*Lantana camara* L.). *Journal of Environmental Science & Engineering*, *57*(3), 203–213.

Anuradha, J., Abbasi, T., & Abbasi, S. A. (2010). 'Green' synthesis of gold nanoparticles with aqueous extracts of neem (*Azadirachtaindica*). *Research Journal of Biotechnology*, *5*, 75–79.

Anuradha, J., Abbasi, T., & Abbasi, S. A. (2011). Rapid and Reproducible 'Green' synthesis of Silver Nanoparticles of consistent shape and size using *Azadirachtaindica*. *Research Journal of Biotechnology*, *6*, 69–70.

Arunachalam, K. D., Annamalai, S. K., Arunachalam, A. M., & Kennedy, S. (2013). Green Synthesis of crystalline silver nanoparticles using Indigoferaaspalathoides-medicinal plant extract for wound healing applications. *Asian Journal of Chemistry*, *25*, S311–S314.

Auffan, M., & Bottero, J. (2009). Chemical stability of metallic nanoparticles: A parameter controlling their potential cellular toxicity in vitro. *Environ. Pollut.*, *157*(4), 1127–1133. doi:10.1016/j.envpol.2008.10.002 PMID:19013699

Baker, S., & Satish, S. (2012). Endophytes: Toward a Vision in Synthesis of Nanoparticles for Future Therapeutic Agents. Int. J. Bio-Inorg. Hybd. *Nanomat.*, *1*, 1–11.

Bokare V., Murugesan K., Kim J.-H., Kim E.-J., & Chang, Y.-S. (2012). Integrated hybrid treatment for the remediation of 2,3,7,8-tetrachlorodibenzo-p-dioxin. *Sci. Total Environ.*, *435-436*, 563-566.

Bokare, V., Murugesan, K., Kim, Y.-M., Jeon, J.-R., Kim, E.-J., & Chang, Y. S. (2010). Degradation of triclosan by an integrated nano-bio redox process. *Bioresource Technology*, *101*(16), 6354–6360. doi:10.1016/j.biortech.2010.03.062 PMID:20381343

Carata, E., Panzarini, E., & Dini, L. (2017). Environmental Nanoremediation and Electron Microscopies. In Nanotechnologies for Environmental Remediation. Springer International Publishing AG.

Cecchin, I., Reddy, K. R., Thome, A., Tessaro, E. F., & Schnaid, F. (2017). Nanobioremediation: Integration of nanoparticles and bioremediation for sustainable remediation of chlorinated organic contaminants in soils. *International Biodeterioration & Biodegradation*, *119*, 419–428. doi:10.1016/j.ibiod.2016.09.027

Firdhouse, M. J., & Lalitha, P. (2015). Biocidal potential of biosynthesized silver nanoparticles against fungal threats. *Journal of Nanostructure in Chemistry*, *5*(1), 25–33. doi:10.1007/s40097-014-0126-x

Gajendran, B., Chinnasamy, A., Durai, P., Raman, J., & Ramar, M. (2014). Biosynthesis and characterization of silver nanoparticles from *Daturainoxia* and its apoptotic effct on human breast cancer cell line MCF7. *Materials Letters*, *122*, 98–102. doi:10.1016/j.matlet.2014.02.003

Gardea-Torresdey, J. L., Dokken, K., Tiemann, K. J., Parsons, J. G., Ramos, J., Pingitore, N. E., & Gamez, G. (2002). Infrared and X-ray absorption spectroscopic studies on the mechanism of chromium(III) binding to alfalfa biomass. *Microchemical Journal*, *71*(2-3), 157–166. doi:10.1016/S0026-265X(02)00007-3

Gordon, T., & Margel, S. (2011). Synthesis and characterization of zinc/iron oxide composite nanoparticles and their antibacterial properties. *Colloids Surf. Physicochem Eng. Asp.*, *374*(1-3), 1–8. doi:10.1016/j.colsurfa.2010.10.015

Karn, B., Kuiken, T., & Otto, M. (2009). Nanotechnology and in situ remediation: A review of the benefits and potential risks. *Environmental Health Perspectives*, *117*(12), 1823–1831. doi:10.1289/ehp.0900793 PMID:20049198

Kirschling, T. L., & Tilton, R. D. (2010). Impact of nanoscale zero valent iron on geochemistry and microbial populations in trichloroethylene contaminated aquifer materials. *Environmental Science & Technology*, *44*(9), 3474–3480. doi:10.1021/es903744f PMID:20350000

Le, T. T., Nguyen, K.-H., Jeon, J.-R., Francis, A. J., & Chang, Y.-S. (2015). Nano/bio treatment of polychlorinated biphenyls with evaluation of comparative toxicity. *Journal of Hazardous Materials*, *287*, 335–341. doi:10.1016/j.jhazmat.2015.02.001 PMID:25679799

Lefevre, E., Bossa, N., Wiesner, M., & Gunsch, C. (2016). A review of the environmental implications of in situ remediation by nanoscale zero valent iron (nZVI): Behavior, transport and impacts on microbial communities. *The Science of the Total Environment*, *15*, 889–901. doi:10.1016/j.scitotenv.2016.02.003 PMID:26897610

Li, H., & Jiang, G. (2009). Effects of waterborne nano-iron on medaka (oryziaslatipes): Antioxidant enzymatic activity, lipid peroxidation and histopathology. *Ecotoxicology and Environmental Safety*, *72*(3), 684–692. doi:10.1016/j.ecoenv.2008.09.027 PMID:19058851

Mubayi, A., Chatterji, S., & Watal, G. (2017). Green Fabrication of Metal Nanoparticles. In Intelligent Nanomaterials (2nd ed.). Scrivener Publishing LLC.

Nguyen, T.H.D., Vardhanabhuti, B., Lin, M., & Mustapha, A. (2017). *Food Control 2017. Antibacterial properties of selenium nanoparticles and their toxicity to Caco-2 cells*. Academic Press.

Nisha, S.N., Aysha, O.S., & Rahaman, J.S.N. (2014). Lemon peels mediated synthesis of silver nanoparticles and its antidermatophytic activity. *SpectrochimicaActa Part A: Molecular and Biomolecular Spectroscopy, 124*, 194–198.

Nithya, R., & Ragunathan, R. (2010). *Proc. On, Decolorization of the dye Congo red by Aspergillus nigernanoparticle, Nation*. Semi. Anna Univer., Trichy.

Tratnyek & Johnson. (2006). Nanotechnologies for environmental cleanup. Nanatoday, 1(2), 44-48.

Pradeep, T. (2007). *Nano: The Essentials*. New Delhi: Tata-McGraw Hill.

Qu, X., Alvarez, P. J. J., & Li, Q. (2013). Applications of Nanotechnology in Water and Wastewater Treatment. *Water Research*, *47*(12), 3931–3946. doi:10.1016/j.watres.2012.09.058 PMID:23571110

Rajakumar, G., & Rahuman, A. A. (2011). Larvicidal activity of synthesized silver nanoparticles using Ecliptaprostrata leaf extract against fiariasis and malaria vectors. *Acta Tropica*, *118*(3), 196–203. doi:10.1016/j.actatropica.2011.03.003 PMID:21419749

Reddy, K. R. (2010). Nanotechnology for site remediation: Dehalogenation of organic pollutants in soils and groundwater by nanoscale iron particles. *Proceedings of the 6th International Congress on Environmental Geotecnics*, *1*, 163-180.

Reddy, K. R., & Adams, J. A. (2010). *Sustainable Remediation of Contaminated Sites*. New York, NY: Momentum Press.

Reddy, K. R., Khodadoust, A. P., & Darko-Kagya, K. (2014). Transport and reactivity of lactate-modified nanoscale iron particles for remediation of DNT in subsurface soils. *Journal of Environmental Engineering*, *140*(12), 04014042. doi:10.1061/(ASCE)EE.1943-7870.0000870

Reeves, R. D., & Baker, A. J. M. (2000). Metal accumulating plants. In I. Raskin & B. D. Ensley (Eds.), *Phytoextraction of toxic metals-using plants to clean the environment* (pp. 193–229). New York, NY: Wiley.

Sakulchaicharoen, N., O'Carroll, D. M., & Herrera, J. E. (2010). Enhanced stability and dechlorination activity of pre-synthesis stabilized nanoscaleFePd particles. *Journal of Contaminant Hydrology*, *118*(3-4), 117–127. doi:10.1016/j.jconhyd.2010.09.004 PMID:20934234

Shankar, S. S., Ahmad, A., Parsricha, R., & Sastry, M. (2003). Bioreduction of chloroaurate ions by geranium leaves and its endophytic fungus yields gold nanoparticles of different shapes. *Journal of Materials Chemistry*, *13*(7), 1822–1826. doi:10.1039/b303808b

Singh, P., Kim, Y. J., Zhang, D., & Yang, D.-C. (2016). Biological Synthesis of Nanoparticles from Plants and Microorganisms. *Trends in Biotechnology*, *34*(7), 588–599. doi:10.1016/j.tibtech.2016.02.006 PMID:26944794

Singh, R., Manickam, N., Mudiam, M. K. R., Murthy, R. C., & Misra, V. (2013). An integrated (nano-bio) technique for degradation of -HCH contaminated soil. *Journal of Hazardous Materials*, *35*, 258–259. PMID:23692681

Thome, A., Reddy, K. R., Reginatto, C., & Cecchin, I. (2015). Review of nanotechnology forsoil and groundwater remediation: Brazilian perspectives. *Water, Air, and Soil Pollution*, *226*(4), 1–20. doi:10.1007/s11270-014-2243-z

Xi, Y., Mallavarapu, M., & Naidu, R. (2010). Reduction and adsorption of Pb 2 þ in aqueous solution by nano zero-valent iron: A SEM, TEM and XPS study. *Materials Research Bulletin*, *45*(10), 1361–1367. doi:10.1016/j.materresbull.2010.06.046

Xiu, Z., & Alvarez, P. J. J. (2010). Effects of nano-scale zero-valent iron particles on a mixed culture dechlorinating trichloroethylene. *Bioresource Technology*, *101*(4), 1141–1146. doi:10.1016/j.biortech.2009.09.057 PMID:19819128

Yan, W., Lien, H.-L., Koel, B. E., & Zhang, W.-X. (2013). Iron nanoparticles for environmental clean-up: Recent developments and future outlook. *Environmental Science. Processes & Impacts*, *15*(1), 63–77. doi:10.1039/C2EM30691C PMID:24592428

APPENDIX

Figure 1. Reported living systems utilized for in vivo and/or in vitro nanoparticles synthesis (Anuradha, 2013)

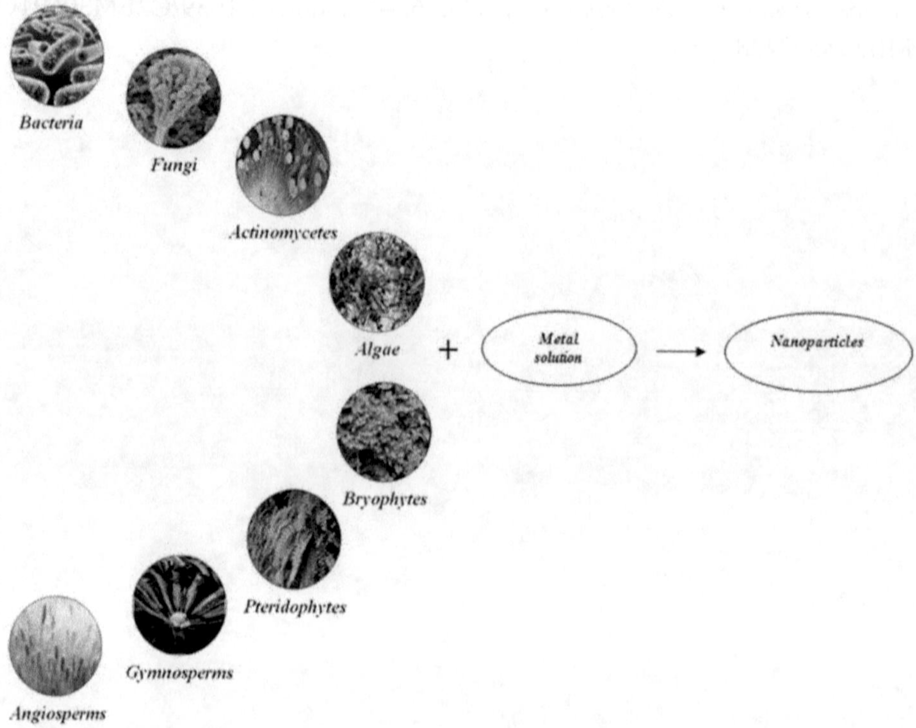

Figure 2. A generalized flowchart illustrating the steps involved in the extracellular synthesis of nanoparticles (adopted from Anuradha, 2013)

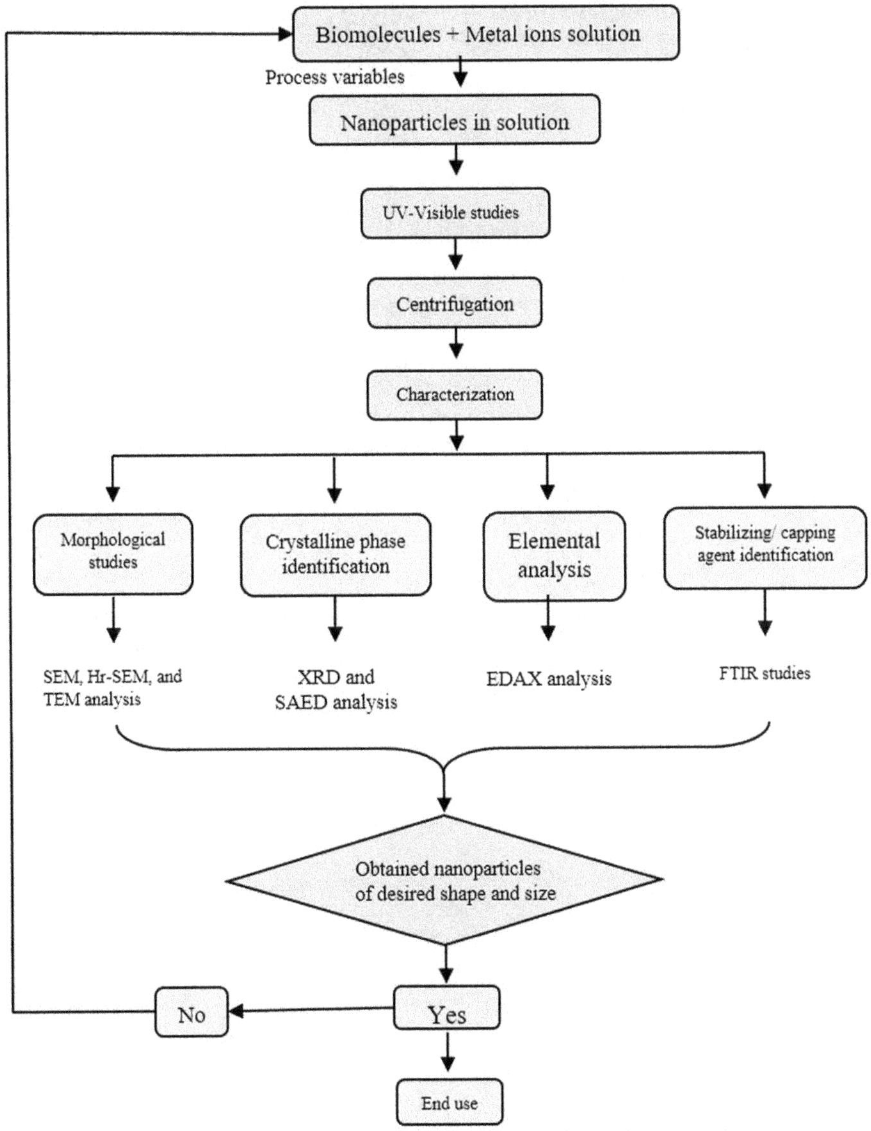

Figure 3. Mechanism of nanoparticles formation (Anuradha et al., 2015a)

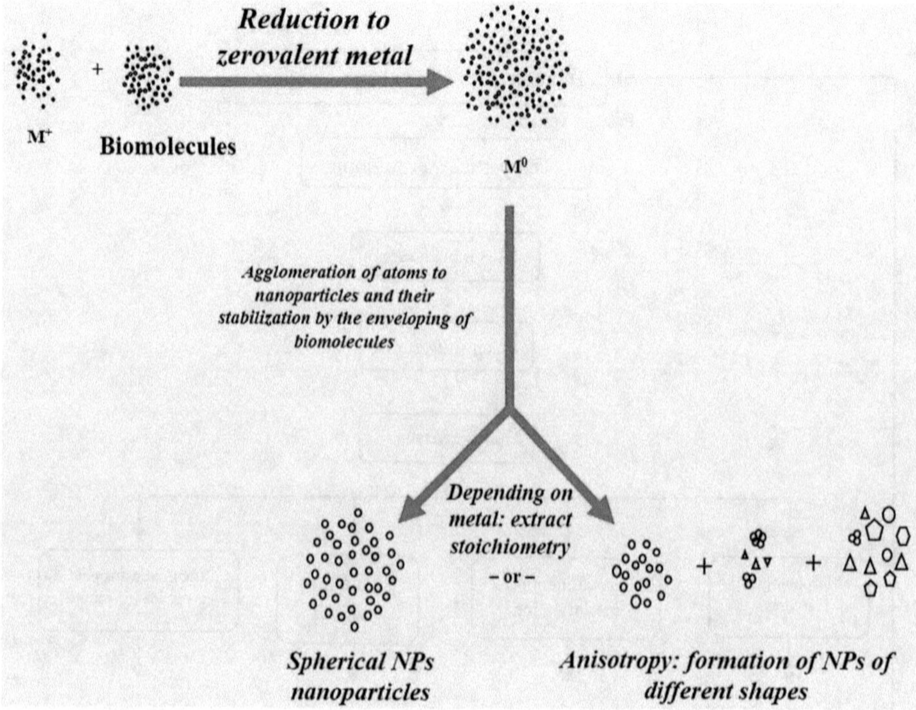

Figure 4. Illustrates the domains of biogenic nanoparticles application

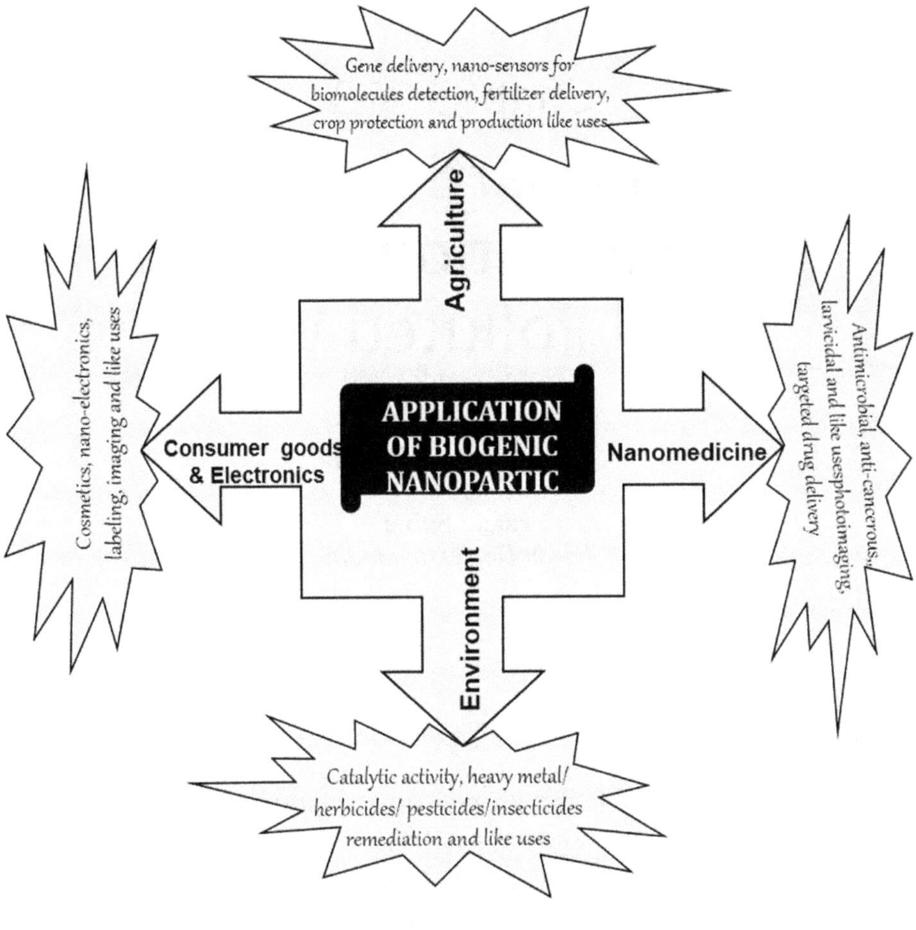

Table 1. Summary of nanoparticles and microorganisms involved in on-site nanobioremediation

Nanoparticles used	Microorganisms used	Contaminant	% of NR	% of NBR	Reference
Pd/Fe	*Trametesversicolor*	Triclosan	100	100	Bokare*et al.*, 2010
Pd/Fe	*Sphingomonaswittichii*	Dioxin	100	100	Bokare*et al.*, 2012
Pd/Fe	Sphingomonas sp.	Lindane	99	99	Singh *et al.*, 2013
nZVI	*Burkholderiaxenovorans*	Aroclor1248 (PCB)	89	90	Le *et al* ., 2015

Chapter 13
Nanoparticles for Bioremediation of Heavy Metal Polluted Water

Rajiv Dutta
Sharda University, India

Pragati Sahai
Sharda University, India

ABSTRACT

The process of bioremediation can be intrinsic or natural attenuation, where the process of remediation happens on its own, or it can be extrinsic or bio-stimulated when it is incited with help of some growth productive conditions like addition of fertilizers or nanoparticles, the smallest active particles on the earth. Nanoparticles are charged entities with low activation energy and exhibiting quantum effects making the chemical reaction between the nanoparticle and surrounding feasible in lesser time, and they also exhibit surface plasmon resonance that helps in identification of toxic material in surrounding. Apart from these properties, their different shapes and sizes help in designing the environmental cleanup process as per the suitable conditions and requirements. Polluted water treatment is being done with the help of nanoparticles due to their property of being highly profitable as adsorbents and for filtration purposes.

DOI: 10.4018/978-1-5225-4162-2.ch013

Copyright © 2018, IGI Global. Copying or distributing in print or electronic forms without written permission of IGI Global is prohibited.

INTRODUCTION

The way the heavy metals are being used in various applications and factory processes like mining and processing of minerals, food products, dyes etc., it produces large amount of toxic waste that is being accumulated leading to the generation of large quantities of effluent containing high levels of toxicity. These effluents are discharged in open water resources like rivers etc. where it harms the marine ecology. Sometimes these toxic elements leach into the groundwater system thereby causing groundwater contamination. It is this water that is used for irrigation and drinking and the presence of different organic and heavy metal contaminants in groundwater that creates a large environmental, public health and economic impact. There is an abundant need of safe drinking water worldwide. In places like Somalia, the kids of tender age travel about ten kilometres every day for drinking water that is not even clean and safe. So a major cause of death in many parts of the world, particularly in young children, the elderly, or those with compromised immune systems is still the water borne diseases. Over the years with changing environment, the epidemiology of the disease is continuously changing with development of new or more resistant microbes. Therefore it becomes very important that these microbes are screened and identified so that the proper remedial process can be designed. Hence in present scenario, the remediation has to be at two levels: a) Remediation of Environment and b) Remedial measures for effective treatment of water borne diseases. For remediation of environment the existing applications include solvent extraction, activated carbon adsorption, biological degradation and common chemical oxidation, but these processes are not cost effective and are time consuming. Similarly for effective remedial measures for treatment of disease the effective microbial testing is done by time consuming culture methods that are not completely accurate. At times the most resistant strains are difficult to screen. However, newer enzymatic, immunological and genetic methods are being developed to replace and/or support classical approaches to microbial detection but it is also an expensive process.

The word Bioremediation is made up of two words; Bio means various biological agents such as bacteria, fungi, protists or their enzymes or living component active principle and Remediate means to solve the problem thus Bioremediation means the process by which various biological agents, such as bacteria, fungi, protists, or their enzyme are used to degrade the environmental contaminants into less toxic forms or in other terms it is the process of cleaning the environment with help of biological agents (Dillewijn et. al, 2007). There is an increase in the application of this process over conventional ways of cleaning because of following benefits

- It is economical,
- No requirements of supplementary nutrient,

- It has screening and selectivity to specific metals,
- It is highly competent as it can work under different parameters,
- It causes reduction of chemical and biological sludge,
- The biosorbent can be regenerated,
- The metal recovery is also possible.

The process of Bioremediation can be Intrinsic or Natural attenuation where the process of remediation happens on its own or it can be extrinsic or bio-stimulated when it is incited with help of some growth productive conditions like addition of fertilizers or Nanoparticles. The innovations in Nanotechnology and Nanosciences are having a significant impact on biological applications like biodiagnostics, biomedical and bioremediation because the Nanotechnology is the study of ways of manipulating the surface to bring down it to nanoscale that then acts as highly reactive materials with smallest size and largest surface area. It is this property that offers the potential of novel nanomaterials for treatment of surface water, groundwater and wastewater which is not only contaminated by toxic metal ions but also by organic soluble impurities and inorganic soluble particles, and microbes. For instance the use of nanoscale iron particles in bioremediation over conventional clean up processes.

Physiological Basis of Nanoparticles in Remediation

The Nanoparticles are those particles whose size is as small as measured on nano-scale of 10^9 making them the smallest active particles with large surface. Therefore any material brought to nano-scale will exhibit increase in surface area per unit mass of a material causing increase in reactivity with neighboring materials. Secondly nanoparticles are charged entities with low activation energy and exhibiting quantum effects making the chemical reaction between the nanoparticles and surrounding feasible in lesser time and they also exhibit surface plasma on resonance that helps in identification of toxic material in surrounding. Apart from above properties, their different shapes and sizes help in designing the environmental cleanup process as per the suitable conditions and requirements. Table 1 (Tambe Patil, 2015) shows different Nanoparticles and their synthesis with examples. The polluted water treatment in past many years is being done with the help of nanoparticles because of its property of being highly profitable as adsorbents and for filtration purpose It is for all these characteristics the nanoparticles are being extensively used in remediation program for environmental cleanup (Lunge, Singh, & Sinha, 2014). Secondly due to their unique activity toward recalcitrant contaminants and their use in more than one application they are largely being used for microbial testing of most resistant strains by the use of number of nanoparticle-based assays and nano-devices. Thirdly many nano-materials are under active research and development and the study of

Table 1. Different nanomaterial and their synthesis

Nanomaterial (NM)	Synthesis Methods	Examples
Metal nanoparticles (NP)	Photochemical Electrochemical Biochemical Thermochemical	Pt, Rh, Pd, Ir, Ag, Au, Cu, Co, Ni, FeNi, Cu3Au, CoNi, CdTe, CdSe, ZnS.
Carbon NM	Chemical Vapor deposition Arc Discharge Lase Ablation	Cylindrical nanotude, Fullerenes
Metal Oxide Nanoparticle	Hydrothermal Solvothermal Sol-gel Reverse micelles Methods Electrochemical deposition	Zno, Fe2O3, Fe3O4, MgO, BaCO3, BaSO4, TiO2
Polymer NM	Electrochemical polymerization	Nanowire of polypyrrole, Polyaniline, Poly (3, 4-ethylenedioxidethiophane) dendrimers (PAMAM)
Nano-composite	Innovative Methods	Nanocomposite of polyethylene oxide and polyethyleneimine, CNT epoxy composite include hydrocarbon polymer, conjugated polymer composites, CNTs with polycarbonates, Fluoropolymers, PEG, polyester polyemides, etc.
Bio-nanomaterial	Biological	Virus, Plasmids, Proteins NP

literature about current research on different nano-materials like nanostructured catalytic membranes, nano-sorbents, nano-catalysts and bioactive nanoparticles and their application in water treatment, purification and disinfection along with the knowledge regarding toxicological effects of engineered nano-materials on humans and the environment is under review and experimentation.

APPLICATION OF NANOMATERIAL IN BIOREMEDIATION

The Nanomaterial has various eco-friendly applications in environmental sciences in terms of cleaning polluted water and providing clean water (Lui et. al, 2014) for both large scale and portable applications (Figure 1).

It also helps in detection of environmental contaminants and can further be used for clean-up process by various oxidation reduction reactions. It is also helpful in bioremediation processes that can be *in situ* (bioremediation treats the toxic material in the location in which it is found) where fewer amounts of Nanoparticles can

Figure 1. Application of nanomaterial bioremediation

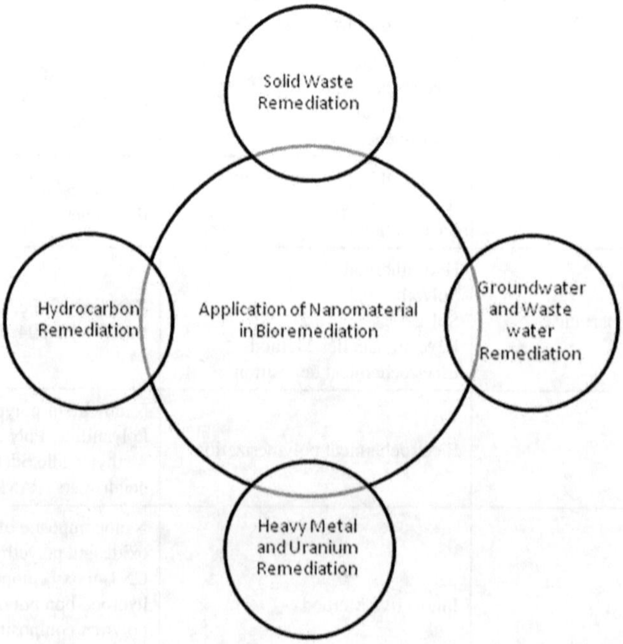

treat large volume or area or *ex situ* (bioremediation processes require excavation of contaminated material or toxic substances before they can be treated) where the amount of Nanoparticles is calibrated as per the unit sample thus making latter faster, more competent, wider range of application on all types of soils and all types of contaminants but expensive than the former type of remediation that is slow, cheaper and cannot be calibrated.

DIFFERENT SOURCES OF WATER POLLUTION AND NEED FOR WATER TREATMENT

It has been reported that only 1% of water is only available as human resource and even this available 1% larger is inadequate for drinking as per WHO report of 2015 (Grey et. al, 2013; WHO, 2015). Secondly this 1% is also largely affected due to human activities and large population explode that is causing high environmental concern for one and all (Adeleye et. al, 2016). There are many different sources of water pollution that can be due to direct or indirect contaminants that can be organic, inorganic, radioactive or acid/base. In case of direct contaminant sources or point sources, the effluent from refineries, factories, waste water plants etc. containing waste fluids of different toxic quality are directly discharged into urban water supplies whereas in indirect contaminant sources or non-point sources include contaminant that are originated somewhere else but ultimately leach into the water (Figure 2). The groundwater pollution is example of pollution due to indirect contaminant sources major because of residue of human agricultural practices like fertilizers and pesticides that leach into the groundwater and improperly disposed industrial wastes (Li, Ma, Kuijp, Quan, & Huang, 2014). The pollution in water cycle is also due to indirect contaminants like gaseous emissions from automobiles, factories and even bakeries that finally cause the water pollution.

Classes of Water Pollutants

The water pollutants can be biological or non-biological and that divides them into different classes. The first most deadly class is of disease causing pollutants that compromises of microbes (bacteria and viruses), parasitic worms (guinea worm) and protozoans (amoeba) that cause some of the deadliest water borne diseases. These pollutants enter sewage systems and untreated wastes. The second class of pollutants involve all the oxygen demanding wastes that leads to utilization of dissolved oxygen in water and causing an increase in COD and BOD levels. The decrease in COD and BOD levels lead to accumulation of toxic contents and depletion in oxygen levels causing death of biological community in water like fishes etc. this further poison the water as these dead matters decay and causes poising of water. The third class of pollutants comprises inorganic pollutants that are water soluble and suspended sediments that absorb the light and lead to depletion in water's diffused light levels that are necessary for marine ecology. The accumulation of such pollutants further makes the water poisonous and unfit for drinking. The fourth category is of dissolve organic pollutants like water soluble nitrate and phosphate fertilizers that causes excessive algal growth, often referred as algal blooms that decreases the DO level, thereby making it unfit for marine ecosystem. This leads to death of fishes etc.

Figure 2. Sources of water pollution

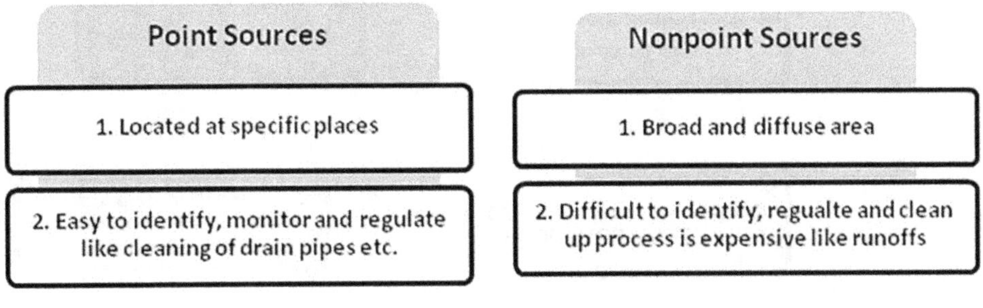

and poisoning of water (Figure 3) (Little Pro, n.d.). The last category comprises of radioactive compounds that are discharged directly into the water bodies or the various nuclear tests done underwater causes vast destruction of water bodies as they are water soluble and drinking such water can lead to death, cancer, various birth and genetic defects.

Need for Water Treatment

Heavy metals pollution is becoming one of the most serious environment problems globally (Fujita et. al, 2014). It is a prominent threat for population in dense countries particularly for China and India (Gowd, Reddy, & Govil, 2010). Its presence in low concentration of heavy metals in various water resources could be harmful to human health. The treatment of heavy metals is so important due to their property of accumulation and deposition in the environment. Thus in order to achieve environmental detoxification and for maintaining the ecological balance, numerous ways of water treatment are employed. Some of these ways are conventional ways of separation like membrane filtration, evaporation, precipitation, adsorption, ion exchange, reverse osmosis, electrochemical treatments and flotation etc. and some are biological and chemical like oxidation, biosorption and biochemical degradation (Dunwellgroup, n.d.). In all these processes the recent advancement is the use of nanoparticle for water purification, for instance, in adsorption process the use of magnetic particles to remove of bacterial pathogens from water is an efficient technique under conventional ways of separation.

Figure 3. Water treatment plant of urban city

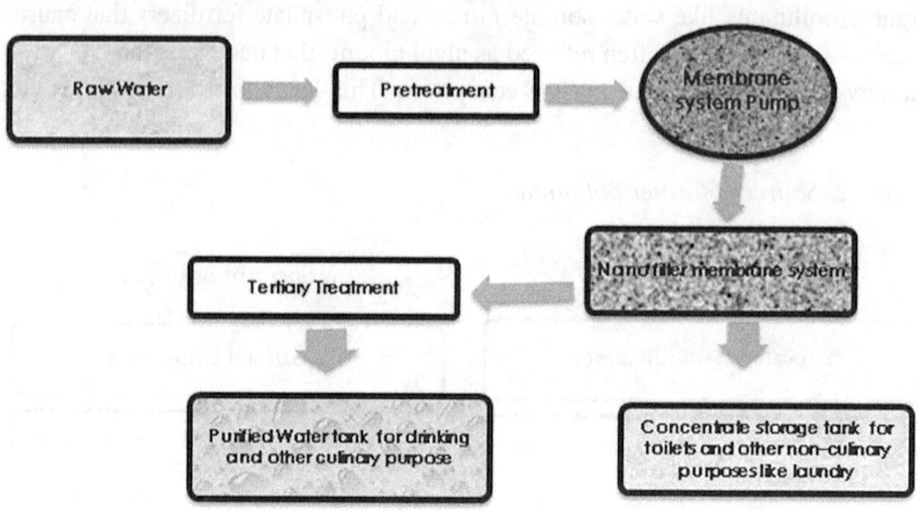

NANOPARTICLES AND BIOREMEDIATION OF WATER

Role of Nanoparticles and Process of Nanofilteration

The recent development in nanotechnology led production of various nanostructures that were used in water treatment. These structures were: magnetic nanoparticles, photo catalytic TiO_2 nanoparticles, CNT, cyclodextrin impregnated in Silver nanoparticles, especillized nano membrane filters, Silica nanoparticles, nanocomposites, Iron-Zeolite nanostructures and C-Fe nanoparticles. These nanostructures were proven to be more efficient in removal of heavy metal contamination, dangerous microbes and pathogens, suspended sediments of organic and inorganic nature and chemical impurities. It has been reported that by using the magnetic nanoparticles as adsorbant like that of magnetite, an iron oxide nanoparticle in water treatment helped in removal of heavy metal contamination very quickly and efficiently (Egirani, Latif, Poyi, Wessey, & Acharjee, 2017). The Metal Oxide Nanoparticles (Dave & Chopda, 2014) are another best remediation possibility for removal of heavy metals from water. Similarly, the process of Nanofilteration (Tambe Patil, 2015) is an important part of bioremediation through nanoparticles that are synthesised so that maximum absorption can be achieved for maximum removal of toxic components from water by various process that are better than the convention ones. The modern methods include: sol-gel method of separation, mechanical grinding, Pyrolysis not by heat but by laser, wet chemical methods, separation by specific binding to a designed template impregnated with nanoparticles and co-precipitation in solution gradient.

Benefits of Nanofilteration

There are various benefits of Nanofilteration as in:

- Low operation and energy cost
- Being chemical free the treated water is non-aggressive and safe
- Lesser discharge than reverse osmosis and lower amount of waste water
- Effective softening of water and removal of hardness
- Reduction of following components:
    - Total dissolved solids and turbidity
    - Volatile organic compounds
    - Phenols, colours, tannins
    - Pesticides
    - Heavy metals and salts

The Nanofilteration Assembly

The Nano-filteration assembly is a membrane system placed after the membrane system pump. This Pump pumps the waste water/ effluent after primary treatment or clarification. Therefore the secondary treated water/ secondary effluent is only passed through the Nano-filteration assembly (Anjum, Al-Makishah, & Barakat, 2017; Anjum, Oves, Kumar, & Barakat, 2017). The nano-filteration membrane assembly may vary as per the treatment process, like for treatment of water from factories the nano-filteration membrane is placed directly with the fed batch bioreactor or feed tank as the effluent not only contains larger amount of water but also many other factory products and rich microbial culture that are helpful in further bioremediation. Thus the effluent water through Nano-filteration membrane system filters into permeate and concentrate, that is returned back to the feed tank and is mixed with secondary effluent as shown in ray diagram. But the treatment in water plant in city is bit different since the waste content is less in it, soil takes up the raw water and passes it through the Nano-filteration membrane system through pump and after Nano-filteration the water is further treated for purification known as tertiary treatment of water Henceforth after filtration the permeate is used as purified water for drinking and the concentrate is used as nondrinking water for toilet and laundry purposes.

The Nano-filteration is a part of secondary treatment of waste water because if untreated water is send directly to Nano-filteration membrane system then it would block the filters thereby effecting the absorption and filtration rate secondly this will lead to spoiling of the membrane making the process least cost effective so it is suggestive that pre-treated water after primary treatment should be sent for Nano-filteration and not before that to make the process fast, cost effective and with high reduction rate (Swamy, Singh, & Sarathy, 2011).

DIFFERENT NANOPARTICLES AND THEIR USE IN REMOVAL OF DIFFERENT HEAVY METALS

Removal of heavy metals from industrial wastewater involves several techniques have been used including precipitation, ion exchange, membrane filtration and adsorption (Fu, Xie, Tang, Wang, & Jiang, 2012). The production of oxide nanoparticles from metals like iron, aluminium, titanium, selenium, manganese, magnesium and zirconium helped in effective removal of heavy metals from water. These metal oxide nanoparticles because of their large surface area and high reactivity due to size quantization, acted as effective adsorbent to the heavy metal impurities (Bhattacharya, Saha, Mukhopadhyay, Chattopadhyay, & Chand, 2013; Sounthararajah, Loganathan, Kandasamy, & Vigneswaran, 2015).

However almost all approaches rely on the following factors:

1. The functional chemistry between the nanoparticle and different reactive groups
2. Impregnating the different support materials with nanostructure for effective capturing of heavy metals.

Nanoadsorbants/ Nanosorbants

The role of nanoadsorbants in treatment has already discussed earlier. In the aqueous environment, the factors affecting the adsorption process are high surface area, adsorption activity, chemical activity, location of atoms on surface, lacking of internal diffusion resistance and higher surface binding energy (Khajeh, Laurent, & Dastafkan, 2013). Smaller size of the nanoparticles increases the surface area that enhances the chemical activity and adsorption capacity of nanoparticles for the adsorption of metals on their surface. The process of adsorption depends on adsorption coefficient K_d and precipitation partitioning of pollutant i.e. heavy metals or organic pollutants precipitate under equilibrium condition. Moreover, for persistent inorganic pollutants redox-reaction is favoured to start the ionic structure transformation (Gupta et. al, 2015). Nanoadsorbents possess two main properties: innate surface and external functionality. Their physical, chemical and material properties are also related to their extrinsic surface structure, apparent size and intrinsic composition. Nanoparticles used as Adsorbent for the removal of heavy metals should be nontoxic, high adsorption capacity, have the ability to adsorb pollutants in less concentration (ppb), adsorbed pollutants that can be easily removed from adsorbent surface and can be recycled for numerous time (Gupta et. al, 2015). Factors controlling the Nano-adsorbents properties are size, surface chemistry as in Grapheme Oxide (Zare-Dorabei, Ferdowsi, Barzin, & Tadjarodi, 2016; Santhosh et. al, 2016; Azamat, Sattary, Khataee, & Joo, 2015; Dong, Zhang, Wang, Liu, & Jin, 2015; Vu et. al, 2017), agglomeration state according to which adsorption of heavy metals is high due to the high concentration of metals, which decreases as time increases due to blockage of active sites (Shirsath & Shirivastava, 2015) as reported in Pb(II) and Cr(III) from wastewater where the equilibrium time for maximum adsorption was 120 min (Lingamdinne, Koduru, Choi, Chang, & Yang, 2016), then further shape and fractal dimensions, chemical composition, crystal structure and solubility for example the surface modification of Fe_2O_3 (Figure 4) increases the adsorption capacity (Lin, Wang, Luo, & Xu, 2015). Moreover, modification in nanoparticles can be carried out by some reagent in enhancing its properties for metals ions pre concentration as in Fe_2O_3, if the particle is surface modified then

Figure 4. Example of nanoabsorbant Fe203-NP showing absorption of fluoride in water

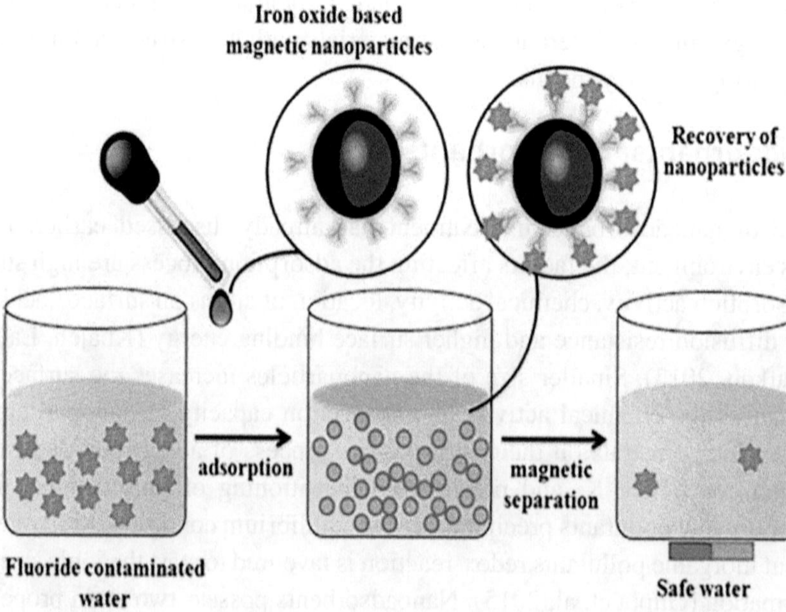

binding capacity further increases (Gopalakrishnan, Krishnan, Thangavel, Venugopal, & Kim, 2015). These modification techniques increase the adsorption capacity of Nanoadsorbents (Li et. al, 2016).

Nanocatalyst

The nanocatalysts, especially those of inorganic materials such as semiconductors and metal oxides, are gaining considerable attention of the researchers in application of wastewater treatment. Various kinds of nanocatalysts are employed for wastewater treatment such as photocatalysts (Dutta, Maji, & Adhikary, 2014), electrocatalysts (Dutta, Maji, & Adhikary, 2014), and Fenton based catalysts (Kurian & Nair, 2015) for improving chemical oxidation of organic pollutants (Faccini, 2015) and antimicrobial actions (Lin, Thirumavalavan, Jiang, & Lee, 2014). Nanoparticle application based on interaction of light energy with metallic nanoparticles is known as Photocatalytic interactions which are of great interest due to their broad and effective functionality for various pollutants. Usually these photo-catalysts are comprised of semiconductor metals that can degrade variety of persistent organic pollutants in wastewater such as dyes, detergents, pesticides and volatile organic compound (Wang et. al, 2015). Photocatalysis has been proven as a promising technique for purification and treatment of various kinds of wastewater. In addition, it has an efficient ability to inactive the pathogenic organism such as bacteria, in the wastewater. As discussed in the above section, TiO_2 is extensively used photocatalyst and reported for its high antimicrobial power (Yang, Geng, & Chen, 2015).

Nanomaterials as Electro-Catalysts as Part of Bioremediation

The process of electro catalysis in microbial fuel cell is an emerging topic of discussion for wastewater treatment and direct electricity generation. In microbial fuel cell, electro-catalyst plays a detrimental role in working of a fuel cell (Habibi & Mohammadyari, 2015). The use of Nanomaterial as electro-catalysts can improve the performance of fuel cell by achieving larger surface area and uniform distribution of catalyst in the reaction media. Tremendous research has been conducted on development of carbon supported nano-electrocatalysts for application in fuel cells. It is reported the Pt nanocatalyst supported by carbon black XC72 showed a potential up to 6.2 mA cm^{-2} of current density in glucose oxidation electro-catalysis reaction (Habibi & Mohammadyari, 2015). Moreover, Pt. electrocatalysts also showed high potential for ethanol oxidation reaction in fuel cell in number of studies (Hu, Yin, Deng, & Hu, 2015).

Although Pt can be used as electro-catalyst, yet, it has various disadvantages which limit its application. For instance, Pt is a precious metal and it has limited availability, and high cost narrowed the interest to use it as catalyst. In addition, during electrocatalysis Pt may restrict the reaction due to poisoning of intermediate compounds. However, these problems can be overcome by replacing Pt with Pd nanoparticles. For instance in ethanol fuel cell Pd nano-catalyst can reduce the cost of anodes due its abundance in the earth. Recently, the researchers have prepared the Pd nanoparticles and functionalized carbon nano-materials by a facile spontaneous redox process (Hu, Yin, Deng, & Hu, 2015). The Pd Nanoparticles with functionalized single wall carbon nanotubes showed the high electro-catalytic activity, and thus could be used as supporting nano-carbon material for synthesis of nano-composite.

COMBINATION OF NANOTECHNOLOGY AND BIOREMEDIATION

The recent advancements in nanotechnologies and nano-sciences has largely helped in resolving the problems of water quality by using various nano-materials as discussed earlier like nano-catalysts, nano-adsorbent, nanotubes, nanostructured catalytic membranes, nano-powder and micro-molecules. All these nanoparticles and colloids had a high impact on water quality in treatment process, however the research studies showed that integration of biological wastewater treatment process with advance nanotechnology resulted in efficient water purification system (Abou-Shanab, Ji, Kim, Paeng, & Jeon, 2013).

Algal Membrane Bioreactor (A-MBR) With Nanoparticles

Algal treatment of waste water has already been proven as one of the promising techniques in terms of energy production and water purification as many algae species effectively grow in wastewater due to the availability of micronutrients (trace metals and vitamins like cyanocobalamin, thiamin) and macronutrients (salts of NO^{-3}, PO^{4}_{3} with Ca, Na, K and NH^{4+}) that are essential for their growth (Barakat, Al-Hutailah, Hashim, Qayyum, & Kuhn, 2013). Thus the reactors are made up of solutions that are a result of mixture of these chemical salts (Nutrients) and water and algal biomass. Nutrient solutions (along with light and carbon dioxide) provide the materials required for algal growth. The result is nutrients removal from wastewater and algal biomass for energy production. For algal biomass harvesting there are various techniques like sedimentation, air flotation, and centrifugation in support with chemical flocculation but these techniques are unaffordable on large scale due to high cost. Among advance techniques, the most advantageous approach for algae cultivation and biomass harvesting is membrane technology in which high-density algae cultivation is simply done through membrane bioreactor (Abou-Shanab, Ji, Kim, Paeng, & Jeon, 2013).

The benefits of membrane technology are that no addition chemicals such as coagulants are required for membrane filtration, thus facilitate the water reuse after filtration and simplify the algal biomass separation (Abou-Shanab, Ji, Kim, Paeng, & Jeon, 2013). Moreover, better algal biomass recovery can be achieved without cell damage and low energy required for algae harvesting as compared to conventional methods (Abou-Shanab, Ji, Kim, Paeng, & Jeon, 2013). Membranes made of polysulfone (PSF), polyvinylidene fluoride (PVDF) and polyethersulfone (PES) are largely used due to their chemical and physical stability but the only problem is the membrane fouling due to hydrophobic mechanism between membrane materials and microbial cells. For increasing hydrophilicity and diminishing membrane fouling there are many techniques like plasma treatment, surface coating and incorporation of Nanomaterials. Research studies revealed that nanoparticles improve the hydrophilicity and reduce membrane fouling. The researchers modified the polyvinyl top layer of the reverse osmosis membrane through coating with TiO_2 nanoparticles and minimize fouling through self-cleaning mechanism under ultraviolet radiation (Martínez et. al, 2013). Besides self-cleaning, TiO_2 particle photocatalysis has also been studied for pollution control due to their high surface area and hydrophilic properties (Moghimifar, Raisi, & Aroujalian, 2014). Due to these characteristics, the hydrophobicity and fouling can be reduced through the incorporation of these nanoparticles with membranes (Yan, Han, Gao, Xue, & Chen, 2015). An investigation was conducted by Hu *et al.* (2015) in which they fabricated the PVDF hollow fibre membranes with TiO_2 by phase inversion method on a custom designed single head spinning machine and this fabricated PVDF/TiO_2 nanocomposite membranes were then tested on algal membrane bioreactor (A-MBR) for wastewater treatment. Results showed that maximum nutrient removal was achieved and up to 75% of phosphorus and nitrogen was removed by AMBRs. In addition, the PVDF/TiO_2 membranes in the A-MBR also increased hydrophilic characteristic of PVDF and reduced membrane fouling.

Pre-Treatment of Aerobic Digestion With Nanoparticles

Traditionally municipal wastewater has been treated with biological process (activated sludge process) but industrial wastewater is more challenging due to the presence of toxic and less biodegradable pollutants (Koók, Rózsenberszki, Nemestóthy, Bélafi-Bakó, & Bakonyi, 2016). The demand for high quality, less toxic and nutrients proportion in treated water results in intensive assessment of the emerging technologies. One of these technologies includes zero valent iron (ZVI) nanoparticles, which have not been used on large scale for water treatment. This nanoparticle can degrade organic contaminants and has been used for the remediation of chlorinated organic compounds from ground water. Integration of biological processes (aerobic degradation) with ZVI nanoparticle treatment is promising in efficient biodegradation of the organic pollutants from wastewater on large scale. In the combination, half of the contaminants are degraded by nanoparticles, which are further easy for biodegradation under aerobic conditions. Partial degradation and reduction in organic compounds occurred due to bimetallic structures form micro and nanoscale cells as electron donor with metallic iron. Further, the products from ZVI reactor are then more biodegradable by aerobic microorganisms. For example in case of degradation of petroleum hydrocarbons and chlorinated solvents, the first section consists of granular iron in which chlorinated ethenes reduced through dechlorination while in second part the aerobic biodegradation of remaining chlorinated compounds and any other hydrocarbons is promoted by the addition of dissolved oxygen. Similarly, it has also been reported that Azo dye and nitro aromatic compounds were treated through coupling of biological process with ZVI nanoparticles (Pan et. al, 2017). A bench scale experiment was conducted to examine the impact of ZVI on biological treatment. Experimental setup was comprised of pre-treatment by ZVI reactor followed by a sequencing batch bioreactor. The influent water to the reactor was comprised of large amounts of persistent organics like halogenated hydrocarbons, petroleum hydrocarbons, nitro aromatic compounds, heavy metals and dyes. The findings showed that BOD was reduced from 235 to 7.7 mg/L (96.5% removal) while COD removal efficiency was 86%. Likewise ammonia removal efficiency was 92% whereas nitrogen removal efficiency was increased from 35 to 52.2%. From the study it was concluded that ZVI pre-treatment followed by biological treatment is the promising potential for industrial wastewater purification (Koók, Rózsenberszki, Nemestóthy, Bélafi-Bakó, & Bakonyi, 2016).

Improvement of Microbial Fuel Cells Efficiency Through Nanotechnology

The attraction toward microbial fuel cell (MFC) technology increased due to double benefits of treating water as well as energy production from inorganic and organic compounds in wastewater by using microbes as biocatalysts (Pan et. al, 2017). The power generated by MFCs can potentially cut down the electricity requirement for a conventional treatment process while complex organic molecules such as butyrate, acetate and propionate are broken down to H_2O and CO_2 by bacteria. The invention of microbial fuel cell by utilization of waste materials to produce electricity is a real breakthrough. The MFC was constructed as a dual chamber cell in which municipal waste was used as substrate in liquid fraction and mesophilic anaerobic inoculum as anaerobic sludge. The surplus production of energy depended on two factors: a) the organic content of the waste and b) the concentration of the fresh inoculum in the anaerobic sludge. It was found that the two factors were inversely related as in if the cell was re-inoculated with fresh inoculum when substrate had low organic content, produced lesser energy and vice versa. Furthermore it was reported that the results highest energy yield was obtained when COD levels were low stating the excellent biodegradability or anaerobic digestion of the substrate (Logroño et. al, 2017). On similar grounds aerobic (in presence of atmospheric and diffused CO_2 single chamber microbial cell was designed using cathodes bound with algae organic dyes polluted water from textile industry acted as substrate in this MFC and the biodegradation of this dye generated electricity. The roles of CO2 both atmospheric and diffused lead to good algal growth that increases the rate of degradation of complex dyes and higher yield of electricity. The cell was effective in treatment of textile waste water along with removal of heavy metals and foul smell because of lowering of COD levels but the single chamber microbial fuel cell had low bio-electrochemical efficiency than double chamber MFC, however the rate of degradation was higher in single chamber MFC than in double chamber that makes them a cost-effective alternative to the costly Pt. electrodes (Cheng et. al, 2017).

However, major barriers in commercialization of this technology were the overall low performance as compared to other fuel cell technologies and high costs of its components. In order to make it economically feasible it is required either to cut down the cost of membrane maintenance or to use cheaper membrane and cathode catalyst. Moreover, it is also necessary to select an appropriate material to improve the performance of MFCs. This includes modification of electrode surface to increase its conductivity, as cathode surface must have high catalytic activity for oxygen reduction reaction. The best option to increase certain properties is the use of nanoparticles as these particles are biologically more active than other particles with similar composition due to its high surface area that effectively interact with biological system. Currently the performance of MFCs was improved by using cheap nano-composite materials like nanostructured carbon in electrodes. These electrodes are mechanically stable with larger surface area, high conductive and better electrochemical catalytic activity. The efficiency of MFC directly depends on the two factors: a) the concentration and diversity of microbial consortia aggregated on electrode (Anode) surface of the MFC and b) the rate of biodegradability of organic content. Various enhancements are being done in increasing the efficiency of MFC. Recent advancement in this field was addition of highly reactive epoxy specie in the waste water and immobilisation of aerobic granular sludge on anode surface of the MFC. It was found that the generation of electricity vastly increased and COD level were greatly reduced. Thus stating that the direct contact of activated sludge with electrode led to increase in bio-electrochemical reactions as the biodegradation was from two sides: a) by microbial film on the anode and b) by microbial consortia in the sludge. The process was also charged because of epoxy reductants present in waste water. These reductants increased the bio-electrochemical process that in led produced higher energy yield (Mahanta & Valiyaveettil, 2013).

Integration of Microbes With Electrospunnanofibrous Webs (Nfw) for Water Purification

Electro-spinning is a recent technology for the production of nanoweb/nanofiber due to its unique properties and cost effectiveness. Environmental application of electrospunnanofibrous webs is its large surface area along with nan-oscale porosity that makes them capable for membrane and filter materials (San, Celebioglu, Tümtaş, Uyar, & Tekinay, 2014). The electrospun nanofibrous integration with microbes can increase the potential of purification and filtration. Research has been conducted on the integration of microbe-electrospun nano-fibre containing algae or bacteria showed momentous impacts for environmental applications (Eroglu et. al, 2012). A study group generated a hybrid system in which algal cells were immobilized on electrospun chitosan nano-fibre mats (ECNMs) for the removal of nitrates from wastewater (Sarioglu, Yasa, Celebioglu, Uyar, & Tekinay, 2013). From the study they summarized that immobilization of the microbial cells is advantageous than free cells due to less space occupancy, easy to handle and low volume required for growth medium. Moreover they also found that ECNMs are water insoluble, non-toxic to algal growth and effectively immobilized the microbial cells. For wastewater treatment they found that 87% of nitrates were removed due to physiochemical adsorption by chitosan and nutrient consumption by algae. In other study the immobilization of ammonium oxidizing bacteria Acinetobacter *calcoaceticus* in cellulose acetate nano-fibre webs that are formed by process of electro-spinning. This system was found to be effective in removal of ammonium from wastewater (Padil, Wacławek, & Černík, 2016). The recent advancement in this spinning process of formation of nano-fibre webs was use of tree gums as a part of green approach. The nano-fibre webs so formed were gum stabilized nanoparticles and functionally charged membranes that were then used for treatment process in different applications (Sarioglu, San Keskin, Celebioglu, Tekinay, & Uyar, 2017).

Likewise a research group also carried out an experiment for decolorization of methylene blue (MB) dye by selecting electrospun CA-NFW for immobilization of dye decolorization bacteria species i.e. Pseudomonas *aeruginosa*, Clavibacter *michiganensis*, Aeromonas *eucrenophila* in wastewater. From the study, they noted that efficient decolorization of MB (95%) was attained in 24 h. The fact behind decolorization of MB dye was the biodegradation carried out by bacteria. Moreover the reusability of the nano-fibrous biocomposite was also examined and it was found that bacteria immobilized NFW possess the capacity to decolorized dye at the end of 4th cycle. Consequently due to its reusability, simple and porous characteristics this nano-fibrous biocomposite can be utilized for treating industrial wastewater. As immobilized bacteria NFW require less space and volume for growth medium compared to free bacteria and hence it is more economical, secondly the biofilm formed also possesses high resistance to metal toxicity, salinity and harsh environmental conditions.

The reduction of methylene blue dye has been done by preparing a unique bio-nano-composite web that had *Pseudomonas aeruginosa* encapsulated in polyvinyl alcohol and Polyethylene oxide matrix by process of electro spinning. These bio nano-fibrous webs were found to be effective in degradation of methylene blue dye in waste water. However the bio-nano-fibrous web of microbe and Polyethylene oxide had higher rate of remediation than that of microbe and PVA. Secondly it was also found that these nano-fibrous webs kept bacteria viable for longer duration and can even be preserved at 4^0C in these webs. This finding creates a platform where the treatment of water can be onsite by directly using these webs as starting inoculum for bioremediation (Zarghami, Akbari, Latifi, & Amani, 2016).

Dendrimers and Its Role in Bioremediation

The dendrimers are those classes of nanoparticles that have their name from their appearance as branches of the tree (made of two Greek words: dendri-branches+ meros-part of the tree). As the name suggests they are highly branched and monodisperse nanoparticles of bigger size because of aggregation. It was this property that was used to prepare bio-nano-dendrimers of chitosan that used to remove heavy metals like Pb^{2+}. The rate of removal was directly proportional to the number of dendrimers as they acted as adsorbents to these heavy metals. The researcher groups apart from developing chitosan adsorbents also enhanced the Nano-filteration process by using porous ceramic filters that were embedded with dendrimers of polyethyleneimine or polypropyleneimine or in green process it was impregnated with β-cyclodextrin for filteration of organic pollutants as they had high mechanical strength along with large surface area (Liu, Jin, Ouyang, & Wang, 2017).

Nanocrystals and Carbon Nanotubes in Bioremediation

The carbon based nano-materials such as nano-crystals and carbon nanotube(s) enable new technologies to identify and solve a broad range of environmental applications because of their exceptional and tenable properties as adsorbents, highly reactive membranes, being efficient as in depth filters, as antimicrobial agents and as sensors for detecting changes in environment. Apart from these they also hold importance in renewable energy technologies and pollution prevention remedial measures. There are many carbon nanotubes developed from single walled to multi walled to both that is hybrid carbon nanotubes. All the three forms had different degradation rate toward ethylene benzene, a commonest pollutant of water due to oil spills. The rate of degradability of ethylene benzene was highest in single walled CNT followed by multi walled CNT and hybrid CNT that have almost similar rate. This shows that single walled CNT are more useful in environmental clean-up. The same group also performed experiment studies by using carboxylated cellulose nano-crystals are for bioremediation. The composite adsorbent (CCN-PEI) composed of carboxylated cellulose nanocrystals (CCN) and polyethyleneimine (PEI) was prepared through an amidation reaction between the carboxyl groups of the cellulose and the amine groups of the Polyethyleneimine. The adsorption performance of the CCN-PEI was tested by removing Cr (VI) ions from aqueous solutions. Regeneration tests indicated that CCN-PEI showed good durability and good efficiency for repeated Cr (VI) adsorptions. Based on the results obtained in this work, it can be concluded that CCN-PEI is a potentially effective adsorbent for removing Cr (VI) ions from aqueous solutions (Jiang, Tan, Ning, & Shi, 2016).

Single Enzyme NP in Bioremediation

In any chemical reaction a catalyst is that molecule that accelerates the reaction without getting converted themselves. Enzymes are such catalytic proteins. Recently these enzymes are being produced extracellular and immobilized in nanoparticles that further enhance the catalytic process. Not only this this conjugate also increased the shelf life of enzymes by masking them and preventing them from oxidation. This increases the stability and lifetime of enzymes as it is protected by nanoparticle coating that prevents it from oxidising agents as seen in trypsin and peroxidase enzyme when the two were bound to magnetic nanoparticle, their shelf life and stability increased from hours to week. Secondly if the enzymes are bound to magnetic nanoparticle then procurement of enzyme becomes easy by magnetic separation in presence of magnetic field, making this process cost effective as well (Hong et. al, 2017). Similarly for the encapsulation of biomolecules in inorganic materials, a unique enzyme-silicate conjugate material that consists of a self-assembled molecularly thin silicate layer on the surface of each individual enzyme molecule has been recently developed, called as single enzyme caged nanoparticles (SECNs). Due to their nearly molecular size, SECNs can also be adsorbed into mesoporous silica materials to yield robust and easily-recyclable enzymatic systems that can be used in a number of potential bio-catalytic applications such as diagnostics, biosensors, bio transformations, biofuel production, bioremediation and CO_2 capture and also for removal of dyes from water (Kumari & Singh, 2016).

Engineered Polymeric NPS In Bioremediation of Hydrophobic Pollutants

The process of size quantification and nucleation are used in designing nanoparticle polymers. Usually polymerisation is a natural activity in nanoparticles that forms nano-aggregates like dendrimers but with the help of existing knowledge this polymerisation can be manipulated as desired. Such engineered nano-polymers were found to be highly effective in removal of organic contaminant like polycyclic aromatic hydrocarbons by process of adsorption. When these nanoparticles form composites with PAHs then they reduce the bioavailability of the organic compound for utilization by microbial community in soil or elsewhere. Thus the COD level is drastically reduced. Similarly in another case these engineered nano-polymers increased the bioavailability of Phenanthrene by enhancing its solubility, thus making it easy for bacteria to degrade. The various engineered nano-polymers are PEG modified Polyurethane acrylate nanoparticles that are used in adsorption of

Phen. Thus it provided the basis of *In situ bioeremediation* of water contaminated with Phen by changing its solvency and making it bioavailable for degradation by microbes (Şengör et. al, 2016).

BIOGENIC URANITE AND URANIUM REMEDIATION

The contamination of water by radioactive substances is the most dangerous contamination as the effect goes in generations as child defects or genetic damage. Therefore nanotechnologists found natural nanoparticles from Uranium known as Uranite. This biogenic Uranite is of special interest to geoscientist who feel that it can replace the chemical Uranium oxide, being similar in chemical and structure along with its own natural properties of nanoparticles that provide it with tiny size and large surface area. This makes Uranite more effective, highly reactive and environment friendly and can be used in remediation of Uranium contamination (Xiao, Chu, Tian, Thring, & Cui, 2016).

Bioremediation With Phytoremediation

Many countries are facing environmental pollution with heavy metal and in the last two decades a great effort has been made to minimize the pollution sources and to remediate the polluted soil and water resources with help of plants via process of Phytoremediation. In this process the specialized structures in plants called Mycorrhiza act as adsorbents and accumulate the contaminants usually heavy metals like Lead, Copper, Zinc and Nickle. It was noted that if in dried waste heap the plants are grown, then an increase in the concentration of above heavy metals were found. The theory behind is that when these plants grew on the organic contents of the waste they started accumulating these metals in their parts that was proven by simple chemical reactions and flame test. It was also found that the concentration of these heavy metals were more inside the plant than in the soil outside, proving the plants to be good accumulators. Still in many parts of the world the ecosystem is restored by plantation in that area. Some of the commonest plants that are used in Phytoremediation of heavy metals are Gundelia, Centaurea, Reseda, Scariola, Eleagnum and Noaea. Based on the study with these six plants it was found that *Noaea mucronata* is the best lead accumulator and good accumulator for other three whereas *Reseda lutea* was best Iron accumulator. The best accumulator like *Noaea mucronata* can do complete biodegradation in just three days (Ferroudj et. al, 2013).

CONCLUSION

The bioremediation processes that involve integration of biological process with advance nanotechnology for wastewater treatment resulted in efficient water purification. It has been studied that through nanoparticles integration the efficiency of each biological process i.e. activated membrane bioreactor, microbial fuel cell, activated sludge process has been positively enhanced than the process alone. In literatures, it has also been reported that process efficiency in terms of pollutants removal like nutrients removal has been increased up to 98.5% (Long, Xiao, & Cao, 2017), dye decolourization up to 95% (Pavithra & Shanthakumar, 2017), and BOD and COD reduction up to 96 and 86% respectively (Sadegh et. al, 2017). Similarly the different nanoparticles like CNT, TiO_2, nanofibers and ZVI have been successfully studied for wastewater treatment. Hence, this area of research can provide efficient and eco-friendly ways for wastewater reclamation but there are limitations in adopting these technologies where the major constrains is the requirement of high level of technical expertise, specific biological agents for treating each pollutants (nutrients, dyes, organic compounds) and equilibrium level of each microbial and nanoparticles assembling in each technology on large scale apart from being time-consuming like pre-treatment with nanoparticles followed by biological process, algae cultivation for wastewater reclamation, and microbial immobilization on nanofibers mats etc. are time consuming processes to be done prior to treatment.

REFERENCES

Abou-Shanab, R. A., Ji, M. K., Kim, H. C., Paeng, K. J., & Jeon, B. H. (2013). Microalgal species growing on piggery wastewater as a valuable candidate for nutrient removal and biodiesel production. *Journal of Environmental Management*, *115*, 257–264. doi:10.1016/j.jenvman.2012.11.022 PMID:23270891

Adeleye, A. S., Conway, J. R., Garner, K., Huang, Y., Su, Y., & Keller, A. A. (2016). Engineered nanomaterials for water treatment and remediation: Costs, benefits, and applicability. *Chemical Engineering Journal*, *286*, 640–662. doi:10.1016/j.cej.2015.10.105

Anjum, M., Al-Makishah, N. H., & Barakat, M. A. (2016). Wastewater sludge stabilization using pre-treatment methods. *Process Safety and Environmental Protection*, *102*, 615–632. doi:10.1016/j.psep.2016.05.022

Anjum, M., Oves, M., Kumar, R., & Barakat, M. A. (2017). Fabrication of ZnO-ZnS@ polyaniline nanohybrid for enhanced photocatalytic degradation of 2-chlorophenol and microbial contaminants in wastewater. *International Biodeterioration & Biodegradation*, *119*, 66–77. doi:10.1016/j.ibiod.2016.10.018

Azamat, J., Sattary, B. S., Khataee, A., & Joo, S. W. (2015). Removal of a hazardous heavy metal from aqueous solution using functionalized graphene and boron nitride nanosheets: Insights from simulations. *Journal of Molecular Graphics & Modelling*, *61*, 13–20. doi:10.1016/j.jmgm.2015.06.012 PMID:26186492

Barakat, M. A., Al-Hutailah, R. I., Hashim, M. H., Qayyum, E., & Kuhn, J. N. (2013). Titania-supported silver-based bimetallic nanoparticles as photocatalysts. *Environmental Science and Pollution Research International*, *20*(6), 3751–3759. doi:10.1007/s11356-012-1306-1 PMID:23161501

Bhattacharya, S., Saha, I., Mukhopadhyay, A., Chattopadhyay, D., & Chand, U. (2013). Role of nanotechnology in water treatment and purification: Potential applications and implications. *Int J Chem Sci Technol*, *3*(3), 59–64.

Cao, Z., Li, S., Zhang, J., & Zhang, H. (2017). An electro-microbial membrane system with anti-fouling function for phenol wastewater treatment. *Journal of Chemical Technology and Biotechnology (Oxford, Oxfordshire)*, *92*(3), 693–699. doi:10.1002/jctb.5056

Cecchin, I., Reddy, K. R., Thomé, A., Tessaro, E. F., & Schnaid, F. (2017). Nanobioremediation: Integration of nanoparticles and bioremediation for sustainable remediation of chlorinated organic contaminants in soils. *International Biodeterioration & Biodegradation*, *119*, 419–428. doi:10.1016/j.ibiod.2016.09.027

Chandrakar, R. K., Baghel, R. N., Chandra, V. K., & Chandra, B. P. (2015). Synthesis, characterization and photoluminescence studies of Mn doped ZnS nanoparticles. *Superlattices and Microstructures*, *86*, 256–269. doi:10.1016/j.spmi.2015.07.043

Cheng, K., Hu, J., Hou, H., Liu, B., Chen, Q., Pan, K., & Yang, C. (2017). Aerobic granular sludge inoculated microbial fuel cells for enhanced epoxy reactive diluent wastewater treatment. *Bioresource Technology*, *229*, 126–133. doi:10.1016/j.biortech.2016.12.115 PMID:28110229

Dave, P. N., & Chopda, L. V. (2014). Application of iron oxide nanomaterials for the removal of heavy metals. *Journal of Nanotechnology*.

Dong, Z., Zhang, F., Wang, D., Liu, X., & Jin, J. (2015). Polydopamine-mediated surface-functionalization of graphene oxide for heavy metal ions removal. *Journal of Solid State Chemistry*, *224*, 88–93. doi:10.1016/j.jssc.2014.06.030

Dunwellgroup. (n.d.). *App08_Flow*. Retrieved March 15, 2017, from http://www.dunwellgroup.com/ProductsServices/DVMAT/VSEP/Applications/App08_Flow.jpg

Dutta, A. K., Maji, S. K., & Adhikary, B. (2014). γ-Fe 2 O 3 nanoparticles: An easily recoverable effective photo-catalyst for the degradation of rose bengal and methylene blue dyes in the waste-water treatment plant. *Materials Research Bulletin*, *49*, 28–34. doi:10.1016/j.materresbull.2013.08.024

Egirani, D. E., Latif, M. T., Poyi, N. R., Wessey, N., & Acharjee, S. (2017). Synthesis and characterization of kaolinite coated with cu-oxide and its effect on the removal of aqueous mercury (ii) ions: part ii. *International Research Journal of Chemistry and Chemical Sciences, 4*(1), 55061.

Eroglu, E., Agarwal, V., Bradshaw, M., Chen, X., Smith, S. M., Raston, C. L., & Iyer, K. S. (2012). Nitrate removal from liquid effluents using microalgae immobilized on chitosan nanofiber mats. *Green Chemistry*, *14*(10), 2682–2685. doi:10.1039/c2gc35970g

Faccini, M., Borja, G., Boerrigter, M., Martín, D. M., Crespiera, S. M., Vázquez-Campos, S., & Amantia, D. (2015). Electrospun carbon nanofiber membranes for filtration of nanoparticles from water. *Journal of Nanomaterials*, *2015*, 2. doi:10.1155/2015/247471

Ferroudj, N., Nzimoto, J., Davidson, A., Talbot, D., Briot, E., Dupuis, V., & Abramson, S. (2013). Maghemite nanoparticles and maghemite/silica nanocomposite microspheres as magnetic Fenton catalysts for the removal of water pollutants. *Applied Catalysis B: Environmental*, *136*, 9–18. doi:10.1016/j.apcatb.2013.01.046

Fu, F., Xie, L., Tang, B., Wang, Q., & Jiang, S. (2012). Application of a novel strategy—Advanced Fenton-chemical precipitation to the treatment of strong stability chelated heavy metal containing wastewater. *Chemical Engineering Journal*, *189*, 283–287. doi:10.1016/j.cej.2012.02.073

Fujita, M., Ide, Y., Sato, D., Kench, P. S., Kuwahara, Y., Yokoki, H., & Kayanne, H. (2014). Heavy metal contamination of coastal lagoon sediments: Fongafale Islet, Funafuti Atoll, Tuvalu. *Chemosphere*, *95*, 628–634. doi:10.1016/j.chemosphere.2013.10.023 PMID:24200049

Gao, C., Liu, L., & Yang, F. (2017). Development of a novel proton exchange membrane-free integrated MFC system with electric membrane bioreactor and air contact oxidation bed for efficient and energy-saving wastewater treatment. *Bioresource Technology*, *238*, 472–483. doi:10.1016/j.biortech.2017.04.086 PMID:28475989

Gopalakrishnan, A., Krishnan, R., Thangavel, S., Venugopal, G., & Kim, S. J. (2015). Removal of heavy metal ions from pharma-effluents using graphene-oxide nanosorbents and study of their adsorption kinetics. *Journal of Industrial and Engineering Chemistry*, *30*, 14–19. doi:10.1016/j.jiec.2015.06.005

Gowd, S. S., Reddy, M. R., & Govil, P. K. (2010). Assessment of heavy metal contamination in soils at Jajmau (Kanpur) and Unnao industrial areas of the Ganga Plain, Uttar Pradesh, India. *Journal of Hazardous Materials*, *174*(1), 113–121. doi:10.1016/j.jhazmat.2009.09.024 PMID:19837511

Grey, D., Garrick, D., Blackmore, D., Kelman, J., Muller, M., & Sadoff, C. (2013). Water security in one blue planet: twenty-first century policy challenges for science. *Phil. Trans. R. Soc. A, 371*(2002), 20120406.

Gupta, V. K., Tyagi, I., Sadegh, H., Shahryari-Ghoshekandi, R., Makhlouf, A. S. H., & Maazinejad, B. (2015). Nanoparticles as adsorbent; a positive approach for removal of noxious metal ions: A review. *Sci. Technol. Dev*, *34*(3), 195–214. doi:10.3923/std.2015.195.214

Habibi, B., & Mohammadyari, S. (2015). Facile synthesis of Pd nanoparticles on nano carbon supports and their application as an electrocatalyst for oxidation of ethanol in alkaline media: The effect of support. *International Journal of Hydrogen Energy*, *40*(34), 10833–10846. doi:10.1016/j.ijhydene.2015.07.021

Hong, S. G., Kim, B. C., Na, H. B., Lee, J., Youn, J., Chung, S. W., ... Kim, S. H. (2017). Single enzyme nanoparticles armored by a thin silicate network: Single enzyme caged nanoparticles. *Chemical Engineering Journal*, *322*, 510–515. doi:10.1016/j.cej.2017.04.022

Hu, W., Yin, J., Deng, B., & Hu, Z. (2015). Application of nano TiO 2 modified hollow fiber membranes in algal membrane bioreactors for high-density algae cultivation and wastewater polishing. *Bioresource Technology*, *193*, 135–141. doi:10.1016/j.biortech.2015.06.070 PMID:26125613

Jiang, B., Tan, L., Ning, S., & Shi, S. (2016). A novel integration system of magnetically immobilized cells and a pair of graphite plate-stainless iron mesh electrodes for the bioremediation of coking wastewater. *Bioresource Technology*, *216*, 684–690. doi:10.1016/j.biortech.2016.06.009 PMID:27289060

Khajeh, M., Laurent, S., & Dastafkan, K. (2013). Nanoadsorbents: Classification, preparation, and applications (with emphasis on aqueous media). *Chemical Reviews*, *113*(10), 7728–7768. doi:10.1021/cr400086v PMID:23869773

Koók, L., Rózsenberszki, T., Nemestóthy, N., Bélafi-Bakó, K., & Bakonyi, P. (2016). Bioelectrochemical treatment of municipal waste liquor in microbial fuel cells for energy valorization. *Journal of Cleaner Production*, *112*, 4406–4412. doi:10.1016/j.jclepro.2015.06.116

Kumari, B., & Singh, D. P. (2016). A review on multifaceted application of nanoparticles in the field of bioremediation of petroleum hydrocarbons. *Ecological Engineering*, *97*, 98–105. doi:10.1016/j.ecoleng.2016.08.006

Kurian, M., & Nair, D. S. (2015). Heterogeneous Fenton behavior of nano nickel zinc ferrite catalysts in the degradation of 4-chlorophenol from water under neutral conditions. *Journal of Water Process Engineering*, *8*, e37–e49. doi:10.1016/j.jwpe.2014.10.011

Li, W., Zhang, Y., Tian, G., Xie, S., Xu, Q., Wang, L., ... Bu, Y. (2016). Fabrication of graphene-modified nano-sized red phosphorus for enhanced photocatalytic performance. *Journal of Molecular Catalysis A Chemical*, *423*, 356–364. doi:10.1016/j.molcata.2016.07.039

Li, Z., Ma, Z., Kuijp, T. J. D., Yuan, Z., & Huang, L. (2014). A review of soil heavy metal pollution from mines in China: Pollution and health risk assessment. *The Science of the Total Environment*, *468-469*, 843–853. doi:10.1016/j.scitotenv.2013.08.090 PMID:24076505

Lin, L., Wang, H., Luo, H., & Xu, P. (2015). Enhanced photocatalysis using side-glowing optical fibers coated with Fe-doped TiO 2 nanocomposite thin films. *Journal of Photochemistry and Photobiology A Chemistry*, *307*, 88–98. doi:10.1016/j.jphotochem.2015.04.010

Lin, S. T., Thirumavalavan, M., Jiang, T. Y., & Lee, J. F. (2014). Synthesis of ZnO/Zn nano photocatalyst using modified polysaccharides for photodegradation of dyes. *Carbohydrate Polymers*, *105*, 1–9. doi:10.1016/j.carbpol.2014.01.017 PMID:24708945

Lingamdinne, L. P., Koduru, J. R., Choi, Y. L., Chang, Y. Y., & Yang, J. K. (2016). Studies on removal of Pb (II) and Cr (III) using graphene oxide based inverse spinel nickel ferrite nano-composite as sorbent. *Hydrometallurgy*, *165*, 64–72. doi:10.1016/j.hydromet.2015.11.005

Little Pro. (n.d.). *How to Calculate German Water Hazard Class*. Retrieved March 15, 2017, from http://www.chemsafetypro.com/Topics/Other/How_to_Calculate_German_Water_Hazard_Class_WGK.html

Liu, C., Jin, R. N., Ouyang, X. K., & Wang, Y. G. (2017). Adsorption behavior of carboxylated cellulose nanocrystal—polyethyleneimine composite for removal of Cr (VI) ions. *Applied Surface Science*, *408*, 77–87. doi:10.1016/j.apsusc.2017.02.265

Liu, M., Wang, Z., Zong, S., Chen, H., Zhu, D., Wu, L., & Cui, Y. (2014). SERS detection and removal of mercury (II)/silver (I) using oligonucleotide-functionalized core/shell magnetic silica sphere@ Au nanoparticles. *ACS Applied Materials & Interfaces*, *6*(10), 7371–7379. doi:10.1021/am5006282 PMID:24738775

Logroño, W., Pérez, M., Urquizo, G., Kadier, A., Echeverría, M., Recalde, C., & Rákhely, G. (2017). Single chamber microbial fuel cell (SCMFC) with a cathodic microalgal biofilm: A preliminary assessment of the generation of bioelectricity and biodegradation of real dye textile wastewater. *Chemosphere*, *176*, 378–388. doi:10.1016/j.chemosphere.2017.02.099 PMID:28278426

Long, Y., Xiao, L., & Cao, Q. (2017). Co-polymerization of catechol and polyethylenimine on magnetic nanoparticles for efficient selective removal of anionic dyes from water. *Powder Technology*, *310*, 24–34. doi:10.1016/j.powtec.2017.01.013

López-Serrano, A., Olivas, R. M., Landaluze, J. S., & Cámara, C. (2014). Nanoparticles: A global vision. Characterization, separation, and quantification methods. Potential environmental and health impact. *Analytical Methods*, *6*(1), 38–56. doi:10.1039/C3AY40517F

Lunge, S. S., Singh, S., & Sinha, A. (2014). Magnetic nanoparticle: Synthesis and environmental applications. *International Conference on Chemical, Civil and Environmental Engineering*, 7-9.

Ma, H., Wang, H., & Na, C. (2015). Microwave-assisted optimization of platinum-nickel nanoalloys for catalytic water treatment. *Applied Catalysis B: Environmental*, *163*, 198–204. doi:10.1016/j.apcatb.2014.07.062

Mahanta, N., & Valiyaveettil, S. (2013). Functionalized poly (vinyl alcohol) based nanofibers for the removal of arsenic from water. *RSC Advances*, *3*(8), 2776–2783. doi:10.1039/c2ra22768a

Martínez, F., López-Muñoz, M. J., Aguado, J., Melero, J. A., Arsuaga, J., Sotto, A., & Cerro, L. (2013). Coupling membrane separation and photocatalytic oxidation processes for the degradation of pharmaceutical pollutants. *Water Research*, *47*(15), 5647–5658. doi:10.1016/j.watres.2013.06.045 PMID:23863375

Moghimifar, V., Raisi, A., & Aroujalian, A. (2014). Surface modification of polyethersulfone ultrafiltration membranes by corona plasma-assisted coating TiO 2 nanoparticles. *Journal of Membrane Science*, *461*, 69–80. doi:10.1016/j.memsci.2014.02.012

Norvill, Z. N., Toledo-Cervantes, A., Blanco, S., Shilton, A., Guieysse, B., & Muñoz, R. (2017). Photodegradation and sorption govern tetracycline removal during wastewater treatment in algal ponds. *Bioresource Technology*, *232*, 35–43. doi:10.1016/j.biortech.2017.02.011 PMID:28214443

Padil, V. V. T., Wacławek, S., & Černík, M. (2016). Green Synthesis: Nanoparticles and Nanofibres Based on Tree Gums for Environmental Applications. *Ecological Chemistry and Engineering. S*, *23*(4), 533–557. doi:10.1515/eces-2016-0038

Pan, F., Zhong, X., Xia, D., Yin, X., Li, F., Zhao, D., & Liu, W. (2017). Nanoscale zero-valent iron/persulfate enhanced upflow anaerobic sludge blanket reactor for dye removal: Insight into microbial metabolism and microbial community. *Scientific Reports*, 7. PMID:28300176

Pavithra, S., & Shanthakumar, S. (2017). Removal of COD, BOD and Color from Municipal Solid Waste Leachate using Silica and Iron nano particles-A Comparative Study. *Global NEST Journal*, 19.

Sadegh, H., Ali, G. A., Gupta, V. K., Makhlouf, A. S. H., Shahryari-ghoshekandi, R., Nadagouda, M. N., & Megiel, E. (2017). The role of nanomaterials as effective adsorbents and their applications in wastewater treatment. *Journal of Nanostructure in Chemistry*, 1-14.

San, N. O., Celebioglu, A., Tümtaş, Y., Uyar, T., & Tekinay, T. (2014). Reusable bacteria immobilized electrospun nanofibrous webs for decolorization of methylene blue dye in wastewater treatment. *RSC Advances*, *4*(61), 32249–32255. doi:10.1039/C4RA04250F

Santhosh, C., Velmurugan, V., Jacob, G., Jeong, S. K., Grace, A. N., & Bhatnagar, A. (2016). Role of nanomaterials in water treatment applications: A review. *Chemical Engineering Journal*, *306*, 1116–1137. doi:10.1016/j.cej.2016.08.053

Sarioglu, O. F., San Keskin, N. O., Celebioglu, A., Tekinay, T., & Uyar, T. (2017). Bacteria encapsulated electrospun nanofibrous webs for remediation of methylene blue dye in water. *Colloids and Surfaces. B, Biointerfaces*, *152*, 245–251. doi:10.1016/j.colsurfb.2017.01.034 PMID:28119219

Sarioglu, O. F., Yasa, O., Celebioglu, A., Uyar, T., & Tekinay, T. (2013). Efficient ammonium removal from aquatic environments by Acinetobacter calcoaceticus STB1 immobilized on an electrospun cellulose acetate nanofibrous web. *Green Chemistry*, *15*(9), 2566–2572. doi:10.1039/c3gc40885j

Şengör, S. S., Singh, G., Dohnalkova, A., Spycher, N., Ginn, T. R., Peyton, B. M., & Sani, R. K. (2016). Impact of different environmental conditions on the aggregation of biogenic U (IV) nanoparticles synthesized by Desulfovibrio alaskensis G20. *Biometals*, *29*(6), 965–980. doi:10.1007/s10534-016-9969-6 PMID:27623995

Sharma, N., Bhatnagar, P., Chatterjee, S., John, P. J., & Soni, I. P. (2017). Indian Journal of Pharmaceutical and Biological Research (IJPBR). *The Indian Journal of Pharmacy*, *5*(1), 17–24.

Shirsath, D. S., & Shirivastava, V. S. (2015). Adsorptive removal of heavy metals by magnetic nanoadsorbent: An equilibrium and thermodynamic study. *Applied Nanoscience*, *5*(8), 927–935. doi:10.1007/s13204-014-0390-6

Sounthararajah, D. P., Loganathan, P., Kandasamy, J., & Vigneswaran, S. (2015). Adsorptive removal of heavy metals from water using sodium titanate nanofibres loaded onto GAC in fixed-bed columns. *Journal of Hazardous Materials*, *287*, 306–316. doi:10.1016/j.jhazmat.2015.01.067 PMID:25668299

Swamy, N. K., Singh, P., & Sarathy, I. P. (2011). Precipitation of phenols from paper industry wastewater using ferric chloride. *Rasayan Journal of Chemistry*, *4*, 452–456.

Tambe Patil, B. B. (2015). Wastewater Treatment Using Nanoparticles. *J. Advances in Chemical Engineering*, *5*, 131.

Van Dillewijn, P., Caballero, A., Paz, J. A., González-Pérez, M. M., Oliva, J. M., & Ramos, J. L. (2007). Bioremediation of 2, 4, 6-trinitrotoluene under field conditions. *Environmental Science & Technology*, *41*(4), 1378–1383. doi:10.1021/es062165z PMID:17593745

Vu, H. C., Dwivedi, A. D., Le, T. T., Seo, S. H., Kim, E. J., & Chang, Y. S. (2017). Magnetite graphene oxide encapsulated in alginate beads for enhanced adsorption of Cr (VI) and As (V) from aqueous solutions: Role of crosslinking metal cations in pH control. *Chemical Engineering Journal*, *307*, 220–229. doi:10.1016/j.cej.2016.08.058

Wang, H., Yuan, X., Wu, Y., Zeng, G., Chen, X., Leng, L., ... Li, H. (2015). Facile synthesis of amino-functionalized titanium metal-organic frameworks and their superior visible-light photocatalytic activity for Cr (VI) reduction. *Journal of Hazardous Materials*, *286*, 187–194. doi:10.1016/j.jhazmat.2014.11.039 PMID:25585267

WHO. (2015). *Drinking-water: Fact sheet No. 391*. Retrieved from http://www.who.int/mediacentre/factsheets/fs391/en/

Xiao, J., Chu, S., Tian, G., Thring, R. W., & Cui, L. (2016). An Eco-tank system containing microbes and different aquatic plant species for the bioremediation of N, N-dimethylformamide polluted river waters. *Journal of Hazardous Materials*, *320*, 564–570. doi:10.1016/j.jhazmat.2016.07.037 PMID:27501878

Yan, J., Han, L., Gao, W., Xue, S., & Chen, M. (2015). Biochar supported nanoscale zerovalent iron composite used as persulfate activator for removing trichloroethylene. *Bioresource Technology*, *175*, 269–274. doi:10.1016/j.biortech.2014.10.103 PMID:25459832

Yang, B., Geng, P., & Chen, G. (2015). One-dimensional structured IrO 2 nanorods modified membrane for electrochemical anti-fouling infiltration of oily wastewater. *Separation and Purification Technology*, *156*, 931–941. doi:10.1016/j.seppur.2015.10.040

Zare-Dorabei, R., Ferdowsi, S. M., Barzin, A., & Tadjarodi, A. (2016). Highly efficient simultaneous ultrasonic-assisted adsorption of Pb (II), Cd (II), Ni (II) and Cu (II) ions from aqueous solutions by graphene oxide modified with 2, 2′-dipyridylamine: Central composite design optimization. *Ultrasonics Sonochemistry*, *32*, 265–276. doi:10.1016/j.ultsonch.2016.03.020 PMID:27150770

Zarghami, Z., Akbari, A., Latifi, A. M., & Amani, M. A. (2016). Design of a new integrated chitosan-PAMAM dendrimer biosorbent for heavy metals removing and study of its adsorption kinetics and thermodynamics. *Bioresource Technology*, *205*, 230–238. doi:10.1016/j.biortech.2016.01.052 PMID:26836608

Chapter 14
Effective Removal of Heavy Metals From Aqueous Solution by Nano-Composites:
Bio Remediation Using Nano Technology

Ashok K. Rathoure
Independent Researcher, India

R. Sanjeevi
NIMS University, India

J. Anuradha
NIMS University, India

Dushyant Singh Chauhan
NIMS University, India

Sandeep Tripathi
NIMS University, India

ABSTRACT

Heavy metal contamination in industrial effluents presents a serious threat to the environment and human health because of their toxicity, non-biodegradability, carcinogenicity, and bioaccumulation in living organisms. Recently, the preparation and application of iron oxides, especially magnetite nanoparticles, for metals removal have been investigated due to their nano size, magnetic separation, and the ease of synthesis, coating, and modification. However, magnetic nanoparticles lose some magnetization due to air oxidation. Magnetite nanoparticles coating with inorganic shell, like silica and carbon, have been reported and were capable of improving chemical stability. The effects of pH, contact time, and initial concentrations on the removal of heavy metals should be studied using nano-composites in water. In this chapter, the authors present a technical review on different nano-composites used for bioremediation and their limitations.

DOI: 10.4018/978-1-5225-4162-2.ch014

Copyright © 2018, IGI Global. Copying or distributing in print or electronic forms without written permission of IGI Global is prohibited.

INTRODUCTION

Rapidity in urbanization and industrialization results in releaseand discharge of hazardous heavy metal ions and synthetic organic dyes into the water resources. The industrial process responsible for the release of the carcinogenic and mutagenic substances into the environment includes, mining, tannery, paper and pulp, textile, leather, metal plating, pesticide, fertilizer, batteries manufacturing and like. The discharge of these substances primarily leads to elusive and pervasive threat to the environment and the world occupants. Distinguishing feature of these substances attributes to stability, non-biodegradability and higher toxicity. Most significantly, the heavy metal that are causing serious environmental issues and that endangers the human survival in the world comprises the metal ions such as arsenic ($As^{3+/5+}$), lead (Pb^{2+}), cadmium (Cd^{2+}), chromium ($Cr^{3+/6+}$), cobalt (Co^{2+}), copper (Cu^{2+}), mercury ($Hg^{+/2+}$), nickel (Ni^{+}), and zinc (Zn^{2+}) (Mahmud et al., 2016). In order to maintain the environmental sustainability and effluent standard levels of these inorganic toxic moieties, an efficient treatment strategies need to be adopted before being discharged into the water bodies.

Several managerial techniques that are extensively studied for the purpose includes physical, chemical and biological methods for mitigating the metal ion concentration from the wastewater includes adsorption, coagulation/ flocculation, chemical precipitation, membrane filtration, electrochemical, and ion exchange methods. Of these, adsorption is the far reaching and promising technology for treating huge quantity of wastewater at lower concentration. Thus synthesis of advanced adsorbent characterized with higher efficacy in complete removal of heavy metal is of resent research interest and is essential for achieving sustainable water resource. With emergence of nano-bioremediation, several opportunities are in progress in remediating the contaminated sites with facile, rapid, efficient, green and low-cost technology (Yang et al., 2008; Yan et al., 2014; Zhu et al., 2014). In the recent past, use of natural polymer contained nano-composites, layered nanomaterials and nanofiltration membranes (De et al., 2011; Sarkar et al., 2015; Zhu et al., 2015) finds wider application in handling of heavy metals from the contaminated wastewater.

Amid of number of nano-composite materials used for remediating the heavy metal in water pollution control, zeolites, activated carbon, zero valent iron, metal oxides and like nanomaterials find an important role due to the fact of lower toxicity, larger surface area, and degradability with ageing. The heavy metal released into the environment percolates into soil and causes threat to ground water and surface water. Exposure of heavy metals or its ingestion can cause serious consequences. In general, the human physiology attempts to eliminate the toxic substances in all possible excision human mechanisms or otherwise gets accumulated in various storage sites. These toxicants might lead to serious threat to the human health.

Arsenic (As III/V)

Arsenic (As) metal ion is widely distributed and is naturally available as the common constituent of the earth's crust. It is widely available in insecticides, fungicides and herbicides. Contamination of As in ground and surface water is very critical. Chronic exposure of arsenic poisoning causes skin or lung cancer and even bladder cancer. The permissible limit for As in drinking water is 0.01 mg L^{-1} (WHO 2006).

Lead (Pb II)

Lead is widely distributed in the rocks and soils of the earth's crust. Its acute Pb poisoning in human beings interferes in heme synthesis, leading to hematological damage. Environmental exposure of Pb causes mental retardation, especially in children. The permissible limit for Pb in drinking water is 0.05 mg L^{-1} (WHO 2006).

Cadmium (Cd II)

Cadmium is a divalent metal, chemically similar to Zn and Hg. It is widely distributed in crystal rocks. The chronic Cd toxicity causes *Itai–Itai* disease. Acute or chronic Cd poisoning does causes hyper blood pressure, kidney malfunction, destruction of testicular tissue and red blood cells. The permissible limit for Pb in safe drinking water is 0.005 mg L^{-1} (WHO 2006).

Chromium (Cr III/VI)

The major Chromium (Cr) is extensively released from the tanning, leather and electroplating industry. Contamination of Cr (III) and Cr (VI) into the aqueous environment is highly toxic and modifies the human physiology. Gradual accumulation of Cr causes severe health issues including skin irritation and even to lung cancer. The permissible limit of Cr (VI) for drinking water is 0.05 mg L^{-1} (WHO 2006).

Cobalt (Co II)

Cobalt (Co) has major importance on human health, excessive Co is being released from nuclear power plants, petrochemical, electroplating, mining industries and like contaminates surface and groundwater. Continual exposure leads to several physical and mental problems with the symptoms, including pneumonia, kidney congestion, skin degeneration and weight loss. The permissible limit of drinking water is 0.01 mg L^{-1} (WHO 2006).

Copper (Cu II)

Similar to Zn, copper (Cu) is an important element for human beings and its excessive ingestion causes serious toxicity, such as nausea, diarrhea, and stomach ache and may even lead to death. The permissible limit of drinking water is 1.5 mg L^{-1} (WHO 2006).

Mercury (Hg I/II)

Mercury (Hg) is one of the toxic metals in the environment. It is discharged from the mining activities, agricultural activities, dental amalgams, laboratory instruments *etc* and is well-known for the *'Minamata disease'*. Prolonged exposure of Hg causes damage in central nervous system and at high concentrations causes chest pain, impairment of pulmonary and kidney function. The permissible limit of drinking water is 0.001 mg L^{-1} (WHO 2006).

Nickel (Ni)

The major sources of nickel contamination in water come from industrial process such as electroplating, batteries manufacturing, mining, metal finishing and forging. Ni ions are non-biodegradable toxic heavy metals and may cause dermatitis and allergic sensitization, lung and kidney problems and are a known human carcinogen. According to the WHO guidelines, the maximum permissible concentration of Ni in drinking water it should be less than 0.1 mg/L.

Zinc (Zn II)

Trace amounts of some metal ions, such as zinc (Zn), copper(Cu) and cobalt (Co), are required by organisms as cofactors for enzymatic processes. However, an excess of these metal ions will cause serious problems in living organisms due to their higher toxicity, carcinogenic and bioaccumulation. Zn is one of the most common pollutants for surface and groundwater as it has versatile uses. Again, due to its non-biodegradability and acute toxicity, Zn-containing liquid and solid wastes are considered as hazardous wastes. An excessive amount of Zn exposure can cause well-known health problems such as stomach cramps, skin irritations, vomiting, nausea and anemia. The WHO recommends the maximum acceptable concentration of Zn ions in drinking water at 5.0 mg/L.

EFFICACY OF NANO-COMPOSITES IN THE REMOVAL OF HEAVY METAL IONS FROM THE WASTEWATER

Heavy metal ions existing in various aqueous streams/ wastewater from multiple sources are currently one of the most important environmental concerns. Therefore, it is necessary to remove heavy metal ions from contaminated wastewater prior to its discharge to the environment to protect the aquatic lives and human beings. Although several methods have been used for many years to remove heavy metal ions from wastewater, achieving the most effective treatment is still challenging for environmental scientists/engineers. Physical and chemical precipitation, ion exchange, reverse osmosis, membrane filtration, electrochemical treatment, solvent extraction and adsorption processes are the widely used methods for removing heavy metal ions from aqueous streams/wastewater. To date, various adsorption methods have been considered as effective and widely used methods due to their simplicity and easy operational conditions. In addition, adsorption is mostly reversible, thus the adsorbents can be easily regenerated using a suitable desorption process.

Several inorganic and organic adsorbents have been utilized for the adsorption method, including zeolites, montmorillonite, clay minerals, trivalent and tetravalent metal phosphates, biosorbents, activated carbon, polymer-based adsorbent and polymer-inorganic hybrid adsorbents.

Zero Valent Iron Nano Composite

The recapitulated studies on utilization of nanoparticles include the adsorption of various pollutants like heavy metals and dyes. This adsorption technique combined with magnetic separation has far reaching benefits in water treatment and environmental clean-up. Mechanism of in-situ technique involving nZVI in treatment of polluted soil and aquifers include,

1. Injection of nanoscale ZVI into contaminated site;
2. Injection of nZVI with high mobility inducing nZVI plume;
3. Spiking of nanoparticles onto the topsoil to adsorb/degrade the organic and inorganic pollutants.

Nano-remediation that is influenced with contaminant degradation, rather than adsorption, is of greater significance, particularly in treatment of organic contaminants. In *ex-situ* technology, biogenic nanoparticles such as TiO_2, ZnO, nZVI and noble metal nanoparticles in particular, has potential role in remediating the organic contaminants. The investigation on the bimetallic gold-palladium NPs serves as an active catalyst, by enhancing the degradability of major water pollutants like trichlorothane (TCE) and oxidation of mercury. Gold nanoparticles have also been used for designing gas sensors for detecting pollution levels of carbon monoxide (CO) and oxides of nitrogen (NOx). Metal oxide nanoparticles like ZnO, TiO_2 and CeO_2 have stimulated the growth of both root and shoot of plant species such as soybean, wheat, corn and alfalfa, suggesting the nano-biotechnology showing significant impact on phytoremediation. TiO_2 at nanoscale increased the reactivity of several enzymes and promoted nitrate adsorption, conversion of inorganic into organic nitrogen.

In another study, silver nanoparticles at nano form incubated with Congo red containing *Aspergillus niger* broth, showed more than 85.8% decolorization within 24 hours (Nithya *et al.*, 2010). The reduction and adsorption of lead (Pb) and hexachlorohexanes (HCHs) from contaminated water was found to be remediated with 99.9% and 95% of removal using nanoscale ZVI (Xi *et al.*, 2010; Daniel *et al.*, 2008). Summary of nanobioremediation method employed for decontamination of pollutants is presented below. The process of nanobioremediation initiates with reduction of contaminants levels and then promotes the biodegradability of contaminants below the risk level.

In addition to soil, water/ wastewater treatment, air pollution monitoring and treatment is under progress with the advancement in nanotechnology. Carbon nanotubes are found to be efficient than activated carbon on absorbing dioxin to a multifold due to the increased surface area in nano form. Technological development in the field of nano-biotechnological approach in nano-remediation are still in their infant stage, since the size, shape and surface chemistry are not yet optimal. The use of these particles in nano-remediation will inevitably lead to release of nano-materials into the environment. But mankind could be exposed to these nanoparticles through inhalation, ingestion, dermal contact, and/ or injection. In order to confirm the safety of employing fabricated nanostructures, in depth research is required to know nano toxicity on the biological systems.

Table 1. Nanoparticles and Microorganisms involved in on-site Nano-bioremediation

Nanoparticles Used	Microorganisms Used	Contaminant	% of NR	% of NBR	Reference
Pd/Fe	*Trametes versicolor*	Triclosan	100	100	Bokare *et al.*, 2010
Pd/Fe	*Sphingomonas wittichii*	Dioxin	100	100	Bokare *et al.*, 2012
Pd/Fe	Sphingomonas sp.	Lindane	99	99	Singh *et al.*, 2013
nZVI	*Burkholderia xenovorans*	Aroclor1248 (PCB)	89	90	Le *et al* ., 2015

Fe_3O_4 Nano Composites

Various forms, size and shapes of PPy/Fe_3O_4 nano-composites have recently been used for the adsorption of heavy metal ions from wastewater or an aqueous solution. A Fe_3O_4 coated PPy magnetic nano-composite has been prepared via the in situ polymerization of pyrrole monomer for the removal of highly toxic Cr (VI). The batch experimental results showed upto 100% adsorption with a 200 mg L 1 Cr (VI) aqueous solution at pH 2. The adsorption results showed that the Cr (VI) removal efficiency using the nano-composite decreased with an increase in pH. XPS studies also suggested that ion exchange and reduction on the surface of the nano-composite may be the possible mechanism for Cr (VI) removal by the PPy/Fe_3O_4 nano-composite. Another study showed that a similar PPy/Fe_3O_4 nano-composite has been synthesized via an in situ chemical oxidative polymerization technique with 9 h of preparation time. A fixed-bed column with continuous flow was also selected at a slow rate of 3 mL min 1 with 100 mg L 1 of Cr (VI). The bed mass, initial Cr (VI) concentration, pH and flow rate. The efficiency of the PPy/Fe_3O_4 nanocomposite in environmental water to remove Cr (VI) ions was effective to give below acceptable levels of Cr (VI) upon processing 5.04 L of water with an initial 76.59 mg L 1 Cr (VI) concentration using only 2 g of adsorbent.

Graphene Nanocomposites

Due to optimal transformation of life, the demand for higher performance applications of nanoadsorbent materials has dramatically increased in the last few years. Different nanoparticles, such as zeolites, Fe_2O_3, TiO_2, carbon nanotubes (CNT), graphene oxide nanosheets or nanoplatelets, have been employed to characterize the nanocomposites at different stages of polymerization. Graphene is a two-dimensional monoatomic thin sheet with a large lateral dimensional sp2-hybridized carbon nanostructure and has unique properties of high tensile strength and Young's modulus, good electrical and thermal conductivities, and high aspect ratio. The adsorption capacity of the PPy/GO composite nanosheets is about two times as large as that of conventional PPy nanoparticles. Again, in another study, a polypyrrolegraphene oxide nanocomposite (PPy–GO NC) was synthesized via an in situ polymerization of Py monomer in GO dispersion and used for Cr (VI) ion adsorption from aqueous solutions using both batch and packed-bed column methods. The results showed that in the batch mode at 25^0C and pH 2, the maximum sorption capacity was 625 mg.

The synthesis of polymer nano-composites has been extensively investigated by many researchers, however, the end of life fate of polymer nano-composites is still largely unknown. It is expected that at the end of their service life, these polymer nano-composites will most likely end up in soil and water systems where microorganisms will interact and, perhaps, even biodegrade them. In this study, we investigate the ability of wastewater microorganisms to biodegrade nano-composite films containing different graphene oxide (GO) loads (0% to 0.6%, (w/w%)) embedded in a model biopolymer *i.e.* chitosan. The ability of wastewater microorganisms to grow and form biofilms on the surface of the nano-composite films was determined by live and dead staining assisted with confocal laser scanning microscopy. The capability of wastewater biofilms to biodegrade nano-composites was assessed through nano-composite film weight losses, Fourier transformed infrared (FTIR) and scanning electron microscopy (SEM) analyses. Results showed that microorganisms present in the activated sludge can grow on the surface of the nano-composites and biodegrade the polymer surrounding the graphene oxide nanoplatelets. As the biopolymer gets degraded, there is increasing exposure of GO on the surface, which yields microbial inactivation and biofilm growth inhibition. Fan et al., (2017) have determined the emergence of the sharp edges of GO on the surface of the nano-composite through atomic force microscopy (AFM), as well as the production of reactive oxygen species (ROS) with the Ellman's assay before and after biodegradation of the nano-composites. Their results show that as GO surfaces the nano-composite film during biodegradation, there is increasing production of ROS, which explains the increasing inactivation of the microorganisms.

The nanosized needle-like MnO_2 was obtained by Jeong et al., (2016) using a graphene substrate. The prepared composite exhibited an ideal supercapacitive behavior. A capacitance retention of 94% was achieved with a 4 h deposition time (an initial capacitance of 574 mF/cm^2 at a scan rate of 20 mV/s) and the retention declined with further deposition time. The results demonstrate enhanced contact between the electrode and electrolyte and improved power density as an electrochemical capacitor.

Epoxy-Based Nano-Composites

The epoxy based nanocoposites are generally prepared by dispersion of known amount of nanoparticles in one of the non-reactive parts of the resin and mix with reactive part of the resin and finally cure. Depending on the physical and chemical properties of nanoparticles, and epoxy resin the processing technique is selected. Most of the commercial metal, metal oxide, and carbon based nanoparticles are used as received. Whereas some of the high density and highly agglomerated/entangled nanoparticles such as WO_3 and CNTs require a special modification for better dispersion to obtain optimal improved properties. In one of our recent studies a high density WO_3 nano-particles were subjected to sized reduction process for better dispersion and improve mechanical properties. The porous WO_3 (pore size 2- 5 nm) nanoparticles were synthesized using a high intensity ultrasound irradiation of commercially available

WO_3 nanoparticles (80 nm) in ethanol. The nano-composites of WO_3/SC-15 epoxy were prepared by infusion of 1wt%, 2wt% and 3wt% of porous WO_3 nanoparticles into SC-15 epoxy resin by using a non-contact mixing technique. In this technique the material container is set at 45 degrees angle inside the mixer and revolves and rotates at high acceleration with the speed of ~ 2000 rpm, dual centrifugal forces were given to the container that keep pressing materials to outward and down along with the slope of inner wall of the container and accomplish powerful mixing and removing air pockets formed in the reaction mixture simultaneously. This technique is non-contact and non-reactive unlike ultrasound and other mixing techniques.

CONCLUSION

Adsorption plays an important role in removing heavy metal s from effluent. Nano-composites are the most popular adsorbent material and has been used widely in waste water treatment for decades. However, commercial nano-composites have drawbacks in terms of heavy metal removal efficiency and cost problems. So the overall idea is to find suitable cost-effective alternatives to replace traditional method. Under these circumstances, new adsorbent materials, especially cost-effective adsorbent materials, need to be developed. The low-cost adsorbent materials usually have the problem of low adsorption capacity; thus cheap and effective modification methods should be introduced to improve the adsorption capacity while cont rolling the cost of activation at the same time. Low-cost materials have been divided into four categories based on the nature of the materials. The common characteristics of these materials are abundance either naturally occurring or low-value by-products. By analyzing the composition and developing low-cost treatment methods, they become value-added materials with good adsorbent properties.

REFERENCES

Bokare, V., Murugesan, K., Kim, J.-H., Kim, E.-J., & Chang, Y.-S. (2012). Integrated hybrid treatment for the remediation of 2,3,7,8-tetrachlorodibenzo-p-dioxin. *Sci. Total Environ., 435-436*, 563-566.

Bokare, V., Murugesan, K., Kim, Y.-M., Jeon, J.-R., Kim, E.-J., & Chang, Y. S. (2010). Degradation of triclosan by an integrated nano-bio redox process. *Bioresource Technology, 101*(16), 6354–6360. doi:10.1016/j.biortech.2010.03.062 PMID:20381343

Fusalba, A., & Belanger, D. (1999). Electropolymerization of Polypyrrole and Polyaniline–Polypyrrole from Organic Acidic Medium. *The Journal of Physical Chemistry B*, *103*(42), 9044–9054. doi:10.1021/jp9916790

Geim, A. K., & Novoselov, K. S. (2007). The rise of Graphene. *Nature Materials*, *6*(3), 183–191. doi:10.1038/nmat1849 PMID:17330084

Le, T. T., Nguyen, K.-H., Jeon, J.-R., Francis, A. J., & Chang, Y.-S. (2015). Nano/bio treatment of polychlorinated biphenyls with evaluation of comparative toxicity. *Journal of Hazardous Materials*, *287*, 335–341. doi:10.1016/j.jhazmat.2015.02.001 PMID:25679799

Mahmud, H. N. M. E., Huq, A. K. O., & Yahya, R. B. (2017). The removal of heavy metal ions from wastewater/aqueous solution using polypyrrole-based adsorbents: A review. *RSC Advances*, *6*(18), 14778–14791. doi:10.1039/C5RA24358K

Neto, C., Guinea, F., Peres, N., Novoselov, K. S., & Geim, A. K. (2009). The electronic properties of grapheme. *Reviews of Modern Physics*, *81*(1), 109–162. doi:10.1103/RevModPhys.81.109

Nithya, R., & Ragunathan, R. (2010). *Decolorization of the dye Congo red by Aspergillus nigernanoparticle, Nation. Semi*. Anna Univer., Trichy.

Singh, P., Kim, Y. J., Zhang, D., & Yang, D.-C. (2016). Biological Synthesis of Nanoparticles from Plants and Microorganisms. *Trends in Biotechnology*, *34*(7), 588–599. doi:10.1016/j.tibtech.2016.02.006 PMID:26944794

Singh, R., Manickam, N., Mudiam, M. K. R., Murthy, R. C., & Misra, V. (2013). An integrated (nano-bio) technique for degradation of -HCH contaminated soil. *Journal of Hazardous Materials*, *35*, 258–259. PMID:23692681

WHO. (2006). *Guidelines for drinking-water quality: recommendations*. World Health Organization.

Xi, Y., Mallavarapu, M., & Naidu, R. (2010). Reduction and adsorption of Pb 2 þ in aqueous solution by nano zero-valent iron: A SEM, TEM and XPS study. *Materials Research Bulletin*, *45*(10), 1361–1367. doi:10.1016/j.materresbull.2010.06.046

Yoonessi, M., & Gaier, J. R. (2010). Highly Conductive Multifunctional Graphene Polycarbonate Nanocomposites. *ACS Nano*, *4*(12), 7211–7220. doi:10.1021/nn1019626 PMID:21082818

Zhang, S., Zeng, M., Xu, W., Li, J., Li, J., Xu, J., & Wang, X. (2013). Polyaniline nanorods dotted on graphene oxide nanosheets as a novel super adsorbent for Cr (vi). *Dalton Transactions (Cambridge, England)*, *42*(22), 7854. doi:10.1039/c3dt50149c

Chapter 15
Nanobiotechnology for Bioremediation:
Recent Trends

Swati Jagdale
Maharashtra Institute of Pharmacy, India

Asawaree Hable
Maharashtra Institute of Pharmacy, India

Aniruddha Chabukswar
Maharashtra Institute of Pharmacy, India

ABSTRACT

Nanobiotechnology is the combination of nanotechnology and biotechnology. Bioremediation is defined as a waste management technique that involves use of organisms to remove or neutralize pollutants from a contaminated site. This involves no use of chemicals and it allows waste to get recycled; this process has evolved as one of the most important methods. Remediation of contaminants using existing conventional technology is neither effective nor efficient in environmental cleanup. Techniques developed from nanobiotechnology can detect, control, and remediate pollutants by acting as sensors. Use of nanomaterials has less toxic effects and this will not only reduce consumption time but also reduce cost. Nanoparticles developed by using microbial systems provide green nanotechnology and help in keeping the environment clean. Green nanotechnology needs to develop to make cleaner environment with great ecological balance on earth. This chapter deals with traditional methods of bioremediation, methods in nanobiotechnology, and future aspects.

DOI: 10.4018/978-1-5225-4162-2.ch015

Copyright © 2018, IGI Global. Copying or distributing in print or electronic forms without written permission of IGI Global is prohibited.

INTRODUCTION

Currently all over globe there is rapid development going on in terms of industrialization, urbanization and modern agricultural practices. Due to these reasons pollution is on rise. Pollution in soil, ground water and air is increasing tremendously. Elimination or minimization of pollutants is a very big challenge now. Exploitation of natural resources is being done for fulfillment of human needs. This results in degradation of soil, water resources and reduced air quality which leads to imbalance of ecological system. Pollution can be defined as the presence of pollutants in the environment which causes instability, disorder, harm and discomfort to the ecosystem. Present treatment technologies are although efficient causes several problems which make remediation processes more complex. Among these technologies, bioremediation has been prominently practiced as an efficient cost-effective technology for controlling hazardous pollutants like heavy metals in the soil and water. This chapter reviews the treatment technologies currently available for removing heavy metals and applications of nanotechnology in water treatment. A novel method of nano-bioremediation is effective and more significant for heavy metal removal in all the aspects. The drawback of bioremediation can be possibly avoided by the application of nanotechnology (Mehndiratta et al., 2013, Qu et al., 2013, Bihari, 2013).

BIO-REMEDIATION

As a consequence of human activities, huge quantities of organic and inorganic compounds are discharged into the environment every year. In some cases the discharge is deliberate and regulated (e.g. industrial waste) while in some they are accidental (e.g. spill of chemicals or oil).Majority cases these compounds are toxic and persistent in the environment. The accumulation of toxic compounds in excess levels results in contamination of soil, air and groundwater. The quality of life on earth is dependent on the overall quality of environment on earth. In early years, there was a great abundance of land and other natural resources. In today's world, because of our negligent and careless behavior in use of natural resources it is getting contaminated. Now, problems associated with contaminated sites are increasing in many countries. Contamination of land/ soil/air/water is a result of domestic and industrial activities like production and disposal of hazardous pollutants. Now contamination of natural resources is recognized as a potential threat to human health. The research on minimization on contamination provides not only remedies to reduce the risk of adverse health or environmental effects caused by contamination but also redevelopment of resources. Many conventional methods are available for remediation such as digging up of soil and removal in landfill or cap contaminated areas. But these types of conventional methods have some drawbacks as significant risks in the excavation, handling and transport of hazardous material. The conventional methods are very difficult to work and expensive (Vidali, 2001, Bihari., 2013).

A newer approach is to destroy or transform the pollutants into innocuous substances. These new approaches include use of high-temperature incineration and chemical decomposition (e.g. dechlorination, UV oxidation). These methods are effective for minimizing the contaminants but have few drawbacks like complexity, cost, small-scale application, lack of acceptance, high exposure to contaminants for workers and nearby residents. Now the new approach is bioremediation which offers the destruction or reduction of contaminants using natural biological activity. The bioremediation techniques are widely accepted as it utilizes relatively low-cost, low-technology techniques, high public acceptance and can be easily carried out on site. Although these methods are not very complex technically, they are applicable for limited range of contaminants, involve lengthy process and expertise is required for design and implementation. There should be through assessment of a site for a successful bioremediation plan. Research in this field is rapidly increasing as bioremediation is a good alternative over conventional clean-up technologies. Bioremediation has been used worldwide with variation of success rate. With the information and experience available, there are improvements in these techniques. Thus these techniques have a great potential for treatment of contamination. (Pandey et al., 2012, Bihari, 2013).

Treatment Technologies for Removal of Pollutants/Contaminants

There are three types of technologies available for remediation - physical, chemical and biological (Kensa, 2011, Vidali, 2001, Pandey et al., 2012, Savage, and Diallo, 2005). The overview of these methods is as below

1. **Physical Treatment Methods:** It includes techniques like precipitation, ion exchange, electro-winning, electro-coagulation, cementation, membrane filtration and electro-dialysis.
    a. **Precipitation:** In this technique metals ions are precipitated from contaminated sites with use of anions. This process is expensive and ineffective for metal ions in low concentration.
    b. **Ion Exchange**: It consists of ion exchange material which exchanges cations or anions. This method is used to recover heavy metals from industrial waste. It is relatively expensive, not selective in nature and very sensitive to pH variations.

c. **Electro-Winning:** It is used for removal and recovery of heavy metals. It is mainly used in mining and metallurgic industry.
d. **Electro-Coagulation:** It involves use of electric current/external voltage to remove contaminants from solution.
e. **Cementation:** It involves precipitation of pollutants with use of electrochemical approach.
f. **Membrane Filtration:** It involves separation of pollutants by using semi permeable membrane with pressure as the driving force. The drawbacks of this method is that it is expensive, require continuous monitoring for pressure difference, water treatment and fouling due to the co precipitation of ions present in water.
g. **Electro-Dialysis**: It uses the process reverse osmosis (RO) with electric field applied across membrane. This method effectively removes heavy metals from water.

2. **Chemical Treatment Methods:** It includes techniques as reduction, chemical washing and chelate flushing. The review is as follows
   a. **Reduction:** The reducing agents are injected into contaminated site for removal of contaminants. Drawback is production of toxic intermediates.
   b. **Chemical Washing:** This method involves use of strong acids for removal of pollutants.
   c. **Chelate Flushing:** This method involves use of active agents. The chelating agents used are ethylene diamine tetraacetic acid (EDTA) and diethylene triamine pentaacetic acid (DTPA).They are expensive and carcinogenic in nature.
3. **Biological Treatment Methods:** It includes techniques as bioremediation, biofiltration and biosorption. Biological treatments are eco-friendly and cost-effective for the removal of contaminants from water and soil. But sometimes the contaminants are toxic to the microorganisms involved in the process. Researchers are working on alternative solution to obtain stable and high rate of bioremediation.
   a. **Bioremediation:** This method utilizes living organisms like fungi, bacteria, algae to detoxify and degrade pollutants. Different bioremediation approaches are represented in Figure 1.
   b. **Biofiltration:** It consists of porous surface covered with microbes. When the contaminated water passed through membrane, complex of organic substance and contaminants forms. Then contaminants get transformed into metabolic products.
   c. **Biosorption:** It is based on the biological origin of the materials. This process does not involve any external energy. The process is efficient and cost effective.

Figure 1. Overview of bioremediation approaches
Source: Joutey, N.T, Bahafid, W., Sayel, H., Ghachtouli. N.E., Agricultural and Biological Sciences, 2013. doi: 10.5772/56194

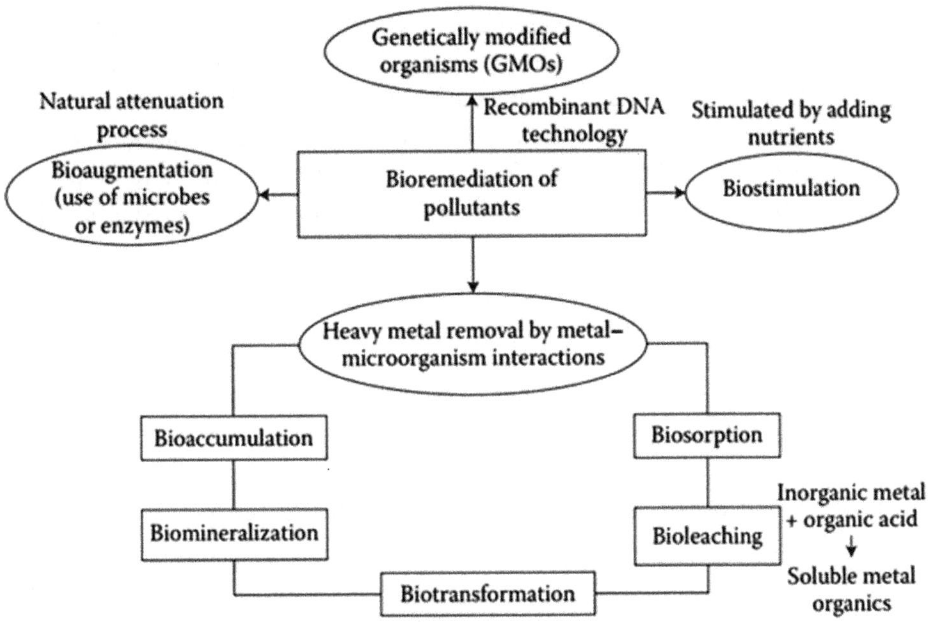

Bioremediation is the process of removal of contamination from waste materials by using organisms. It can be defined as a waste management technique that involves the use of organisms to remove or neutralize pollutants from a contaminated site. As there is no use of chemicals and it allows waste to get recycled, bioremediation plays a key role now days. There are two major causes for waste contamination i.e. natural and man-made.

Principles of Bioremediation

Environmental biotechnology is now a well-known field as waste management is a familiar technology. Recent studies are carried out at molecular level to make the process more efficient. Bioremediation involves degradation of organic waste biologically as it uses living organisms for conversion of contaminants into non-toxic products. The detoxification of environmentally hazardous substances is carried out by using bacteria, fungi, plants etc. (Kensa, 2011, Joutey et al., 2013). The microbes used for remediation process are grown and isolated from another site and carried to the contaminated sites. Metabolic reaction between microbes and contaminants takes place. Action of multiple microbes results in biodegradation.

For enhancement of the process, microbes are imported to the contaminated site is nothing but bioaugmentation. For effective bioremediation microbes do attack on contaminants enzymatically, allows microbial growth and activity for conversion of pollutants to harmless products. Bioremediation has limitations like some pollutants are resistant to microbial attack. Some contaminants have slow rate of degradation. As this method involves natural/ biological sources for remediation, process is not expensive. Most of the reactions are carried out under aerobic conditions, so permits degradation (Kensa, 2011, Vidali, 2001). For microbial bioremediation there are few factors which are responsible to make the process more effective (Kensa, 2011).

FACTORS OF BIOREMEDIATION

Bioremediation involves many factors to control and optimize the process. The factors include presence of microbial load, presence of contaminants and many environmental factors like type of contaminated site, nutrients, pH, availability of amount of oxygen etc (Kensa, 2011, Joutey et al., 2013).

1. **Microbial Populations for Bioremediation Processes:** From any environment, microorganisms can be isolated. Microorganisms are adaptable at any environmental conditions like high or low temperatures, in water, in aerobic or anaerobic conditions, with any waste or hazardous compound. Carbon is the only energy source required. With use of this adaptability feature of microbes, they are used to remediate contaminants from the environment. These microbes can be subdivided and summarized as shown in Table 1 (Cecchin, et al., 2016).
    a. **Aerobic (Presence of Oxygen):** These use contaminant as carbon source for energy and degrade hydrocarbons, pesticides. Examples, *Pseudomonas, Alcaligenes, Sphingomonas, Rhodococcus*and *Mycobacterium*.
    b. **Anaerobic (Absence of Oxygen):** They degrade polychlorinated biphenyls (PCBs) in water, dechlorination of the solvent trichloroethylene (TCE) and chloroform.
    c. **Ligninolytic Fungi:** They have the ability to remediate diverse range of environmental pollutants.
    d. **Methylotrophs:** They utilize methane for carbon and energy. They use methane monooxygenase enzyme for aerobic degradation.
    e. The microbes and contaminants are not uniformly spread in the soil, so they do not come easily in contact which is necessary step for degradation.

Table 1. Some contaminants potentially suitable for bioremediation

Class of Contaminants	Specific Examples	Aerobic	Anaerobic	More Potential Sources
Chlorinated solvents	Trichloroethylene Perchloroethylene		+	Drycleaners
Polychlorinated biphenyls	4-Chlorobiphenyl 4,4-Dichlorobiphenyl		+	Electrical manufacturing, Power station
Chlorinated Phenol "BTEX"	Pentachlorophenol Benzene Toluene Ethyl benzene Xylene		+	Timber treatment, Landfills, Oil production and storage, Gas work sites, Airports, Paint manufacture, Port facilities Railway yards, Chemical manufacture
Polyaromatic Hydrocarbons (PAHs)	Naphthalene Antracene Fluorene Pyrene Benzo(a)pyrene	+		Oil production and storage Gas work sites, Coke plants, Engine works, Landfills, Tar production and storage, Boiler ash dump sites, Power stations
Pesticides	Atrazine Carbaryl Carbofuran Coumphos Diazinon Glycophosphate Parathion Propham 2,4-D	+	+	Agriculture, Timber treatment plants, Pesticide manufacture, Recreational areas, Landfills

Some microbes are mobile while contaminants can be mobilized by using some surfactants.

2. **Environmental Factors:** As the number of microbes present in soil are not sufficient for the bioremediation process, their growth and activity should be enhanced with the supply of additional nutrients and oxygen. This is called as biostimulation. The supplied nutrients stimulate microbes for synthesis of enzymes required to degrade pollutants. Carbon is utilized as major source of energy by microbes. The composition of microbial cell is summarized in table 2 (Colberg et al., 1995).
3. **Environmental Requirements:** Optimum conditions required for the decomposition of pollutants are reported in Table 3.

Table 2. Composition of a microbial cell

Element	Percentage	Element	Percentage
Carbon	50	Sodium	1
Nitrogen	14	Calcium	0.5
Oxygen	20	Magnesium	0.5
Hydrogen	8	Chloride	0.5
Phosphorous	3	Iron	0.2
Sulphur	1	All others	0.3
Potassium	1		

The activity and growth pattern of microbes are affected by environmental conditions such as pH, water content, oxygen, temperature. Most of the microbes grow optimally in narrow range of conditions. So, achievement of optimal environment conditions is important. Acidic pH of the soil can be adjusted by using lime. Rate of biochemical reaction depends upon temperature. Increase in temperature increases the rate of reaction while cell death occurs after threshold temperature. Moisture content plays important role for growth of microbes so irrigation is carried out for maintaining water content. Structure of soil, control the supply of nutrients to microbes and this can be improved by addition of salts and organic matter in soil.

Table 3. Environmental conditions affected degradation

Parameters	Condition Required for Microbial Activity	Optimum Value for Oil Degradation
Soil moisture	25-28% of water holding capacity	30-90%
Soil pH	5.5-8.8	6.5-8.0
Oxygen content	Aerobic. Minimum air-filled 10-40%	pore space of 10%
Nutrient content	N and p for microbial growth	C:N:P = 100:10:1
Temperature (°C)	15-45	20-30
Contaminants	Not too toxic	Hydrocarbon 5-10% of dry weight of soil
Heavy metals	Total content 2000 ppm	700 ppm
Type of soil	Low clay or silt content	

C: Carbon, N: Nitrogen, P: Phosphorous

Source: Cecchin, I., Reddy, K., R., Tessaro, E., F., Schnaid, F., 2016

- **Classes of Bioremediation:** There are 2 classes of bioremediation according to the area of the treatment of waste. Classes are in-situ and ex-situ (Kensa, 2011, Kumar et al., 2011).
- **In-Situ:** When contaminated waste is processed at the origin point, it is called as in-situ bioremediation. This has main advantage that it prevent spreading of contamination during transportation.
- **Ex-Situ:** When contaminated waste removes to a treatment area for processing then it is called as ex-situ bioremediation. This process assists to control the bioremediation products and also use of contaminated area.

Types of Bioremediation Techniques

There are more than 9 types of bioremediation based on constituents used to remove contaminants (Kensa, 2011, Kumar et al., 2011). Some common types are

- **Phytoremediation:** Use of plants
- **Bioventing:** Blowing of air through soil
- **Bioleaching:** Use of living organisms to eliminate metals from soil
- **Land Farming:** Use of aeration through contaminated soil and sifting
- **Bioreactor:** Use of specially designed containers to hold the waste
- **Composting:** Natural decomposition of waste
- **Bioaugmentation:** Addition of microbes and organisms in waste
- **Rhizofiltration:** Use of plants for removal of metals in water
- **Biostimulation:** Use of microbes for removal of contamination applied in a medium to the waste.

Advantages of Bioremediation

- Bioremediation is widely accepted process as it is a natural way of waste management.
- It is applicable for variety of contaminants for their complete degradation.
- Some hazardous compounds can be transferred into harmless products.
- The destruction of contaminants is possible instead of conversion of one to another form.
- Bioremediation is comparatively less expensive than other methods.

Disadvantages of Bioremediation (Kumar et al., 2011)

- All compounds cannot be rapidly and completely degraded by bioremediation.
- It cannot be applicable for non-biodegradable compounds.

Table 4. Developmental methods applied in bioremediation

Technique	Examples	Benefits	Applications
In Situ	Biosparging Bioventing Bioaugmentation	Most efficient, Non Invasive, Relative passive, Naturally attenuated process, Treat soil and water	Biodegradative abilities of indigenous microorganisms, Presence of metals and inorganic compounds, Environmental parameters, Biodegradability of pollutants, Chemical solubility, Geological factors, Distribution of pollutants
Ex situ	Land farming (Solid-phase treatment system) Composting (Anaerobic, converts solid organic wastes into humus-like material) Biopiles	Cost efficient, Simple, Inexpensive, self heating Low cost, Rapid reaction rate, Inexpensive, Can be done on site	Surface application, aerobic process, application of organic materials to natural soils followed by irrigation and tilling. To make plants healthier good alternative to land filling or incinerating practical and convenient. Surface application, agricultural to municipal waste
Bioreactors	Slurry reactors Aqueous reactors	Rapid degradation, kinetic optimized environmental parameters, Enhances mass transfer, Effective use of inoculants and surfactant	Bioaugmentat Toxicity of amendments. Toxic concentrations of contaminants
Precipitation or Flocculation	Non-directed physico-chemical complexation reaction between dissolved contaminants and charged cellular components (dead Biomass)	Cost-effective	Removal of heavy Metals
Microfiltration	Microfiltration membranes are used at a constant pressure	Remove dissolved solids rapidly	Waste water treatment; recovery and reuse of more than 90% of original waste water
Electro dialysis	Uses cations and anions exchange membrane pairs	Withstand high temperature and can be reused	Removal of dissolved solids efficiently

Source: Cecchin., I., Reddy, K., R., Tessaro, E., F., Schnaid, F., 2016

- It is not easy to implement the process from pilot to large scale fields.
- It requires long time for complete degradation as compare to other processes.
- It is difficult to degrade the mixture of unevenly spreaded various types of contaminants.
- There are few cases of generation of more toxic biodegradable product than parent compound.

- Process of bioremediation is quite specific in terms of contaminants, nutrients and environment for microbial growth.

NANO-BIO-TECHNOLOGY

Nanobiotechnology is the combination of nanotechnology and biotechnology. The term 'nanotechnology' is the combination of two Greek words *i.e.* nano and technology. Nanotechnology or Nanoscaled Technology is considered to be at a size below 0.1 μm or 100 nm. Nanoscience is the study of concept, properties, and applications of materials at atomic, molecular, and macromolecular scales. Nanoscale ranges between 1 and 100 nm (Qu et al., 2013). Nano materials include nano-particles, nano-rods, quantum dots, carbon tubes, dendrimer, nano-crystals etc. Nanotechnology is an emerging field as there is possibility of advancement in products with new characteristics and functions with extensive potential in a broad range of applications. Applications of nanotechnology are in the fields of medicine, food, electronics, fuel cells, solar cells, space, environmental remediation, fabric etc. (Savage, N., Diallo, M.S., 2005). Biotechnology uses living organisms for manufacturing, modification and development of products for their specific use. Nanobiotechnology describes the multidisciplinary activities related with biosensors, biochemistry, biophysics, medicine etc. (Mehndiratta et al., 2013).

Concepts

Many concepts of bionanotechnology are driven from different fields. Properties of bionanoscience consist of biological, mechanical, optical, electrical and thermal properties. These properties include electromechanical stimulation, capacitors, energy storage/batteries, deformation, adhesion, absorption, failure, luminescence, photochemistry, thermal management, thermo mutability, etc. The detailed structural and mechanical analysis of biological process at nano level is bionanoscience. Most of its core principles are received from nanotechnology. Nanobiotechnology give a detailed summary of overlapping multidisciplinary activities where chemistry, biology, physics, nanomedicine and engineering are associated with each other (Kumar et al., 2011).

Applications

Field of nano-biotechnology is comparatively new. Applications of bionanotechnology are widely spread in various fields. Nanobiotechnology not only provides tools and techniques for the study of biology but also recreates useful biological mechanisms and pathways (Kumar et al., 2011, Fortina et al., 2005).

NANOBIOTECHNOLOGY FOR BIOREMEDIATION

Because of urbanization, pollution in ground water, waste water and other water resources is influencing some serious and long term effect on health of human being. As there is rapid growth of industrialization, urbanization, and modern agricultural practice techniques, pollution of air, water or groundwater and soil is on the hike. In today's era conventional methods for treatment of waste are not suitable because of industrialization. Conventional methods are nonspecific, less efficient and expensive so it cannot be used. The biggest challenge today is to minimize or neutralize contaminants. For fulfillment of human desires for every need, natural resources are being exploited. This exploitation results in humiliation of water, soil and thus environment pollution leads to imbalance of ecological system. Pollution is presence of pollutants in environment which generate ecological disorder and discomfort. The remediation of contaminants by using existing conventional technology is neither effective nor efficient in environment clean up. Use of nanomaterials for bioremediation has less toxic effects and it also improves the microbial action in waste and toxic material. This will not only reduce the consumption time but also cost (Mehndiratta et al., 2013, Yadav et al., 2017).

Currently available technologies for removal of contaminate are physical treatment method, chemical treatment method and biological treatment method. There are few studies for bioremediation which combines some biological methods and other remediation methods. Nanomaterials gained more attention due to their unique physicochemical properties. Research is going on utilizing these properties in the area of environmental science, particularly in bioremediation. Many nanomaterials are being used for waste water remediation. As nanomaterials have size dependant properties, they provide high surface area, more sorption, high reactivity and high dissolution rates. The nano-based technologies consist of nanotechnology associated membranes, nanoadsorbents and nano-photocatalysts. But these types of technologies are successful at lab scale only, not on large scale (Kumar et al., 2011).

Nanotechnology generates material with specified properties to perform particular functions by controlling the material at nanoscale. This is specifically important in environmental pollution issues emerging from the presence of contaminants in solid, liquid or gas form. As the size of nanomaterials is very small, it provides high surface to volume ratio. The properties of nanomaterials leads to smaller, sensitive and accurate pollution tracking devices called as nano-devices/ nanosensors. Also nanomaterials can be designed and developed in such a way that they can interact with contaminants and detoxify them. Nanotechnology not only detects contaminated sites but also remediate them. Thus it minimizes the production of harmful waste material from industrial sector (Mehndiratta et al., 2013, Greque de Morais et al., 2014). Nanotechnology launches various novel methods for remediation of contaminants and purification of resources like water, air and soil. Currently nanomaterials discovered for bioremediation comprise of metallic and bi-metallic, semiconductor and magnetic nanoparticles, dendrimers etc. (Yadav et al., 2017). Nanoparticles have a long history since from naturally occurring organic and inorganic nanoparticles. The historical development of nanoparticles started with Paul Ehrlich and then first attempts by Ursula Scheffel and colleagues. Extensive work by the group of Professor Peter Speiser at the ETH Zürich in the late 1960s and early 1970s are described from a personal point of view. Special attention is given to the years between 1970 and the early 1980s. (Kreuter, 2006). The size of nanoparticles ranges from 1-100 nm and depending upon their size, shape and properties they are classified. Because of their ultrafine particle size they have numerous potential applications in various fields. Some types of nanoparticles used in remediation are explained as below (Heiligtag 2013).

Role of Metal Nanoparticles in Remediation

The contaminated soil and ground water can be remediated by using iron nanoparticles. Iron gets oxidized when it comes in contact with air. It also transforms some contaminants into simple, non-toxic products by oxidizing them (e.g. dioxins, trichloroethylene (TCE) etc.) Iron is non-toxic and present in natural environment in ample amount. Some industries remediate waste with use of iron powder. But due to coating on their surface, oxidation capacity of iron decreases over time period (Rizwan et al., 2014). New solution offered is iron nanoparticles as they are very much reactive than iron powder. Due to their nano-scaled structure, they can react with pollutant over large surface area. The slurry of iron nanoparticles prepared with water and then injected into contaminated site. Nature of nanoparticles remains unchanged by components present in the contaminated site like nutrients, pH and temperature. An iron nanoparticle completely detoxifies and transforms environmental pollutants as they are highly reactive and stable. Figure 2 shows the remediation process by iron nano-particles (Pandey et al., 2015).

Bimetallic nanoparticles (iron/palladium) have been shown to be even more active and stable than iron nanoparticles. These nanoparticles casted on solid supports such as activated carbon or silica for treatment of polluted water and industrial wastes (Rizwan et al., 2014).

Figure 2. Nanoscale iron particles for in situ remediation

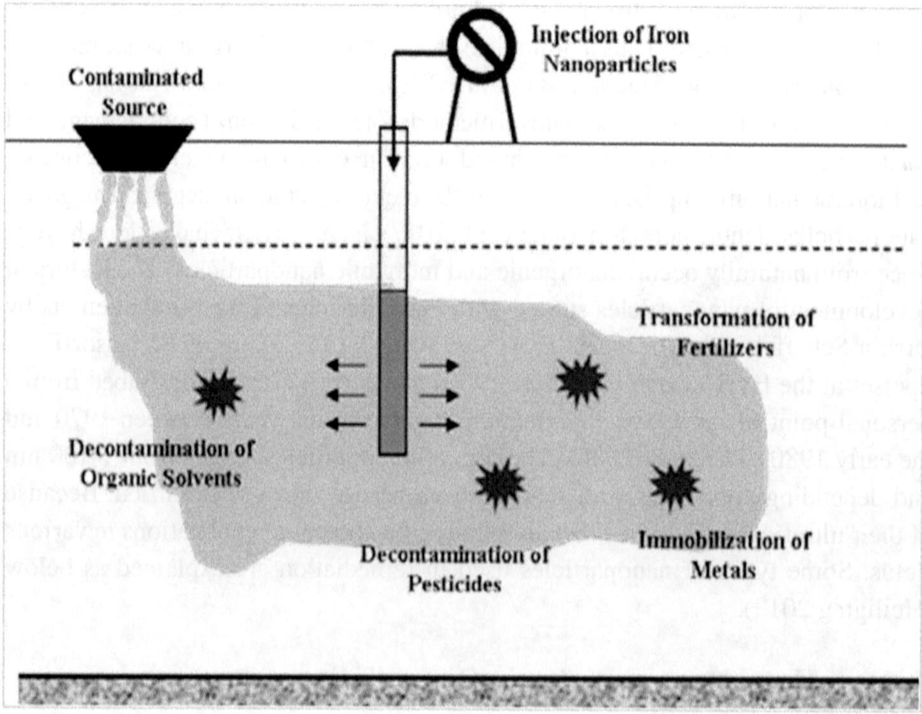

Remediation Using Semiconducting Nanoparticles

Semiconducting nanoparticles of zinc oxide and titanium dioxide are used in remediation by photocatalysis method. As these materials are semiconductor, they formats an electron hole pair after exposure to light. Both types of nanoparticles transfer the charge to organic contaminants and promote their oxidation and converts them into less harmful/ non-toxic products. As the materials are easily available, the remediation method by using semiconducting nanoparticles is inexpensive. Nanosize of particles provide large active surface area for the remediation reaction (Jang et al., 2001, Rizwan et al., 2014).

Remediation Using Dendrimer

Dendrimer are nanomaterials made up of branched polymers with controlled composition. The metal pollutants can be removed by using dendrimer containing chelating agents. They act as cages for metal contaminants and act by binding of pollutants to surface or solubilizing them in suitable solvents. (Chen, C. Z. S., & Cooper, S., 2002, Frechet, J. M. J., Tomalia, D. A. (Eds.), 2001).

Remediation using Magnetic Nanoparticles

the use of magnetic nanoparticles are effective in removing the arsenic from polluted water sources. Arsenic easily gets stuck with the rust which is nothing but iron oxide. This can be removed from water with use of magnet. Rust of nano-sized are prepared for improvement of remediation efficiency as they provide small size and large surface area. In comparison with other conventional techniques used to remediate contaminated water sources, this is very simple, inexpensive techniques with no usage of electricity (Bystrzejewski et al., 2009, Kanel et al., 2006, Rizwan et al., 2014).

Remediation Using Aerogels and Solid Absorbents

The oil spills in seawater is of major concern for the environment. There are many remediation plans available currently to clean up oil. Current strategies consist of microbe culture, nutrients or enzyme additives. The basic purpose of addition of these is to fasten natural technology to degrade the oil components. (Kumar, S., Gopinath, P., 2012). Another majorly used method for remediation is aerogels. To amplify interaction of aerogels with oil, they are altered with hydrophobic molecules.

Table 5. List of Pollutants remediated by Nano iron technology

Acid Orange	Dichlorobenzenes	Orange II
Acid Red	Dichloromethane	Pentachlorobenzene
Arsenic	Cis-Dichloroethene	Pentachlorophenol
Bromoform	Trans- Dichloroethene	Perchlorate
Cadmium	1,1- Dichloroethene	PCBs
Carbon tetrachloride	Dichromate	Silver
Chloroform	DDT	Tetrachlorobenzenes
Chloromethane	Hexachlorobenzene	Tetrachloroethene
Chlorobenzene	Lindane	Trichloroethene
Chrysoidine	Mercury	Trichlorobenzenes
Dioxine	Nickel	TNT
Dibromochloromethane	Nitrate	Tropaeolin
Dichlorobromomethane	NDMA	Vinyl Chloride

Source: Md. Rizwan, Man Singh, Chanchal K. Mitra, and Roshan K. Morve,2014

Aerogels perform like sponge, they can soak oil around sixteen times more than the weight of aerogel as the nanostructure provides very large surface area. They are very effective in remediation of oil spills (Hashim et al., 2011).

Nano-Membranes and Nano-Filters

For remediation of contaminated water and air filters, membranes and adsorbents are used. For more effective remediation nano-filter, nano-membrane, nano-adsorbents are designed and developed with some particular properties by using nanotechnology. Nanomaterials are used for manufacturing of these products as they provide minimum size and maximum surface area for the reaction. For some pollutants nano-traps are designed and developed with particular size of pores and reactivity of surface. Researchers are developing nanomembranes made up of iron oxide ceramic which have capability to remediate organic waste from water (Figure 3). (Dave, P. N., Chopda, L. V. 2014).

Nano-filters and nano-membranes can also be modified in such a way that they can trap the pollutants easily and also reacts with them chemically for conversion into less toxic product. (Liu, W. T., 2006).

Super-Hydrophilic Filters

Water remediation is a huge problem in some situations. Use of nano-filters facilitates the water filtration and removes heavy metal pollutants like arsenic. Nano-filter inserted inside plastic bottles for removing pollutants from contaminated water site. LifeSaver® Bottle is commercially available technology. It can filter material up to 15nm size microbes as it consists of super hydrophilic filter (Tofighy et al., 2011, Mondal et al., 2006).

Pollution Prevention

Nanotechnologies offer many novel strategies to minimize the pollution in many processes. These technologies reduces waste from all types of manufacturing processes, reduces use of harmful chemicals, minimizes emission of hazardous gases in fuel combustions and utilizes more biodegradable plastics. These are few approaches that are used to reduce the pollution from the environment. Nanotechnologies not only used for production of advanced materials but also for the development of a method to improve the efficiency of industrial processes (e.g. catalytic process) which detoxify the pollutants and minimizes the pollution (Md. Rizwan et al., 2014),

Figure 3. A ceramic nano-membrane-
Source: Photo courtesy of Professor M. Wiesner, Wiesner Laboratory, Duke University

Eco-Friendly Materials

Nanobiotechnology utilizes more eco-friendly materials like some biodegradable plastic. This is made up of polymers with degradable molecular structure. These materials can replace electrodes from rechargeable batteries. Self cleaning glass is also a new material manufactured with nanocrystals of titanium dioxide. This when exposed to daylight it removes organic dust present on glass and when gets exposure to water gets easily washed out with dirt deposited on it (Liu, 2006, Md. Rizwan et al., 2014).

Fertilizers and Wood Treatment Products

With the use of nanotechnology, wood treatment products and fertilizers are manufactured in such a way that they became more stable and their percolation in environment is less. In one research study, incorporation of biocide for wood treatment in the nanoparticles was carried out. It has been observed that the biocide gets protected from degradation due to nanoshell around it and with the nano size they easily get inside of wood (Md. Rizwan et al., 2014).

Nanocatalysis

A substance which increases the rate of chemical reaction without any chemically alteration is known as catalyst. They have active surface for the occurrence of chemical reaction. Few earth metals act as catalysts but they are expensive e.g. platinum and palladium. As the size of catalysts is reduced to nanoscale level, the size of reactive surface increases and completion of reaction occur with more efficiency. Manufacturing of new nanostructured catalysts is now the area of research interest. Nano-scaled catalysts not only improve quality of air but also treat and reduce the level of contaminants present in water (Mauter and Elimelech 2008).

Catalytic Gold

There are conventional catalysts available like platinum and palladium. But they are very expensive. In demand of alternative catalysts nano-gold is invented. They are found to be more efficient in many studies in controlling the pollution. It filters air and separates carbon monoxide. Recent studies have shown that nano-gold catalysts oxidizes diesel and minimizes the emission of diesel hydrocarbon up to 40% more than available pollution controlling materials. Thus nanotechnology is making its impact in environment bioremediation field (Husseiny et al., 2007, Lafleur et al., 2012).

Nanowire and Nanotube-Based Sensors

Nanowires are generally used for the detection of the antibodies and proteins. A wide range of analytes can be detected by nanowires due to its very small size. The nanowires act as a sensor for detection of pathogen, chemical biological agents in water, soil and air (Liang et al., 2005; Liu, 2006).

Some nanotubes and nanowires act as a sensor for chemical and biological elements. Single walled carbon nanotubes have high sensitivity for gases like NO_2, NH_3 with fast response. The gas molecules directly bind to the surface of carbon nanotubes. These bindings produce increase or decrease in the electric resistance on carbon nano-tube and thus act as sensor for pollutants. On contrary, for achievement of increased chemical reactivity, conventional sensors generally operate at high temperatures. (Hyung et al., 2007).

Cantilever Sensors

A device of silicon arrays, coated with a Nanomaterial layer are called as cantilever sensor (Figure 4). The coatings of nanolayer are sensitive to specific contaminants.

The cantilever sensors are with the thickness in micrometers and is of length around 10-500nm. The quantitative measurement of the contaminants can be detected with cantilever sensor. The contaminant interacts with the nanolayer of sensor, this changes the surface stress and arm of cantilever sensor bends. Cantilever sensor detects contaminants like heavy metals, toxic bacteria, harmful chemicals and pesticides. (Liu, 2006).

Green Manufacturing

Manufacturing processes are always accompanied by the production of diverse waste products, many of which pose a threat to the environment and thus need to be removed and treated. Ideally, manufacturing processes should be designed to minimize material usage and waste production, while ensuring the use of the least amount of energy possible. "Green manufacturing" is a generic name to broadly cover methods and technologies that are directed towards achieving this goal. It includes the development of new chemical and industrial processes (for instance, water-based rather the solvent- based processes), reduction in the use of unsafe

Figure 4. Schematic diagram of a cantilever based biosensor. The yellow molecules bind specifically to the red molecules on the right hand cantilever and are detected by the bending of the cantilever.
Source: Image courtesy of M. Lorentzen (iNANO, University of Aarhus

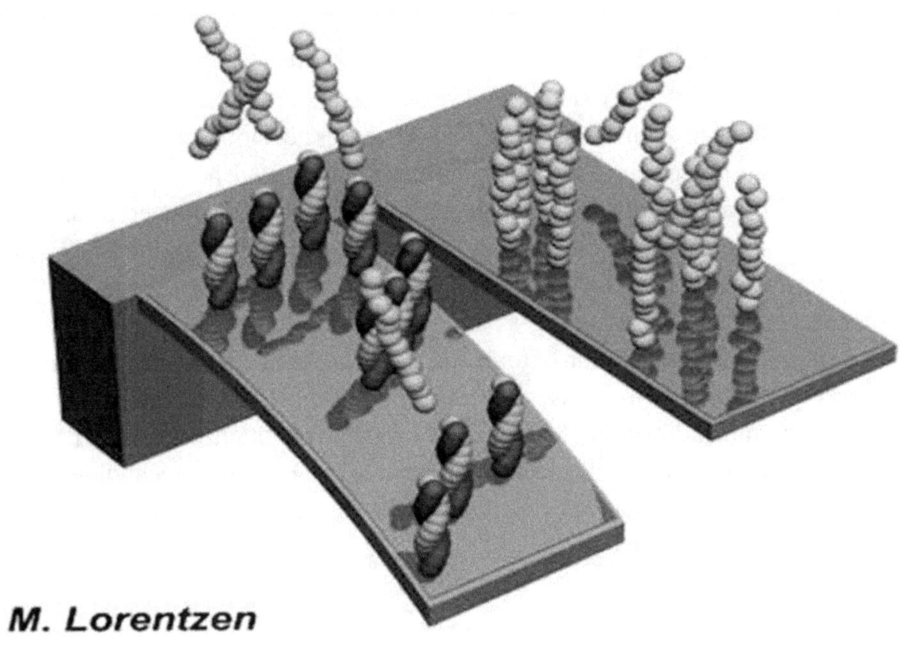

compounds (such as metals), development of "green" chemicals that are more environment-compatible and efficient use of energy. In terms of its application to the reduction of manufacturing waste, nanotechnology can contribute in two ways: by helping the manufacturing to be more controlled and efficient and by using nanomaterials (such as catalysts) that can raise the manufacturing efficiency while reducing or eliminating the use of toxic materials. Overall, nanotechnology has the potential to make industrial processes more efficient in terms of energy usage and material usage, while minimizing the production of toxic waste. Sometimes the application of nanotechnology to "greener" manufacturing processes is referred to as "green nanotechnology". These include, for instance: bottom-up atomic-level synthesis for developing improved catalysts; inserting information into molecules to build new materials (such as DNA) through highly specific synthetic routes; scaling down material usage during chemical reaction by using nanoscale reactors and improving manufacturing to require less energy and less toxic materials (Greque de Morais et al., 2014).

CONCLUSION

Carbon nanotubes, quantum dots work as sensors as well as remediators. Silver and gold nanoparticles work as catalyst for decomposition of pollutants from waste. Nano materials can be used as carrier for pesticides and chemical fertilizers delivery. These carriers deliver the chemicals in equal distribution into soil in required amount so that it can avoid soil pollution effectively. Nanomaterials can limit the usage of chemicals, pesticides, fertilizers with the use of biosynthesis of nanomaterials from microbial system as it will emerge as very novel technology for bioremediation of soil and protection of crop for mankind. Nanoparticles developed by using microbial system provide green nanotechnology for remediation and helps in keeping the environment clean. Microflora in soil used for synthesis of nanomaterials and used as bioremediator further. These types of nanomaterials help in biosorption of heavy metals from contaminated areas (soil or water). Nanotechnology has the capability to increase efficiency of any process in terms of use of material and energy, with minimum amount of generation of toxic wastes. This application of nanotechnology to "greener" production processes is referred to as "green nanotechnology". This helps for bioremediation in natural or organic ways and maintains ecological balance. This gives rise for evolution of green nanotechnology. Green nanotechnology needs to develop with speedy advancement to make cleaner environment with great ecological balance on earth.

REFERENCES

Bihari, Z. (2013). Current Trends in Bioremediation and Biodegradation: Next-Generation Sequencing, Bioremediation & Biodegradation. *Journal of Bioremediation & Biodegradation*, 4–8.

Bystrzejewski, M., Pyrzyńska, K., Huczko, A., & Lange, H. (2009). Carbon-encapsulated magnetic nanoparticlesas separable and mobile sorbents of heavy metal ions from aqueous solutions. *Carbon*, *47*(4), 1189–1206. doi:10.1016/j.carbon.2009.01.007

Cecchin, I., Reddy, K. R., Tessaro, E. F., & Schnaid, F. (2016). Nanobioremediation: Integration of nanoparticles and bioremediation for sustainable remediation of chlorinated organic contaminants in soils. *International Biodeterioration & Biodegradation*, 1–10.

Chen, C. Z. S., & Cooper, S. (2002). Interactions between dendrimerbiocides and bacterial membranes. *Biomaterials*, *23*(16), 3359–3368. doi:10.1016/S0142-9612(02)00036-4 PMID:12099278

Colberg, P. J. S., & Young, L. Y. (1995). Anaerobic degradation of nonhalogenated homocyclic aromatic compounds coupled with Nitrate, Iron, or Sulfate reduction. In *Microbial Transformation and Degradation of Toxic Organic Chemicals* (pp. 307–330). New York: Wiley-Liss.

Dave, P. N., & Chopda, L. V. (2014). Application of Iron Oxide Nanomaterials for the Removal of Heavy Metals, review article. *Journal of Nanotechnology*, 398–412.

Fortina, P., Larry, J., Kricka, L., Surrey, S., & Grodzinski, P. (2005). Nanobiotechnology: The promise and reality of new approaches to molecular recognition. *Trends in Biotechnology*, *23*(4), 168–173. doi:10.1016/j.tibtech.2005.02.007 PMID:15780707

Frechet, J. M. J., & Tomalia, D. A. (Eds.). (2001). *Dendrimers and Other Dendritic Polymers*. New York: Wiley and Sons; doi:10.1002/0470845821

Greque de Morais, M., Martins, V., Steffens, D., & Pranke, P. (2014). Biological Applications of Nanobiotechnology. *Journal of Nanoscience and Nanotechnology*, *14*(1), 1007–1017. doi:10.1166/jnn.2014.8748 PMID:24730317

Gupta, A. K., Deva, D., Sharma, A., & Verma, N. (2010). Fe-grown carbon nanofibers for removal of arsenic (V) in waste water. *Industrial & Engineering Chemistry Research*, *49*(15), 7074–7084. doi:10.1021/ie100392q

Hashim, M. A., Mukhopadhyay, S., Sahu, J. N., & Sengupta, B. (2011). Remediation technologies for heavy metal contaminated groundwater. *Journal of Environmental Management*, *92*(10), 2355–2388. doi:10.1016/j.jenvman.2011.06.009 PMID:21708421

Heiligtag, F., A., Niederberge, M., (2013). *The fascinating world of nanoparticle research*. Academic Press.

Husseiny, M. I., El-Aziz, M. A., Badr, Y., & Mahmoud, M. A. (2007). Biosynthesis of gold nanoparticles using *Pseudomonas aeruginosa. Spectrochimica Acta. Part A: Molecular and Biomolecular Spectroscopy*, *67*(3-4), 3–4. doi:10.1016/j.saa.2006.09.028 PMID:17084659

Hyung, H., Fortner, J. D., Hughes, J. B., & Kim, J. H. (2007). Natural organic matter stabilizes carbon nanotubes in the aqueous phase. *Environmental Science & Technology*, *41*(1), 179–184. doi:10.1021/es061817g PMID:17265945

Jang, H. D., Kim, S. K., & Kim, S. J. (2001). Effect of particle size and phase composition of titanium dioxide nanoparticles on the photocatalytic properties. *Journal of Nanoparticle Research*, *3*(2/3), 141–147. doi:10.1023/A:1017948330363

Joutey, N. T., Bahafid, W., Sayel, H., & Ghachtouli, N. E. (2013). *Overview of bioremediation approaches*. Agricultural and Biological Sciences.

Kanel, S., Greneche, J., & Choi, H. (2006). Arsenic (V) removal from groundwater using nano scale zero-valent iron as a colloidal reactive barrier material. *Environmental Science & Technology*, *40*(6), 2045–2050. doi:10.1021/es0520924 PMID:16570634

Karn, B., Kuiken, T., & Otto. M., (2009). *Nanotechnology and in Situ Remediation: A Review of the Benefits and Potential Risks*. Academic Press.

Kensa, M. (2011). Bioremediation - An Overview. *Jr. of Industrial Pollution Control*, *27*(2), 161–168.

Kreuter, J. (2006). *Nanoparticles—a historical perspective*. Academic Press.

Kumar, A., Bisht, B. S., Joshi, V. D., & Dhewa, T. (2011). Review on Bioremediation of Polluted Environment: A Management Tool. *International Journal of Environmental Sciences*, *1*(6), 1079–1093.

Kumar, S., & Gopinath, P. (2012). Nano-Bioremediation Applications of Nanotechnology for Bioremediation. *Remediation of Heavy Metals in the Environment*, 27-40.

Lafleur, J. P., Senkbeil, S., Jensen, T. G., & Kutter, J. P. (2012). Gold nanoparticle-based optical microfluidic sensors for analysis of environmental pollutants. *Lab on a Chip, 12*(22), 4651–4656. doi:10.1039/c2lc40543a PMID:22824920

Liang, P., Ding, Q., & Song, F. (2005). Application of multiwalled carbon nanotubes as solid phase extraction sorbent for preconcentration of trace copper in water samples. *Journal of Separation Science, 28*(17), 2339–2343. doi:10.1002/jssc.200500154 PMID:16342800

Liu, W. T. (2006). Nanoparticles and Their Biological and Environmental Applications. *Journal of Bioscience and Bioengineering, 102*(1), 1–7. doi:10.1263/jbb.102.1 PMID:16952829

Mauter, M. S., & Elimelech, M. (2008). Environmental applications of carbon-based nanomaterials. *Environmental Science & Technology, 42*(16), 5843–5859. doi:10.1021/es8006904 PMID:18767635

Mehndiratta, P., Jain, A., Srivastava, S., & Gupta, N. (2013). Environmental Pollution and Nanotechnology. *Environment and Pollution, 2*(2), 49–58. doi:10.5539/ep.v2n2p49

Mondal, P., Majumdar, C. B., & Mohanty, B. (2006). Laboratory based approaches for arsenic remediation from contaminated water: Recent developments. *Journal of Hazardous Materials, B137*(1), 464–479. doi:10.1016/j.jhazmat.2006.02.023 PMID:16616812

Pandey, B., & Fulekar, M. H. (2012). Bioremediation technology: A new horizon for environmental clean-up. *Biology and Medicine, 4*(1), 51–59.

Pandey, S., Kumari, M., Singh, S., Bhattacharya, A., Mishra, S., Chauhan, P., & Mishra, A. (2015). *Bioremediation via Nanoparticles: An Innovative Microbial Approach*. Academic Press; doi:10.4018/978-1-4666-8682-3.ch019

Qu, X., Alvarez, P. J. J., & Li, Q. (2013). Applications of nanotechnology in water and wastewater treatment. *Water Research, 47*, 3931–3946.

Rajendran, P., & Gunasekaran, P. (2010). Nanotechnology for Bioremediation of Heavy Metals. *Environmental Bioremediation Technologies*, 211-221.

Md. Rizwan, Singh, M., Chanchal, K., Mitra, & Morve, R. (2014). Ecofriendly Application of Nanomaterials: Nanobioremediation. *Journal of Nanoparticles, 7*.

Rizwan, M., Singh, M., Mitra, C. K., & Morve, R. K. (2014). Ecofriendly application of nanomaterials: Nanobioremediation. *Journal of Nanoparticles*, 1–7.

Savage, N., & Diallo, M. S. (2005). Nanomaterials and water purification: Opportunities and challenges. *Journal of Nanoparticle Research*, *7*(4-5), 331–342. doi:10.1007/s11051-005-7523-5

Tofighy, M. A., & Mohammadi, T. (2011). Adsorption of divalent heavy metal ions from water using carbonNanotube sheets. *Journal of Hazardous Materials*, *185*(1), 140–147. doi:10.1016/j.jhazmat.2010.09.008 PMID:20926186

Vidali, M. (2001). Bioremediation. An overview. *Pure and Applied Chemistry*, *73*(7), 1163–1172. doi:10.1351/pac200173071163

Yadav, K. K., Singh, J. K., Gupta, N., & Kumar, V. (2017). A Review of Nanobioremediation Technologies for Environmental Cleanup: A Novel Biological Approach. *Journal of Materials and Environmental Sciences*, *8*(2), 740–757.

APPENDIX

Table 6. Examples of nanoparticles and nanomaterials for use in water remediation

Nanoparticle/Nanomaterial	Pollutant
Nanocrystalline zeolites	Toluene, NO_2
Carbonaceous nanomaterials CeO_2-Carbon nanotubes (CNTs) Activated carbon fibres (ACFs) CNTs functionalized with polymers CNTs functionalized with Fe Single-walled carbon nanotubes Multi-walled carbon nanotubes	Heavy metal ions Benzene, toluene, ethylbenzene, xylene p-nitrophenol benzene, toluene, dimethylbenzene Heavy metal ions Trihalomethanes (THMs) Heavy metal ions THMs chlorophenols Herbicides Microcystin toxins
Self-assembled monolayer on mesoporous supports (SAMMS) Anion-SAMMS Thiol-SAMMS HOPO-SAMMS	Inorganic ions Heavy metal ions Actinides and lanthanides
Biopolymers	Heavy metal ions
Zero-valent iron nanoparticles (nZVI)	Polychlorinated biphenyls (PCBs) Inorganic ions Chlorinated organic compounds Heavy metal ions
Bimetallic nanoparticles Pd/Fe nanoparticles	PCBs Chlorinated ethane Chlorinated methanes
Ni/Fe nanoparticles Pd/Au nanoparticles	TCE and PCBs Dichlorophenol Trichlorobenzene Chlorinated ethane Brominated organic compounds (BOCs)
Nanocrystalline TiO_2 Nitrogen (N)-doped TiO_2 Fe (III)-doped TiO_2 Supported TiO_2 nanoparticles TiO_2 based p-n junction nanotubes	Heavy metal ions Azo dyes Phenol Aromatic pollutants Toluene

Source: Yadav, K., K., Singh, J., K., Gupta, N., Kumar, V., 2017

Table 7. Examples of the use of nanoparticles in remediation

Process Exploited	Target Compounds	Nanomaterials Used	Some of Novel Properties
Adsorption	Heavy metals, organic compounds, arsenic, phosphate, Cr (IV), mercury, PAHs, DDT, Dioxin	Iron oxides, Carbon-based nanomaterials such as dendrimers and polymers, Carbon nanotubes (CNTs)	High specific surface area and assessable adsorption sites, selective and more adsorption sites, short intra-particle diffusion distance, tunable surface chemistry, easy to reuse
Photocatalysis	Organic pollutants, NOX, VOCs, Azo dye, Congo red dye, 4-chlorophenol and Orange II, PAHs	TiO_2, ZnO, Species of iron oxides (Fe III, Fe_2O_3, Fe_3O_4)	Photocatalytic activity in solar spectrum, low human toxicity, high stability and selectivity, low cost
Redox reactions	Halogenated organic compounds, metals, nitrate, arsenate, oil, PAH, PCB	Nanoscale zero-valent iron (nZVI), nanoscale calcium peroxide	Electron transfers such as photosynthesis, respiration, metabolism and molecular signaling, nature of their redox centers
Disinfection	Diamines, phenols, formaldehyde, hydrogen peroxide, silver ions, halogens, glutaraldehyde, acridines	Nanosilver/titanium dioxide (Ag/TiO_2) and CNTs	Strong antimicrobial activity, low toxicity and cost, high chemical stability ease of use
Membranes	Chlorinated compounds, polyethylene, 1,2-dichlorobenzene, organic and inorganic solutes, halogenated organic solvents	NanoAg/TiO_2/ Zeolites/Magnetite and CNTs	Strong antimicrobial activity, hydrophilicity, low toxicity to humans, high mechanical and chemical stability, high permeability and selectivity, photocatalytic activity

Source: Rizwan, M., Singh, M., Mitra, C., K., Morve, R., K., 2014

Compilation of References

Abbasi, S.A., Abbasi, T., & Anuradha, J. (2012). *A process for synthesis of metal nanoparticles from aquatic weeds*. Off. J. Patent Office. Application No.458/CHE/2012A, p.6184.

Abbasi, S.A., Abbasi, T., Anuradha, J., Neghi, N., Prathiba, S., & Ganaie, S.U. (2011). *Gainful utilization of four otherwise worthless and problematic weeds for silver nanoparticle synthesis*. Off. J. Patent Office. Application No.2175/CHE/2011A, 11869.

Abbasi, T., Anuradha, J., & Abbasi, S. A. (2009). Nanotechnology and its potential in revolutionizing the pollution control scenario. *J. Institution of Public Health Engineers, India*, *10*, 5–12.

Abbasi, T., Anuradha, J., & Abbasi, S. A. (2014). Utilization of the terrestrial weed guduchi (*Tinosporacordifolia*) in clean-green synthesis of gold nanoparticles. *Nanoscience and Technology*, *1*(3), 1–7.

Abbasi, T., Anuradha, J., & Abbasi, S. A. (2016a). Clean and rapid biomimetic synthesis of gold nanoparticles using all parts of the aquatic weed water hyacinth (*Eichhorniacrassipes*). *Journal of the Indian Chemical Society*, *93*, 851–865.

Abbasi, T., Anuradha, J., & Abbasi, S. A. (2016b). Utilization of the pernicious aquatic weed salvinia (*Salviniamolesta* D. S. Mitchell) in generating gold nanoparticles. *Indian Journal of Biotechnology*, *15*, 382–391.

Abbasi, T., Anuradha, J., Ganaie, S. U., & Abbasi, S. A. (2015). Biomimetic synthesis of nanoparticles using aqueous extracts of plants (botanical species). *Journal of Nano Research*, *31*, 138–202. doi:10.4028/www.scientific.net/JNanoR.31.138

Abdel-Aziz, M. H. (2011). Production of copper powder from wastewater containing CuSO 4 and alcoholic additives in a modified stirred tank reactor by cementation. *Hydrometallurgy*, *109*(1), 161–167. doi:10.1016/j.hydromet.2011.06.007

Abdul, M. R., & Schroder, P. (2009). Implications of metal accumulation mechanisms to phytoremediation. *Environ. Sci. Pollut. Resear.*, *16*(2), 162–175. doi:10.1007/s11356-008-0079-z PMID:19067014

Abdulrazak, S., Hussaini, K., & Sani, H. M. (2017). Evaluation of removal efficiency of heavy metals by low-cost activated carbon prepared from African palm fruit. *Applied Water Science*, *7*(6), 3151–3155. doi:10.1007/s13201-016-0460-x

Abedi, M., Salmani, M. H., & Mozaffari, S. A. (2016). Adsorption of Cd ions from aqueous solutions by iron modified pomegranate peel carbons: Kinetic and thermodynamic studies. *International Journal of Environmental Science and Technology*, *13*(8), 2045–2056. doi:10.1007/s13762-016-1002-7

Abou-Shanab, R. A., Ji, M. K., Kim, H. C., Paeng, K. J., & Jeon, B. H. (2013). Microalgal species growing on piggery wastewater as a valuable candidate for nutrient removal and biodiesel production. *Journal of Environmental Management*, *115*, 257–264. doi:10.1016/j.jenvman.2012.11.022 PMID:23270891

Acar, Y. B., & Alshawabkeh, A. N. (1993). Principles of electrokinetic remediation. *Environmental Science & Technology*, *27*(13), 2638–2647. doi:10.1021/es00049a002

Adekola, F. A., Hodonou, D. S. S., & Adegoke, H. I. (2016). Thermodynamic and kinetic studies of biosorption of iron and manganese from aqueous medium using rice husk ash. *Applied Water Science*, *6*(4), 319–330. doi:10.1007/s13201-014-0227-1

Adeleye, A. S., Conway, J. R., Garner, K., Huang, Y., Su, Y., & Keller, A. A. (2016). Engineered nanomaterials for water treatment and remediation: Costs, benefits, and applicability. *Chemical Engineering Journal*, *286*, 640–662. doi:10.1016/j.cej.2015.10.105

Agrawal, B.J., & Patel, B.H. (2000). The reincarnation of Natural Colourants – A Review. *Chemical Weekly*, 139-146.

Agrawal, B. J. (2014). Designing efficient zero liquid discharge system for a textile effluent treatment plant. In *Recent Trends in Water Science and Technology* (pp. 32–38). Anand, India: Charotar Publishing House Pvt. Ltd.

Agrawal, B. J. (2017). *Textile Biological Decolourization and Effluent Remediation Techniques*. In A. K. Rathoure (Ed.), *Bioremediation: Current Research and Applications* (pp. 335–358). IK International Publishing House.

Ahalya, N., Ramachandra, T., & Kanamadi, R. (2003). Biosorption of heavy metals. *Research Journal of Chemistry and Environment*, *7*(4), 71–79.

Ahluwalia, S. S., & Goyal, D. (2007). Microbial and plant derived biomass for removal of heavy metals from wastewater. *Bioresource Technology*, *98*(12), 2243–2257. doi:10.1016/j.biortech.2005.12.006 PMID:16427277

Ahmad, M. S., Yasser, M. M., Sholkamy, E. N., Ali, A. M., & Mehanni, M. M. (2015). Anticancer activity of biostabilized selenium nanorods synthesized by Streptomyces bikiniensis strain Ess_amA-1. *International Journal of Nanomedicine*, *10*, 3389–3401. PMID:26005349

Ahmed, M. (2012). Implications of bacterial resistance against heavy metals in bioremediation: A review. *The IIOAB Journal*, *3*(3), 39–46.

Akbal, F., & Camcı, S. (2011). Copper, chromium and nickel removal from metal plating wastewater by electrocoagulation. *Desalination, 269*(1), 214–222. doi:10.1016/j.desal.2010.11.001

Akcil, A., & Koldas, S. (2006). Acid Mine Drainage (AMD): Causes, treatment and case studies. *Journal of Cleaner Production, 14*(12), 1139–1145. doi:10.1016/j.jclepro.2004.09.006

Aksu. (1992). The biosorption of copper (II) by C. vulgaris and Zramigera. *Environ Technol, 13*, 579-586.

Aksu, Z., Kutsal, T., Gün, S., Haciosmanoglu, N., & Gholaminejad, M. (1991). Investigation of biosorption of Cu (It), Ni (II) and Cr (VI) ions to activated sludge bacteria. *Environmental Technology, 12*(10), 915–921. doi:10.1080/09593339109385086

Alam, M., Allinson, G., Stagnitti, F., Tanaka, A., & Westbrooke, M. (2002). Arsenic contamination in Bangladesh groundwater: A major environmental and social disaster. *International Journal of Environmental Health Research, 12*(3), 235–253. doi:10.1080/0960312021000000998 PMID:12396524

Al-Degs, Y. S., El-Barghouthi, M. I., Issa, A. A., Khraisheh, M. A., & Walker, G. M. (2006). Sorption of Zn (II), Pb (II), and Co (II) using natural sorbents: Equilibrium and kinetic studies. *Water Research, 40*(14), 2645–2658. doi:10.1016/j.watres.2006.05.018 PMID:16839582

Alemayehu, T., & Teklemariam, Z. (2014). Application of natural dyes on textile: a review. *International Journal of Research – Granthaalayah, 2*(2), 61-68.

Ali, A., & Saeed, K. (2015). Decontamination of Cr (VI) and Mn (II) from aqueous media by untreated and chemically treated banana peel: A comparative study. *Desalination and Water Treatment, 53*(13), 3586–3591. doi:10.1080/19443994.2013.876669

Ali, A., Saeed, K., & Mabood, F. (2016). Removal of chromium (VI) from aqueous medium using chemically modified banana peels as efficient low-cost adsorbent. *Alexandria Engineering Journal, 55*(3), 2933–2942. doi:10.1016/j.aej.2016.05.011

Ali, H., Khan, E., & Sajad, M. A. (2013). Phytoremediation of Heavy Metals – Concepts and Applications. *Chemosphere, 91*(7), 869–881. doi:10.1016/j.chemosphere.2013.01.075 PMID:23466085

Ali, I. H., & Alrafai, H. A. (2016). Kinetic, isotherm and thermodynamic studies on biosorption of chromium (VI) by using activated carbon from leaves of Ficus nitida. *Chemistry Central Journal, 10*(1), 36. doi:10.1186/s13065-016-0180-1 PMID:27252777

Ali, I., Asim, M., & Khan, T. A. (2012). Low cost adsorbents for the removal of organic pollutants from wastewater. *Journal of Environmental Management, 113*, 170–183. doi:10.1016/j.jenvman.2012.08.028 PMID:23023039

Ali, N., Hameed, A., & Ahmed, S. (2008). Physicochemical characterization and bioremediation perspective of textile effluent, dyes and metals by indigenous bacteria. *Journal of Hazardous Materials, 164*(1), 322–328. doi:10.1016/j.jhazmat.2008.08.006 PMID:18818017

Al-Khaldi, F. A., Abusharkh, B., Khaled, M., Atieh, M. A., Nasser, M. S., Saleh, T. A., ... Gupta, V. K. (2015). Adsorptive removal of cadmium (II) ions from liquid phase using acid modified carbon-based adsorbents. *Journal of Molecular Liquids*, *204*, 255–263. doi:10.1016/j.molliq.2015.01.033

Allen, R. L. M. (1971). *Colour Chemistry*. London: Thomas Nelson & Sons. doi:10.1007/978-1-4615-6663-2

ALOthman, Z. A., Habila, M. A., Al-Shalan, N. H., Alfadul, S. M., Ali, R., Rashed, I. G. A., & Alfarhan, B. (2016). Adsorptive removal of Cu (II) and Pb (II) onto mixed-waste activated carbon: Kinetic, thermodynamic, and competitive studies and application to real wastewater samples. *Arabian Journal of Geosciences*, *9*(4), 315. doi:10.1007/s12517-016-2350-9

Al-Qahtani, K. M. (2016). Water purification using different waste fruit cortexes for the removal of heavy metals. *Journal of Taibah University for Science*, *10*(5), 700–708. doi:10.1016/j.jtusci.2015.09.001

Al-Qunaibit, M. H. (2004). A kinetic study of uptake of some cationic centities by the alga *Chlorella vulgaris*. *Chemindix*, *6*, 1–12.

Amin, F., Talpur, F. N., Balouch, A., Chandio, Z. A., Surhio, M. A., & Afridi, H. I. (2016). Biosorption of mercury (II) from aqueous solution by fungal biomass Pleurotus eryngii: Isotherm, kinetic, and thermodynamic studies. *Environmental Progress & Sustainable Energy*.

Anaemene, I. (2012). The use of Candida sp. in the biosorption of heavy metals from industrial effluent. *European Journal of Experimental Biology*, *2*(3), 484–488.

Andrews, M. (2010). Transit on iron radar. *Australia's Paydirt*, *1*(178), 60.

Angleviel, F. (2003). Restoring the Economic Balance: The Nickel Stakes in New Caledonia. *The New Pacific Review*, *155*.

Anjum, M., Al-Makishah, N. H., & Barakat, M. A. (2016). Wastewater sludge stabilization using pre-treatment methods. *Process Safety and Environmental Protection*, *102*, 615–632. doi:10.1016/j.psep.2016.05.022

Anjum, M., Oves, M., Kumar, R., & Barakat, M. A. (2017). Fabrication of ZnO-ZnS@ polyaniline nanohybrid for enhanced photocatalytic degradation of 2-chlorophenol and microbial contaminants in wastewater. *International Biodeterioration & Biodegradation*, *119*, 66–77. doi:10.1016/j.ibiod.2016.10.018

Anuradha, J. (2013). *Use of some aquatic and terrestrial weeds in the 'green' synthesis of gold nanoparticles* (PhD thesis). Pondicherry (Central) University.

Anuradha, J., Abbasi, T., & Abbasi, S. A. (2010). 'Green' synthesis of gold nanoparticles with aqueous extracts of neem (*Azadirachtaindica*). *Research Journal of Biotechnology*, *5*, 75–79.

Anuradha, J., Abbasi, T., & Abbasi, S. A. (2011). Rapid and Reproducible 'Green' synthesis of Silver Nanoparticles of consistent shape and size using *Azadirachtaindica*. *Research Journal of Biotechnology*, *6*, 69–70.

Anuradha, J., Abbasi, T., & Abbasi, S. A. (2015a). An eco-friendly method of synthesizing gold nanoparticles using an otherwise worthless weed pistia (*Pistiastratiotes* L.). *Journal of Advanced Research*, *6*(5), 711–720. doi:10.1016/j.jare.2014.03.006 PMID:27563461

Anuradha, J., Abbasi, T., & Abbasi, S. A. (2015b). Rapid and Green synthesis of gold nanoparticles by the use of an otherwise worthless weed lantana (*Lantana camara* L.). *Journal of Environmental Science & Engineering*, *57*(3), 203–213.

Apostoli, P., & Catalani, S. (2011). Metal ions affecting reproduction and development. *Metal Ions in Life Sciences*, *8*, 263–303. PMID:21473384

Arévalo, A., Skewes, A., Stern, C., Floody, R., Goldfarb, R., Nielsen, R., Zúñiga, R. (2002). The giant El Teniente, breccia deposit: hypogene copper distribution and emplacement. *Integrated methods for discovery: Global exploration in the Twenty-first Century*, *9*, 299-332.

Arias, M., Barral, M. T., & Mejuto, J. C. (2002). Enhancement of copper and cadmium adsorption on kaolin by the presence of humic acids. *Chemosphere*, *48*(10), 1081–1088. doi:10.1016/S0045-6535(02)00169-8 PMID:12227513

Arnon, D. I. (1949). Copper enzymes in isolated chloroplast. Polyphenol/oxidase in *Beta vulgaris*. *Plant Physiology*, *24*(1), 1–15. doi:10.1104/pp.24.1.1 PMID:16654194

Aroca, P. (2001). Impacts and development in local economies based on mining: The case of the Chilean II region. *Resources Policy*, *27*(2), 119–134. doi:10.1016/S0301-4207(01)00013-7

Arshadi, M., Amiri, M. J., & Mousavi, S. (2014). Kinetic, equilibrium and thermodynamic investigations of Ni (II), Cd (II), Cu (II) and Co (II) adsorption on barley straw ash. *Water Resources and Industry*, *6*, 1–17. doi:10.1016/j.wri.2014.06.001

Arunachalam, K. D., Annamalai, S. K., Arunachalam, A. M., & Kennedy, S. (2013). Green Synthesis of crystalline silver nanoparticles using Indigoferaaspalathoides-medicinal plant extract for wound healing applications. *Asian Journal of Chemistry*, *25*, S311–S314.

Arunakumara, K. K. I. U., & Xuecheng, Z. (2008). Heavy metal bioaccumulation and toxicity with special reference to microalgae. *Journal of Ocean University of China*, *7*(1), 25–30. doi:10.1007/s11802-008-0060-y

Asberry, H. B., Kuo, C. Y., Gung, C. H., Conte, E. D., & Suen, S. Y. (2014). Characterization of water bamboo husk biosorbents and their application in heavy metal ion trapping. *Microchemical Journal*, *113*, 59–63. doi:10.1016/j.microc.2013.11.011

Association, A. (1984). *Aluminum: Properties and physical metallurgy*. ASM International.

Asubiojo, O. I., & Ajelabi, O. B. (2009). Removal of heavy metals from industrial wastewaters using natural adsorbents. *Toxicological and Environmental Chemistry*, *91*(5), 883–890. doi:10.1080/02772240802614721

Atieh, M. A. (2011). Removal of chromium (VI) from polluted water using carbon nanotubes supported with activated carbon. *Procedia Environmental Sciences*, *4*, 281–293. doi:10.1016/j.proenv.2011.03.033

Auffan, M., & Bottero, J. (2009). Chemical stability of metallic nanoparticles: A parameter controlling their potential cellular toxicity in vitro. *Environ. Pollut.*, *157*(4), 1127–1133. doi:10.1016/j.envpol.2008.10.002 PMID:19013699

Awofolu, O. R., Mbolekwa, Z., Mtshemla, V., & Fatoki, O. S. (2005). Levels of trace metals in water and sediment from Tyume River and its effects on an irrigated farmland. *Water S.A.*, *31*(1), 87. doi:10.4314/wsa.v31i1.5124

Ayangbenro, A. S., & Babalola, O. O. (2017). A New Strategy for Heavy Metal Polluted Environments: A Review of Microbial Biosorbents. *International Journal of Environmental Research and Public Health*, *14*(1), 94–110. doi:10.3390/ijerph14010094 PMID:28106848

Aytar, P., Kay, C. M., Mutlu, M. B. I., & Çabuk, A. (2013). Coal Desulfurization with Acidithiobacillusferrivorans, from Balya Acidic Mine Drainage. *Energy & Fuels*, *27*(6), 3090–3098. doi:10.1021/ef400360t

Azamat, J., Sattary, B. S., Khataee, A., & Joo, S. W. (2015). Removal of a hazardous heavy metal from aqueous solution using functionalized graphene and boron nitride nanosheets: Insights from simulations. *Journal of Molecular Graphics & Modelling*, *61*, 13–20. doi:10.1016/j.jmgm.2015.06.012 PMID:26186492

Aziz, H. A., Adlan, M. N., & Ariffin, K. S. (2008). Heavy metals (Cd, Pb, Zn, Ni, Cu and Cr (III)) removal from water in Malaysia: Post treatment by high quality limestone. *Bioresource Technology*, *99*(6), 1578–1583. doi:10.1016/j.biortech.2007.04.007 PMID:17540556

Badar, U., Ahmed, N., Beswick, A. J., Pattanapipitpaisal, P., & Macaskie, L. E. (2000). Reduction of Chromate by microorganisms isolated from metal contaminated sites of Karachi, Pakistan. *Biotechnology Letters*, *22*(10), 829–836. doi:10.1023/A:1005649113190

Badmus, M. O. A., Audu, T. O. K., & Anyata, B. (2007). Removal of copper from industrial wastewaters by activated carbon prepared from periwinkle shells. *Korean Journal of Chemical Engineering*, *24*(2), 246–252. doi:10.1007/s11814-007-5049-5

Bahobil, A., Bayoumi, R., Atta, H., & El-Sehrawey, M. (2017). Fungal Biosorption for Cadmium and Mercury Heavy Metal Ions Isolated from Some Polluted Localities in KSA. *International Journal of Current Microbiology and Applied Sciences*, *6*(6), 2138–2154. doi:10.20546/ijcmas.2017.606.253

Bai, X. J., Liu, L. J., Zhang, C. H., Ge, Y., & Cheng, W. (2011). Effect of H_2O_2 pretreament on Cd tolerance of different rice cultivars. *Rice Science*, *18*(1), 29–35. doi:10.1016/S1672-6308(11)60005-X

Baker, S., & Satish, S. (2012). Endophytes: Toward a Vision in Synthesis of Nanoparticles for Future Therapeutic Agents. Int. J. Bio-Inorg. Hybd. *Nanomat.*, *1*, 1–11.

Baligar, V. C., Shaffert, R. E., Dos Santos, H. L., Pitta, G. V. E., Bahia Filho, A. F., & De, C. (1993). Soil aluminium effects on uptake influx and transport of nutrients in sorghum genotypes. *Plant and Soil*, *150*(2), 271–277. doi:10.1007/BF00013024

Ballard, J. R., Palin, J. M., Williams, I. S., Campbell, I. H., & Faunes, A. (2001). Two ages of porphyry intrusion resolved for the super-giant Chuquicamata copper deposit of northern Chile by ELA-ICP-MS and SHRIMP. *Geology*, *29*(5), 383–386. doi:10.1130/0091-7613(2001)029<0383:TAOPIR>2.0.CO;2

Ball, E. (1946). Development in sterile culture of stem tips and adjacent regions of *Tropaeolum majus* L. and of *Lupinus albus* L. *American Journal of Botany*, *33*(5), 301–318. doi:10.2307/2437117

Bamforth, S. M., & Singleton, I. (2005). Bioremediation of polycyclic aromatic hydrocarbons: Current knowledge and future directions. *Journal of Chemical Technology and Biotechnology (Oxford, Oxfordshire)*, *80*(7), 723–736. doi:10.1002/jctb.1276

Banerjee, S. S., & Chen, D. H. (2007). Fast removal of copper ions by gum arabic modified magnetic nano-adsorbent. *Journal of Hazardous Materials*, *147*(3), 792–799. doi:10.1016/j.jhazmat.2007.01.079 PMID:17321674

Bani, A., Pavlova, D., Echevarria, G., Mullaj, A., Reeves, R. D., Morel, J. L., & Sulce, S. (2010). Nickel hyperaccumulation by the species of *Alyssum and Thlaspi* (*Brassicaceae*) from the ultramafic soils of the Balkans. *Botanica Serbica*, *34*(1), 3–14.

Barakat, M. A. (2008). Adsorption of heavy metals from aqueous solutions on synthetic zeolite. *Research Journal of Environmental Sciences*, *2*(1), 13–22. doi:10.3923/rjes.2008.13.22

Barakat, M. A. (2011). New trends in removing heavy metals from industrial wastewater. *Arabian Journal of Chemistry*, *4*(4), 361–377. doi:10.1016/j.arabjc.2010.07.019

Barakat, M. A., Al-Hutailah, R. I., Hashim, M. H., Qayyum, E., & Kuhn, J. N. (2013). Titania-supported silver-based bimetallic nanoparticles as photocatalysts. *Environmental Science and Pollution Research International*, *20*(6), 3751–3759. doi:10.1007/s11356-012-1306-1 PMID:23161501

Barregård, L., Svalander, C., Schütz, A., Westberg, G., Sällsten, G., Blohmé, I., ... Haglind, P. (1999). Cadmium, mercury, and lead in kidney cortex of the general Swedish population: A study of biopsies from living kidney donors. *Environmental Health Perspectives*, *107*(11), 867–871. doi:10.1289/ehp.99107867 PMID:10544153

Barsoum, R. S. (2006). Chronic kidney disease in the developing world. *The New England Journal of Medicine*, *354*(10), 997–999. doi:10.1056/NEJMp058318 PMID:16525136

Barton, K. A., Whiteley, H. R., & Yang, N.-S. (1987). Bacillus thurigiensis δ-endotoxin expressed in transgenic Nicotiana tabacum provides resistance to Lepidoptera insects. *Plant Physiology*, *85*(4), 1103–1109. doi:10.1104/pp.85.4.1103 PMID:16665812

Basaldella, E., Vázquez, P. G., Iucolano, F., & Caputo, D. (2007). Caputo Chromium removal from water using LTA zeolites: Effect of pH. *Journal of Colloid and Interface Science*, *313*(2), 574–578. doi:10.1016/j.jcis.2007.04.066 PMID:17553515

Basaran, B., Iscan, M., Bitlisli, B. O., & Aslan, A. (2006). Heavy metal contents of various finished leathers. *Journal of the Society of Leather Technologists and Chemists*, *90*(6), 229–234.

Béchamp, A. (1854). De l'action des protosels de fer sur la nitronaphtaline et la nitrobenzine. nouvelle méthode de formation des bases organiques artificielles de Zinin. *Annales de chimie et de physique*, *42*, 186–196.

Beckers, F., & Rinklebe, J. (2017). Cycling of Mercury in the Environment: Sources, Fate, and Human Health Implications-A Review. *Critical Reviews in Environmental Science and Technology*. (accepted)

Belay, A. A. (2010). Impacts of chromium from tannery effluent and evaluation of alternative treatment options. *Journal of Environmental Protection*, *1*(01), 53–58. doi:10.4236/jep.2010.11007

Bendtsson, B., Asp, H., & Jensn, P. (1994). Uptake and distribution of calcium and phosphorus in beech (*Fagus sylvatica*) as influences by aluminum and nitrogen. *Tree Physiology*, *14*(1), 63–73. doi:10.1093/treephys/14.1.63 PMID:14967634

Benoit, P., Barriuso, E., & Calvet, R. (1998). Biosorption characterization of herbicides, 2, 4-D and atrazine, and two chlorophenols on fungal mycelium. *Chemosphere*, *37*(7), 1271–1282. doi:10.1016/S0045-6535(98)00125-8

Bergmann, L. (1960). Growth and division of single cells of higher plants in vitro. *The Journal of General Physiology*, *43*(4), 841–851. doi:10.1085/jgp.43.4.841 PMID:13799400

Bernhoft, R. A. (2011). Mercury toxicity and treatment: A review of the literature. *Journal of Environmental and Public Health*. PMID:22235210

Beveridge, T. J., & Doyle, R. J. (1989). *Metal ions and bacteria*. Wiley.

Bhargava, A., Carmona, F. F., Bhargava, M., & Srivastava, S. (2012). Approaches for enhanced phytoextraction of heavy metals. *Journal of Environmental Management*, *105*, 103–120. doi:10.1016/j.jenvman.2012.04.002 PMID:22542973

Bhatnagar, A., Hogland, W., Marques, M., & Sillanpaa, M. (2013). An overview of the modification methods of activated carbon for its water treatment applications. *Chemical Engineering Journal*, *219*, 499–511. doi:10.1016/j.cej.2012.12.038

Bhattacharya, S., Saha, I., Mukhopadhyay, A., Chattopadhyay, D., & Chand, U. (2013). Role of nanotechnology in water treatment and purification: Potential applications and implications. *Int J Chem Sci Technol*, *3*(3), 59–64.

Bhatti, K. P., & Noor, A. R. (2009). Isolation and characterization of Chromium, Mercury and Cadmium Resistant bacteria. *Journal of the Chemical Society of Pakistan, 31*(6), 928–936.

Bihari, Z. (2013). Current Trends in Bioremediation and Biodegradation: Next-Generation Sequencing, Bioremediation & Biodegradation. *Journal of Bioremediation & Biodegradation*, 4–8.

Biškup, B., & Subotić, B. (2005). Removal of heavy metal ions from solutions using zeolites. III. Influence of sodium ion concentration in the liquid phase on the kinetics of exchange processes between cadmium ions from solution and sodium ions from zeolite A. *Separation Science and Technology, 39*(4), 925–940. doi:10.1081/SS-120028454

Blaby-Haas, C. E., & Merchant, S. S. (2012). The ins and outs of algal metal transport Biochimica et Biophysica Acta (BBA) -. *Molecular Cell Research., 1823*, 1531–1552.

Bohli, T., & Ouederni, A. (2016). Improvement of oxygen-containing functional groups on olive stones activated carbon by ozone and nitric acid for heavy metals removal from aqueous phase. *Environmental Science and Pollution Research International, 23*(16), 15852–15861. doi:10.1007/s11356-015-4330-0 PMID:25794582

Bohli, T., Ouederni, A., Fiol, N., & Villaescusa, I. (2015). Evaluation of an activated carbon from olive stones used as an adsorbent for heavy metal removal from aqueous phases. *Comptes Rendus. Chimie, 18*(1), 88–99. doi:10.1016/j.crci.2014.05.009

Bokare V., Murugesan K., Kim J.-H., Kim E.-J., & Chang, Y.-S. (2012). Integrated hybrid treatment for the remediation of 2,3,7,8-tetrachlorodibenzo-p-dioxin. *Sci. Total Environ., 435-436*, 563-566.

Bokare, V., Murugesan, K., Kim, J.-H., Kim, E.-J., & Chang, Y.-S. (2012). Integrated hybrid treatment for the remediation of 2,3,7,8-tetrachlorodibenzo-p-dioxin. *Sci. Total Environ., 435-436*, 563-566.

Bokare, V., Murugesan, K., Kim, Y.-M., Jeon, J.-R., Kim, E.-J., & Chang, Y. S. (2010). Degradation of triclosan by an integrated nano-bio redox process. *Bioresource Technology, 101*(16), 6354–6360. doi:10.1016/j.biortech.2010.03.062 PMID:20381343

Bond, A. R., & Levine, R. M. (2001). Noril'sk Nickel and Russian platinum-group metals production. *Post-Soviet Geography and Economics, 42*(2), 77–104.

Bosecker, K. (1997). Bioleaching: Metal solubilization by microorganisms. *FEMS Microbiology Reviews, 20*(3-4), 591–604. doi:10.1111/j.1574-6976.1997.tb00340.x

Bose, J., Babourina, O., & Rengel, Z. (2011). Role of magnesium in alleviation of aluminium toxicity in plants. *Journal of Experimental Botany, 62*(7), 2251–2264. doi:10.1093/jxb/erq456 PMID:21273333

Bose, J., Babourina, O., Shabala, S., & Rengel, Z. (2013). Low-pH and aluminium resistance in *Arabidopsis* correlates with high cytosolic magnesium content and increase magnesium uptake by plant roots. *Plant & Cell Physiology, 54*(7), 1093–1104. doi:10.1093/pcp/pct064 PMID:23620479

Bosnic, M., Buljan, J., & Daniels, R. (2000). Pollutants in tannery effluents. *UNIDO, Vienna, Rev.*

Bouché, C., Brandt, C., Broussaud, A., & Drunick, W. (2005). Advanced control of gold ore grinding plants in South Africa. *Minerals Engineering*, *18*(8), 866–876. doi:10.1016/j.mineng.2005.02.003

Boudrahem, F., Aissani-Benissad, F., & Soualah, A. (2013). Kinetic and equilibrium study of the sorption of lead (II) ions from aqueous phase by activated carbon. *Arabian Journal for Science and Engineering*, *38*(8), 1939–1949. doi:10.1007/s13369-013-0614-1

Bouhamed, F., Elouear, Z., Bouzid, J., & Ouddane, B. (2016). Multi-component adsorption of copper, nickel and zinc from aqueous solutions onto activated carbon prepared from date stones. *Environmental Science and Pollution Research International*, *23*(16), 15801–15806. doi:10.1007/s11356-015-4400-3 PMID:25843824

Brandl, H., & Faramarzi, M. A. (2006). Microbe metal-interactions for the biotechnological treatment of metal-containing solid waste. *Chin Particuol*, *4*(2), 93–97. doi:10.1016/S1672-2515(07)60244-9

Brown, P., Jefcoat, I. A., Parrish, D., Gill, S., & Graham, E. (2000). Evaluation of the adsorptive capacity of peanut hull pellets for heavy metals in solution. *Advances in Environmental Research*, *4*(1), 19–29. doi:10.1016/S1093-0191(00)00004-6

Buat-Menerd, P., & Chesselt, R. (1979). Variable influence of the atmospheric flux on the trace metal chemistry of oceanic suspended matter. *Earth and Planetary Science Letters*, *42*, 398–411.

Bystrzejewski, M., Pyrzyńska, K., Huczko, A., & Lange, H. (2009). Carbon-encapsulated magnetic nanoparticles as separable and mobile sorbents of heavy metal ions from aqueous solutions. *Carbon*, *47*(4), 1201–1204. doi:10.1016/j.carbon.2009.01.007

Cabral, J. P. S. (1992). Selective binding of metal ions to Pseudomonas syringae cells. *Microbios*, *71*, 47–53.

Cain, J. C., & Thorpe, J. F. (1946). *The Synthetic Dyestuffs and Intermediate Products from which they are derived*. London: Charles Griffin & Co.

Cai, X., Li, J., Zhang, Z., Yang, F., Dong, R., & Chen, L. (2013). Novel Pb^{2+} ion imprinted polymers based on ionic interaction via synergy of dual functional monomers for selective solid-phase extraction of Pb^{2+} in water samples. *ACS Applied Materials & Interfaces*, *6*(1), 305–313. doi:10.1021/am4042405 PMID:24344795

Calheiros, C. S. C., Rangel, A. O. S. S., & Castro, P. M. L. (2008). The effects of tannery waste water on the development of different plant species and chromium accumulation in *Phragmites australis*. *Archives of Environmental Contamination and Toxicology*, *55*(3), 404–414. doi:10.1007/s00244-007-9087-0 PMID:18214580

Cao Rocky, X., Lena, M. A., Chen Ming, Q., Singh Satya, P., & Harris Willie, G. (2003). Phosphate Induced metal immobilization in a contaminated site. *Environ. Pollut.*, *122*(1), 19–28. doi:10.1016/S0269-7491(02)00283-X PMID:12535592

Cao, C. Y., Cui, Z. M., Chen, C. Q., Song, W. G., & Cai, W. (2010). Ceria hollow nanospheres produced by a template-free microwave-assisted hydrothermal method for heavy metal ion removal and catalysis. *The Journal of Physical Chemistry C*, *114*(21), 9865–9870. doi:10.1021/jp101553x

Cao, Z., Li, S., Zhang, J., & Zhang, H. (2017). An electro-microbial membrane system with anti-fouling function for phenol wastewater treatment. *Journal of Chemical Technology and Biotechnology (Oxford, Oxfordshire)*, *92*(3), 693–699. doi:10.1002/jctb.5056

Carata, E., Panzarini, E., & Dini, L. (2017). Environmental Nanoremediation and Electron Microscopies. In Nanotechnologies for Environmental Remediation. Springer International Publishing AG.

Carlson, P. S. (1970). Induction and isolation of auxotrophic mutants in somatic cell culture of Nocatiana tabacum. *Science*, *168*(3930), 487–489. doi:10.1126/science.168.3930.487 PMID:17838129

Carral, E., Puente, X., Villares, R., & Carballeira, A. (1995). Background heavy metal levels in estuarine sediments and organisms in Galicia (northwest Spain) as determined by modal analysis. *The Science of the Total Environment*, *172*(2-3), 175–188. doi:10.1016/0048-9697(95)04788-3

Castro, I. M., Fietto, J. L. R., Vieira, R. X., Trópia, M. J. M., Campos, L. M. M., Paniago, E. B., & Brandão, R. L. (2000). Bioleaching of zinc and nickel from silicates using Aspergillus niger cultures. *Hydrometallurgy*, *57*(1), 39–49. doi:10.1016/S0304-386X(00)00088-8

Cecchin, I., Reddy, K. R., Tessaro, E. F., & Schnaid, F. (2016). Nanobioremediation: Integration of nanoparticles and bioremediation for sustainable remediation of chlorinated organic contaminants in soils. *International Biodeterioration & Biodegradation*, 1–10.

Cecchin, I., Reddy, K. R., Thome, A., Tessaro, E. F., & Schnaid, F. (2017). Nanobioremediation: Integration of nanoparticles and bioremediation for sustainable remediation of chlorinated organic contaminants in soils. *International Biodeterioration & Biodegradation*, *119*, 419–428. doi:10.1016/j.ibiod.2016.09.027

Chaidogiannos, G., Petraki, F., Glezos, N., Kennou, S., & Nešpůrek, S. (2009). Low voltage operating OFETs based on solution-processed metal phthalocyanines. *Applied Physics. A, Materials Science & Processing*, *96*(3), 763–767. doi:10.1007/s00339-009-5268-1

Chander, K., Brookes, P. C., & Harding, S. A. (1995). Microbial biomass dynamics following addition of metal-enriched sewage sludges to a sandy loam. *Soil Biology & Biochemistry*, *27*(11), 1409–1421. doi:10.1016/0038-0717(95)00074-O

Chandrakar, R. K., Baghel, R. N., Chandra, V. K., & Chandra, B. P. (2015). Synthesis, characterization and photoluminescence studies of Mn doped ZnS nanoparticles. *Superlattices and Microstructures*, *86*, 256–269. doi:10.1016/j.spmi.2015.07.043

Chatterjee, A., Das, D., Mandal, B. K., Chowdhury, T. R., Samanta, G., & Chakraborti, D. (1995). Arsenic in ground water in six districts of West Bengal, India: The biggest arsenic calamity in the world. I: Arsenic species in drinking water and urine of the affected people. *Analyst (London), 120*(3), 643–650. doi:10.1039/an9952000643

Chaudhary, K., Sumira, J., & Khan, S. (2016). Heavy metal ATPase (*HMA2, HMA3 & HMA4*) genes in hyperaccumulation mechanism of heavy metals. *Plant metal interaction (Emerging Remediation Techniques),* 545-552.

Chavan, R. B. (2001). Indian Textile Industry – Environmental Issues. *Indian Journal of Fibre and Textile Research, 26*, 11–21.

Chen, G.Z., Fray, D.J., & Farthing, T.W. (2000). Direct electrochemical reduction of titanium dioxide to titanium in molten calcium chloride. *Nature, 407*(6802), 361.

Chen, B. C., Lai, H. Y., & Juang, K. W. (2012). Model evaluation of plant metal content and biomass yield for the phytoextraction of heavy metals by switchgrass. *Ecotoxicology and Environmental Safety, 80*, 393–400. doi:10.1016/j.ecoenv.2012.04.011 PMID:22541831

Chen, C. Z. S., & Cooper, S. (2002). Interactions between dendrimerbiocides and bacterial membranes. *Biomaterials, 23*(16), 3359–3368. doi:10.1016/S0142-9612(02)00036-4 PMID:12099278

Chen, C., & Wang, X. (2006). Adsorption of Ni (II) from aqueous solution using oxidized multiwall carbon nanotubes. *Industrial & Engineering Chemistry Research, 45*(26), 9144–9149. doi:10.1021/ie060791z

Chen, G. H. (2004). Electrochemicals technologies in wastewater treatment. *Separation and Purification Technology, 38*(1), 11–41. doi:10.1016/j.seppur.2003.10.006

Cheng, K., Hu, J., Hou, H., Liu, B., Chen, Q., Pan, K., & Yang, C. (2017). Aerobic granular sludge inoculated microbial fuel cells for enhanced epoxy reactive diluent wastewater treatment. *Bioresource Technology, 229*, 126–133. doi:10.1016/j.biortech.2016.12.115 PMID:28110229

Chen, J., Yang, P., Song, D., Yang, S., Zhou, L., Han, L., & Lai, B. (2014). Biosorption of Cr (VI) by carbonized Eupatorium adenophorum and Buckwheat straw: Thermodynamics and mechanism. *Frontiers of Environmental Science & Engineering, 8*(6), 960–966. doi:10.1007/s11783-013-0612-2

Chen, K. C., Wu, J. Y., Liou, D. J., & Hwang, S. C. J. (2003). Decolorization of the textile dyes by newly isolated bacterial strains. *Journal of Biotechnology, 101*(1), 57–68. doi:10.1016/S0168-1656(02)00303-6 PMID:12523970

Chen, Z., Megharaj, M., & Naidu, R. (2007). Speciation of chromium in waste water using ion chromatography inductively coupled plasma mass spectrometry. *Talanta, 72*(2), 394–400. doi:10.1016/j.talanta.2006.10.041 PMID:19071630

Cho, H. H., Wepasnick, K., Smith, B. A., Bangash, F. K., Fairbrother, D. H., & Ball, W. P. (2009). Sorption of aqueous Zn [II] and Cd [II] by multiwall carbon nanotubes: The relative roles of oxygen-containing functional groups and graphenic carbon. *Langmuir*, *26*(2), 967–981. doi:10.1021/la902440u PMID:19894751

Cho, U. H., & Seo, N. H. (2005). Oxidative stress in *Arabidopsis thaliana* exposed to cadmium is due to hydrogen peroxide accumulation. *Plant Science*, *168*(1), 113–120. doi:10.1016/j.plantsci.2004.07.021

Chowdhury, U. K., Biswas, B. K., Chowdhury, T. R., Samanta, G., Mandal, B. K., Basu, G. C., ... Mukherjee, S. K. (2000). Groundwater arsenic contamination in Bangladesh and West Bengal, India. *Environmental Health Perspectives*, *108*(5), 393–397. doi:10.1289/ehp.00108393 PMID:10811564

Chowdhury, Z. Z., Zain, S. M., Khan, R. A., & Islam, M. S. (2012). Preparation and characterizations of activated carbon from kenaf fiber for equilibrium adsorption studies of copper from wastewater. *Korean Journal of Chemical Engineering*, *29*(9), 1187–1195. doi:10.1007/s11814-011-0297-9

Chu, W. (1999). Lead metal removal by recycled alum sludge. *Water Research*, *33*(13), 3019–3025. doi:10.1016/S0043-1354(99)00010-X

Cohen, M. D., Draxler, R. R., Artz, R. S., Blanchard, P., Gustin, M. S., Han, Y.-J., ... Lei, H. (2016). Modeling the global atmospheric transport and deposition of mercury to the Great Lakes. *Elementa*, 4.

Colberg, P. J. S., & Young, L. Y. (1995). Anaerobic degradation of nonhalogenated homocyclic aromatic compounds coupled with Nitrate, Iron, or Sulfate reduction. In *Microbial Transformation and Degradation of Toxic Organic Chemicals* (pp. 307–330). New York: Wiley-Liss.

Cordero, H. C., Trevino, H., Juárez-Aguilar, H. L. G., Castillo, D. I. M., Mendoza, D. I., Montoya, V. H., ... Morán, M. A. M. (2013). Synthesis and adsorption properties of activated carbons from biomass of Prunusdomestica and Jacaranda mimosifolia for the removal of heavy metals and dyes from water. *Industrial Crops and Products*, *42*, 315–323. doi:10.1016/j.indcrop.2012.05.029

Cornu, J. Y., Huguenot, D., Jezequel, K., Lollier, M., & Lebeau, T. (2017). Bioremediation of copper contaminated soils by bacteria. *World Journal of Microbiology & Biotechnology*, *33*(2), 26. doi:10.1007/s11274-016-2191-4 PMID:28044274

Correia, V. M., Stephenson, T., & Judd, S. J. (1994). Characterization of textile wastewaters: A review. *Environmental Technology*, *15*(10), 917–919. doi:10.1080/09593339409385500

Craig, J. R., & Rimstidt, J. D. (1998). Gold production history of the United States. *Ore Geology Reviews*, *13*(6), 407–464. doi:10.1016/S0169-1368(98)00009-2

Cumming, J. R., Eckert, R. T., & Evans, L. S. (1986). Effect of aluminium on ^{32}P uptake and translocation by red spruce seedlings. *Canadian Journal of Forest Research*, *16*(4), 864–867. doi:10.1139/x86-152

Cutillas-Barreiro, L., Ansias-Manso, L., Fernández-Calviño, D., Arias-Estévez, M., Nóvoa-Muñoz, J. C., Fernández-Sanjurjo, M. J., & Núñez-Delgado, A. (2014). Pine bark as bio-adsorbent for Cd, Cu, Ni, Pb and Zn: Batch-type and stirred flow chamber experiments. *Journal of Environmental Management, 144*, 258–264. doi:10.1016/j.jenvman.2014.06.008 PMID:24973614

Da'vila-Jime'nez, M. M., Marı'a, P., Elizalde-Gonza'lez, W., Geyer, J., & Mattusch, R. (2003). Adsorption of metal cations from aqueous solution onto a natural and a model biocomposite. *Physicochem. Eng. Aspects, 219*(1-3), 243–252. doi:10.1016/S0927-7757(03)00052-9

Daims, H., Lebedeva, E. V., Pjevac, P., Han, P., Herbold, C., Albertsen, M., ... Wagner, M. (2015). Complete nitrification by Nitrospira bacteria. *Nature, 528*, 504–509. PMID:26610024

Das, S. K., Das, A. R., & Guha, A. K. (2007). A study on the adsorption mechanism of mercury on Aspergillus versicolor biomass. *Environmental Science & Technology, 41*(24), 8281–8287. doi:10.1021/es070814g PMID:18200852

Dave, P. N., & Chopda, L. V. (2014). Application of Iron Oxide Nanomaterials for the Removal of Heavy Metals, review article. *Journal of Nanotechnology*, 398–412.

Dave, P. N., & Chopda, L. V. (2014). Application of iron oxide nanomaterials for the removal of heavy metals. *Journal of Nanotechnology*.

Davis, A., Kempton, J. H., Nicholson, A., & Yare, B. (1994). Groundwater transport of arsenic and chromium at a historical tannery, Woburn, Massachusetts, USA. *Applied Geochemistry, 9*(5), 569–582. doi:10.1016/0883-2927(94)90019-1

de Vries, J. G., & Elsevier, C. J. (2007). *The Handbook of Homogeneous Hydrogenation* (Vols. 1–3). Weinheim: Wiley-VCH.

Dehghani, M. H., Sanaei, D., Ali, I., & Bhatnagar, A. (2016). Removal of chromium (VI) from aqueous solution using treated waste newspaper as a low-cost adsorbent: Kinetic modeling and isotherm studies. *Journal of Molecular Liquids, 215*, 671–679. doi:10.1016/j.molliq.2015.12.057

Dexbury, A. C., & Yentch, C. S. (1956). Plankton pigment monograph. *Journal of Marine Research, 15*, 93–101.

Dhankhar, R., & Hooda, A. (2011). Fungal biosorption–an alternative to meet the challenges of heavy metal pollution in aqueous solutions. *Environmental Technology, 32*(5), 467–491. doi:10.1080/09593330.2011.572922 PMID:21877528

Dhir, B. (2014). Potential of biological materials for removing heavy metals from wastewater. *Environmental Science and Pollution Research International, 21*(3), 1614–1627. doi:10.1007/s11356-013-2230-8 PMID:24185905

Ding, Z., Hu, X., Morales, V. L., & Gao, B. (2014). Filtration and transport of heavy metals in graphene oxide enabled sand columns. *Chemical Engineering Journal, 257*, 248–252. doi:10.1016/j.cej.2014.07.034

Diniz, C. V., Doyle, F. M., & Ciminelli, V. S. (2002). Effect of pH on the adsorption of selected heavy metal ions from concentrated chloride solutions by the chelating resin Dowex M-4195. *Separation Science and Technology*, *37*(14), 3169–3185. doi:10.1081/SS-120006155

Dixit, R., Malaviya, D., Pandiyan, K., Singh, U. B., & Sahu, A. (2015). Bioremediation of heavy metals from soil and aquatic environment: An overview of principles and criteria of fundamental processes. *Sustainability*, *7*(2), 2189–2212. doi:10.3390/su7022189

Dixon, D., & Hendrix, J. (1993). A general model for leaching of one or more solid reactants from porous ore particles. *Metallurgical and Materials Transactions. B, Process Metallurgy and Materials Processing Science*, *24*(1), 157–169.

Domagalski, J. L., Alpers, C. N., Slotton, D. G., Suchanek, T. H., & Ayers, S. M. (2004). Mercury and methylmercury concentrations and loads in the Cache Creek watershed, California. *The Science of the Total Environment*, *327*(1), 215–237. doi:10.1016/j.scitotenv.2004.01.013 PMID:15172583

Domic, E. M. (2007). A review of the development and current status of copper bioleaching operations in Chile: 25 years of successful commercial implementation. In *Biomining* (pp. 81–95). Springer. doi:10.1007/978-3-540-34911-2_4

Dong, Z., Zhang, F., Wang, D., Liu, X., & Jin, J. (2015). Polydopamine-mediated surface-functionalization of graphene oxide for heavy metal ions removal. *Journal of Solid State Chemistry*, *224*, 88–93. doi:10.1016/j.jssc.2014.06.030

Doty, S. L. (2008). Enhancing phytoremediation through the use of transgenics and endophytes. *The New Phytologist*, *179*(2), 318–333. doi:10.1111/j.1469-8137.2008.02446.x PMID:19086174

Driscoll, C. T., Mason, R. P., Chan, H. M., Jacob, D. J., & Pirrone, N. (2013). Mercury as a global pollutant: Sources, pathways, and effects. *Environmental Science & Technology*, *47*(10), 4967–4983. doi:10.1021/es305071v PMID:23590191

Dua, M., Singh, A., Sethunathan, N., & Johri, A. (2002). Biotechnology and bioremediation: Success and limitations. *Applied Microbiology and Biotechnology*, *59*(2-3), 143–152. doi:10.1007/s00253-002-1024-6 PMID:12111139

Dubey, S., Shri, M., Misra, P., Lakhwani, D., Bag, S. K., Asif, M. H., & Chakrabarty, D. (2014). Heavy metals induce oxidative stress and genome-wide modulation in transcriptome of rice root. *Functional & Integrative Genomics*, *14*(2), 401–417. doi:10.1007/s10142-014-0361-8 PMID:24553786

Dunwellgroup. (n.d.). *App08_Flow*. Retrieved March 15, 2017, from http://www.dunwellgroup.com/ProductsServices/DVMAT/VSEP/Applications/App08_Flow.jpg

Duruibe, J., Ogwuegbu, M., & Egwurugwu, J. (2007). Heavy metal pollution and human biotoxic effects. *International Journal of Physical Sciences*, *2*(5), 112–118.

Dutta, A. K., Maji, S. K., & Adhikary, B. (2014). γ-Fe 2 O 3 nanoparticles: An easily recoverable effective photo-catalyst for the degradation of rose bengal and methylene blue dyes in the waste-water treatment plant. *Materials Research Bulletin*, *49*, 28–34. doi:10.1016/j.materresbull.2013.08.024

Dutta, M., & Mukherjee, S. (2010). An outlook into energy consumption in large scale industries in India: The cases of steel, aluminium and cement. *Energy Policy*, *38*(11), 7286–7298. doi:10.1016/j.enpol.2010.07.056

Dzikiewicz, M. (2000). Activities in nonpoint pollution control in rural areas of Poland. *Ecological Engineering*, *14*(4), 429–434. doi:10.1016/S0925-8574(99)00066-X

Dzionek, A., Wojcieszyńska, D., & Guzik, U. (2016). Natural carriers in bioremediation: A review. *Electronic Journal of Biotechnology*, *23*, 28–36. doi:10.1016/j.ejbt.2016.07.003

Eapen, S., Singh, S., & D'Souza, S. (2007). Advances in development of transgenic plants for remediation of xenobiotic pollutants. *Biotechnology Advances*, *25*(5), 42–451. doi:10.1016/j.biotechadv.2007.05.001 PMID:17553651

Egirani, D. E., Latif, M. T., Poyi, N. R., Wessey, N., & Acharjee, S. (2017). Synthesis and characterization of kaolinite coated with cu-oxide and its effect on the removal of aqueous mercury (ii) ions: part ii. *International Research Journal of Chemistry and Chemical Sciences*, *4*(1), 55061.

Ekmekyapar, F., Aslan, A., Bayhan, Y. K., & Cakici, A. (2006). Biosorption of copper (II) by nonliving lichen biomass of Cladonia rangiformis hoffm. *Journal of Hazardous Materials*, *137*(1), 293–298. doi:10.1016/j.jhazmat.2006.02.003 PMID:16530938

Ellis, K.V. (1989). *Surface water pollution and its control*. Macmillan Press Ltd.

El-Said, A. G., Badawy, N. A., Abdel-Aal, A. Y., & Garamon, S. E. (2011). Optimization parameters for adsorption and desorption of Zn (II) and Se (IV) using rice husk ash: Kinetics and equilibrium. *Ionics*, *17*(3), 263–270. doi:10.1007/s11581-010-0505-3

Elsgaard, L., Petersen, S. O., & Debosz, K. (2001). Effects and risk assessment of linear alkylbenzene sulfonates in agricultural soil. 1. Short-term effects on soil microbiology. *Environmental Toxicology and Chemistry*, *20*(8), 1656–1663. doi:10.1002/etc.5620200806 PMID:11491546

Ercole, V. F., Toro, L., Ficara, G., & Lepidi, A. (1994). Immobilisation of microbial cells for metal adsorption and desorption. In Mineral Bioprocessing II. Snowboard.

Eroglu, E., Agarwal, V., Bradshaw, M., Chen, X., Smith, S. M., Raston, C. L., & Iyer, K. S. (2012). Nitrate removal from liquid effluents using microalgae immobilized on chitosan nanofiber mats. *Green Chemistry*, *14*(10), 2682–2685. doi:10.1039/c2gc35970g

Ersahin, M. E., Ozgun, H., Dereli, R. K., Ozturk, I., Roest, K., & van Lier, J. B. (2012). A review on dynamic membrane filtration: Materials, applications and future perspectives. *Bioresource Technology*, *122*, 196–206. doi:10.1016/j.biortech.2012.03.086 PMID:22534368

Esposito, A., Pagnanelli, F., Lodi, A., Solisio, C., & Veglio, F. (2001). Biosorption of heavy metals by Sphaerotilus natans: An equilibrium study at different pH and biomass concentrations. *Hydrometallurgy*, *60*(2), 129–141. doi:10.1016/S0304-386X(00)00195-X

Evans, D. A., Sharp, W. R., Ammirato, P. V., & Yamada, Y. (1983). *Handbook of Plant Cell Culture*. New York: Macmillan Publishing Company.

Evans, J., & Kvande, H. (2008). Sustainability, climate change, and greenhouse gas emissions reduction: Responsibility, key challenges, and opportunities for the aluminum industry. *JOM Journal of the Minerals. Metals and Materials Society*, *60*(8), 25–31. doi:10.1007/s11837-008-0103-2

Fabrega, J., Luoma, S. N., Tyler, C. R., Galloway, T. S., & Lead, J. R. (2011). Silver nanoparticles: Behaviour and effects in the aquatic environment. *Environment International*, *37*(2), 517–531. doi:10.1016/j.envint.2010.10.012 PMID:21159383

Faccini, M., Borja, G., Boerrigter, M., Martín, D. M., Crespiera, S. M., Vázquez-Campos, S., & Amantia, D. (2015). Electrospun carbon nanofiber membranes for filtration of nanoparticles from water. *Journal of Nanomaterials*, *2015*, 2. doi:10.1155/2015/247471

Falk, J., Green, J., & Mudd, G. (2006). Australia, uranium and nuclear power. *The International Journal of Environmental Studies*, *63*(6), 845–857. doi:10.1080/00207230601047131

Fan, Y., Yang, R., Lei, Z., Liu, N., Lv, J., Zhai, S., ... Wang, L. (2016). Removal of Cr (VI) from aqueous solution by rice husk derived magnetic sorbents. *Korean Journal of Chemical Engineering*, *33*(4), 1416–1424. doi:10.1007/s11814-015-0248-y

Farhan, A. M., Al-Dujaili, A. H., & Awwad, A. M. (2013). Equilibrium and kinetic studies of cadmium (II) and lead (II) ions biosorption onto Ficus carcia leaves. *International Journal of Industrial Chemistry*, *4*(1), 24. doi:10.1186/2228-5547-4-24

Farooq, U., Kozinski, J. A., Khan, M. A., & Athar, M. (2010). Biosorption of heavy metal ions using wheat based biosorbents–a review of the recent literature. *Bioresource Technology*, *101*(14), 5043–5053. doi:10.1016/j.biortech.2010.02.030 PMID:20223652

Fassler, E., Robinson, B. H., Gupta, S. K., & Schulin, R. (2010). Uptake and allocation of plant nutrients and Cd in maize, sunflower and tobacco growing on contaminated soil and the effect of soil conditioners under field conditions. *Nut. Cycle Agro.*, *87*(3), 339–352. doi:10.1007/s10705-009-9342-z

Felestrino, E.B., Santigo, I.F., Freitos, L.S., Rosa, L.H., Ribeiro, S.P., & Moreira, L.M. (2017). Plant growth promoting bacteria associated with *Longsdorffia hypogae* Rhizosphere host biological interfere a neglected model of bacterial prospection. *Frontiers in Microbiol.* doi 10.3389ifmicb.2017.00172

Feng, L., Cao, M., Ma, X., Zhu, Y., & Hu, C. (2012). Superparamagnetic high-surface-area Fe_3O_4 nanoparticles as adsorbents for arsenic removal. *Journal of Hazardous Materials*, *217*, 439–446. doi:10.1016/j.jhazmat.2012.03.073 PMID:22494901

Ferret, C., Cornu, J. Y., Elhabiri, M., Sterckeman, T., Braud, A., Jézéquel, K., ... Geoffroy, V. A. (2015). Effect of pyoverdine supply on cadmium and nickel complexation and phytoavailability in hydroponics. *Environ. Sci. Pollut.*, *22*(3), 2106–2116. doi:10.1007/s11356-014-3487-2 PMID:25167822

Ferret, C., Sterckeman, T., Cornu, J. Y., Gangloff, S., Schalk, I. J., & Geoffroy, V. A. (2014). Siderophore-promoted dissolution of smectite by fluorescent *Pseudomonas*. *Environmental Microbiology Reports*, *6*(5), 459–467. doi:10.1111/1758-2229.12146 PMID:25646536

Ferroudj, N., Nzimoto, J., Davidson, A., Talbot, D., Briot, E., Dupuis, V., & Abramson, S. (2013). Maghemite nanoparticles and maghemite/silica nanocomposite microspheres as magnetic Fenton catalysts for the removal of water pollutants. *Applied Catalysis B: Environmental*, *136*, 9–18. doi:10.1016/j.apcatb.2013.01.046

Feuer, H., Bartlett, R. S., Vincent, B. F. Jr, & Anderson, R. S. (1965). Diborane Reduction of Nitro Salts. A New Synthesis of N-Monosubstituted Hydroxylamines. *The Journal of Organic Chemistry*, *30*(9), 2880–2882. doi:10.1021/jo01020a002

Fierz-David, H. E., & Blangey, L. (1979). *Fundamental Processes in Dye Chemistry* (5th ed.). New York: Interscience Publications.

Filip, Z. (2002). International approach to assessing soil quality by ecologically related biological parameters. *Agriculture, Ecosystems & Environment*, *88*(2), 169–174. doi:10.1016/S0167-8809(01)00254-7

Firdhouse, M. J., & Lalitha, P. (2015). Biocidal potential of biosynthesized silver nanoparticles against fungal threats. *Journal of Nanostructure in Chemistry*, *5*(1), 25–33. doi:10.1007/s40097-014-0126-x

Fliessbach, A., Martens, R., & Reber, H. H. (1994). Soil microbial biomass and microbial activity in soils treated with heavy metal contaminated sewage sludge. *Soil Biology & Biochemistry*, *26*(9), 1201–1205. doi:10.1016/0038-0717(94)90144-9

Fodor, S. P. A. (1997). Massively parallel genomics. *Science*, *270*(5324), 393–394. doi:10.1126/science.277.5324.393

Fortina, P., Larry, J., Kricka, L., Surrey, S., & Grodzinski, P. (2005). Nanobiotechnology: The promise and reality of new approaches to molecular recognition. *Trends in Biotechnology*, *23*(4), 168–173. doi:10.1016/j.tibtech.2005.02.007 PMID:15780707

Fourest, E., & Roux, J. C. (1992). Heavy metal biosorption by fungal mycelial by-products: Mechanism and influence of pH. *Applied Microbiology and Biotechnology*, *37*(3), 399–403. doi:10.1007/BF00211001

Francesca, L. T. (2002). Olive mill solid residues as heavy metal sorbent material: A preliminary study. *Waste Management (New York, N.Y.)*, *22*(8), 901–907. doi:10.1016/S0956-053X(02)00086-7 PMID:12423052

Frechet, J. M. J., & Tomalia, D. A. (Eds.). (2001). *Dendrimers and Other Dendritic Polymers*. New York: Wiley and Sons; doi:10.1002/0470845821

Freeman, H. S., Hinks, D., & Esancy, J. (1996). Genotoxicity of Azo Dyes: Bases and Implications. In A. T. Peters & H. S. Freeman (Eds.), *Physico Chemical Principles of Colour Chemistry* (pp. 254–292). Chapman & Hall. doi:10.1007/978-94-009-0091-2_7

Friis, N., & Myers-Keith, P. (1986). Biosorption of uranium and lead by Streptomyces longwoodensis. *Biotechnology and Bioengineering*, *28*(1), 21–28. doi:10.1002/bit.260280105 PMID:18553838

Fu, F., Xie, L., Tang, B., Wang, Q., & Jiang, S. (2012). Application of a novel strategy—Advanced Fenton-chemical precipitation to the treatment of strong stability chelated heavy metal containing wastewater. *Chemical Engineering Journal*, *189*, 283–287. doi:10.1016/j.cej.2012.02.073

Fujita, M., Ide, Y., Sato, D., Kench, P. S., Kuwahara, Y., Yokoki, H., & Kayanne, H. (2014). Heavy metal contamination of coastal lagoon sediments: Fongafale Islet, Funafuti Atoll, Tuvalu. *Chemosphere*, *95*, 628–634. doi:10.1016/j.chemosphere.2013.10.023 PMID:24200049

Fulekar, M., Singh, A., & Bhaduri, A. M. (2009). Genetic engineering strategies for enhancing phytoremediation of heavy metals. *African Journal of Biotechnology*, *8*, 529–535.

Fusalba, A., & Belanger, D. (1999). Electropolymerization of Polypyrrole and Polyaniline–Polypyrrole from Organic Acidic Medium. *The Journal of Physical Chemistry B*, *103*(42), 9044–9054. doi:10.1021/jp9916790

Fyodorov, G. (2005). *Uranium deposits of the Inkay-Mynkuduk ore field, Kazakhstan. In Developments in uranium resources, production, demand and the environment TECDOC-1425* (pp. 95–112). Vienna: IAEA.

Gadd, G. M. (2000). Bioremedial potential of microbial mechanisms of metal mobilization and immobilization. *Current Opinion in Biotechnology*, *11*(3), 271–279. doi:10.1016/S0958-1669(00)00095-1 PMID:10851150

Gadd, G. M., & De Rome, L. (1988). Biosorption of copper by fungal melanine. *Applied Microbiology and Biotechnology*, *29*(6), 610–617. doi:10.1007/BF00260993

Gadd, G. M., White, C., & De Rome, L. (2003). Heavy metal and radionuclide uptake by fungi and yeasts. In: Hammaini, A., Simultaneous uptake of metals by activated sludge. *Minerals Engineering*, *16*(8), 723–729. doi:10.1016/S0892-6875(03)00166-3

Gadgil, A. (1998). Drinking water in developing countries. *Annual Review of Energy and the Environment*, *23*(1), 253–286. doi:10.1146/annurev.energy.23.1.253

Gaffney, J. S., & Marley, N. (2014). In-depth review of atmospheric mercury: Sources, transformations, and potential sinks. *Energy and Emission Control Technologies*, *2*, 1–21. doi:10.2147/EECT.S37038

Gajendran, B., Chinnasamy, A., Durai, P., Raman, J., & Ramar, M. (2014). Biosynthesis and characterization of silver nanoparticles from *Daturainoxia* and its apoptotic effct on human breast cancer cell line MCF7. *Materials Letters*, *122*, 98–102. doi:10.1016/j.matlet.2014.02.003

Galun, M., Galun, E., Siegel, B. Z., Keller, P., Lehr, H., & Siegel, S. M. (1987). Removal of metal ions from aqueous solutions by Penicillium biomass: Kinetic and uptake parameters. *Water, Air, and Soil Pollution*, *33*(3-4), 359–371. doi:10.1007/BF00294204

Gamalero, E., & Glick, B. R. (2011). Mechanisms used by plant growth-promoting bacteria. In M. K. Maheshwari (Ed.), *Bacteria in Agrobiology: Plant Nutrient Management* (pp. 17–46). Berlin, Heidelberg, Germany: Springer-Verlag. doi:10.1007/978-3-642-21061-7_2

Ganguli, A., & Tripathi, A. (2002). Bioremediation of toxic chromium from electroplating effluent by chromate-reducing Pseudomonas aeruginosa A2Chr in two bioreactors. *Applied Microbiology and Biotechnology*, *58*(3), 416–420. doi:10.1007/s00253-001-0871-x PMID:11935196

Gao, C., Liu, L., & Yang, F. (2017). Development of a novel proton exchange membrane-free integrated MFC system with electric membrane bioreactor and air contact oxidation bed for efficient and energy-saving wastewater treatment. *Bioresource Technology*, *238*, 472–483. doi:10.1016/j.biortech.2017.04.086 PMID:28475989

Gardea-Torresdey, J. L., Dokken, K., Tiemann, K. J., Parsons, J. G., Ramos, J., Pingitore, N. E., & Gamez, G. (2002). Infrared and X-ray absorption spectroscopic studies on the mechanism of chromium(III) binding to alfalfa biomass. *Microchemical Journal*, *71*(2-3), 157–166. doi:10.1016/S0026-265X(02)00007-3

Gardea-Torresdey, J. L., Parsons, J. G., Gomez, E., Peralta-Videa, J., Troiani, H. E., Santiago, P., & Yacaman, M. J. (2002). Formation and growth of Au nanoparticles inside live alfalfa plants. *Nano Letters*, *2*(4), 397–401. doi:10.1021/nl015673+

Gautheret, T. J. (1939). Culture du tissue cambial. *C.R. Acad. Sci.*, *198*, 2195–2196.

Gaya, U. I., Otene, E., & Abdullah, A. H. (2015). Adsorption of aqueous Cd (II) and Pb (II) on activated carbon nanopores prepared by chemical activation of doum palm shell. *SpringerPlus*, *4*(1), 458. doi:10.1186/s40064-015-1256-4 PMID:26339560

Gebrehawaria, G., Hussen, A., & Rao, V. M. (2015). Removal of hexavalent chromium from aqueous solutions using barks of Acacia albida and leaves of Euclea schimperi. *International Journal of Environmental Science and Technology*, *12*(5), 1569–1580. doi:10.1007/s13762-014-0530-2

Geim, A. K., & Novoselov, K. S. (2007). The rise of Graphene. *Nature Materials*, *6*(3), 183–191. doi:10.1038/nmat1849 PMID:17330084

Gerlach, A. S. (1981). *Marine Pollution: Diagnosis and Therapy*. New York: Springer-Verlag Berlin Heidelberg. doi:10.1007/978-3-642-68182-0

Ghaneian, M. T., Bhatnagar, A., Ehrampoush, M. H., Amrollahi, M., Jamshidi, B., Dehvari, M., & Taghavi, M. (2017). Biosorption of hexavalent chromium from aqueous solution onto pomegranate seeds: Kinetic modeling studies. *International Journal of Environmental Science and Technology*, *14*(2), 331–340. doi:10.1007/s13762-016-1216-8

Ghoreishi, S. M., & Haghighi, R. (2003). Chemical catalytic reaction and biological oxidation for treatment of non-biodegradable textile effluent. *Chemical Engineering Journal*, *95*(1-3), 163–169. doi:10.1016/S1385-8947(03)00100-1

Giller, K. E., Witter, E., & Mcgrath, S. P. (1998). Toxicity of heavy metals to microorganisms and microbial processes in agricultural soils: A review. *Soil Biology & Biochemistry, 30*(10-11), 1389–1414. doi:10.1016/S0038-0717(97)00270-8

Girgis, B. S., & Hendawy, A. N. A. (1997). Capacity of activated carbon from date pits in the removal of organic pollutants and heavy metals. *1st Inter. conference on chem. edu.*

Girisha, S. T., & Ragavendra, V. B. (2009). Accumulation of heavy metals in leafy vegetables grown in urban areas by using sewage water and its effect. *Arc. Phytopathol. Plant Protect., 42*(10), 956–959. doi:10.1080/03235400701543806

Giusti, L. (2009). A review of waste management practices and their impact on human health. *Waste Management (New York, N.Y.), 29*(8), 2227–2239. doi:10.1016/j.wasman.2009.03.028 PMID:19401266

Godbold, D. L., & Hüttermann, A. (1985). Effect of zinc, cadmium and mercury on root elongation of Picea abies (Karst.) seedlings, and the significance of these metals to forest die-back. *Environmental Pollution. Series A. Ecological and Biological, 38*(4), 375–381. doi:10.1016/0143-1471(85)90108-4

Goel, P. (2006). *Water pollution: causes, effects and control.* New Age International.

Golding, B., & Golding, S. D. (2017). *Copper and Coal Resources. In Metals, Energy and Sustainability* (pp. 21–35). Springer. doi:10.1007/978-3-319-51175-7_2

Gomes, M. M. S., Cambraia, J., Sant'anna, R., & Estevão, M. M. (1985). Aluminium effects on uptake and translocation of nitrogen in sorghum (*Sorghum bicolor, L. Moench*). *Journal of Plant Nutrition, 8*(6), 457–465. doi:10.1080/01904168509363360

González, F., Romera, E., Ballester, A., Blázquez, L., Muñoz, J. Á., & García-Balboa, C. (2011). Algal Biosorption and Biosorbents. In P. Kotrba, M. Mackova, & T. Macek (Eds.), *Microbial Biosorption of Metals* (pp. 159–178). Heidelberg, Germany: Springer Dordrech. doi:10.1007/978-94-007-0443-5_7

González, I., Águila, E., & Galán, E. (2007). Environ. Partitioning, bioavailability and origin of heavy metals from the Nador Lagoon sediments (Morocco) as a basis for their management. *Geol., 52*, 1581.

Gopalakrishnan, A., Krishnan, R., Thangavel, S., Venugopal, G., & Kim, S. J. (2015). Removal of heavy metal ions from pharma-effluents using graphene-oxide nanosorbents and study of their adsorption kinetics. *Journal of Industrial and Engineering Chemistry, 30*, 14–19. doi:10.1016/j.jiec.2015.06.005

Gordon, P. E., & Gregory, P. (1983). *Organic Chemistry in Colour.* Berlin: Springer-Verlag.

Gordon, T., & Margel, S. (2011). Synthesis and characterization of zinc/iron oxide composite nanoparticles and their antibacterial properties. *Colloids Surf. Physicochem Eng. Asp., 374*(1-3), 1–8. doi:10.1016/j.colsurfa.2010.10.015

Gotovac, S., Yang, C. M., Hattori, Y., Takahashi, K., Kanoh, H., & Kaneko, K. (2007). Adsorption of polyaromatic hydrocarbons on single wall carbon nanotubes of different functionalities and diameters. *Journal of Colloid and Interface Science*, *314*(1), 18–24. doi:10.1016/j.jcis.2007.04.080 PMID:17604043

Gowd, S. S., Reddy, M. R., & Govil, P. K. (2010). Assessment of heavy metal contamination in soils at Jajmau (Kanpur) and Unnao industrial areas of the Ganga Plain, Uttar Pradesh, India. *Journal of Hazardous Materials*, *174*(1), 113–121. doi:10.1016/j.jhazmat.2009.09.024 PMID:19837511

Gratao, P. L., Polle, A., Lea, P. J., & Azevedo, R. A. (2005). Making the life of heavy metal-stressed plant a little easier. *Functional Plant Biology*, *32*(6), 481–494. doi:10.1071/FP05016

Gray, N. F. (1999). *Water Technology*. New York: John Wiley & Sons.

Greque de Morais, M., Martins, V., Steffens, D., & Pranke, P. (2014). Biological Applications of Nanobiotechnology. *Journal of Nanoscience and Nanotechnology*, *14*(1), 1007–1017. doi:10.1166/jnn.2014.8748 PMID:24730317

Grey, D., Garrick, D., Blackmore, D., Kelman, J., Muller, M., & Sadoff, C. (2013). Water security in one blue planet: twenty-first century policy challenges for science. *Phil. Trans. R. Soc. A*, *371*(2002), 20120406.

Grison, C. (2015). Combining phytoextraction and ecocatalysis: A novel concept for greener chemistry, an opportunity for remediation. *Environ. Sci. Pollut. R*, *22*(8), 5589–5591. doi:10.1007/s11356-014-3169-0 PMID:24946705

Grybos, M., Billard, P., Desobry-Banon, S., Michot, L. J., Lenain, J. F., & Mustin, C. (2011). Bio-dissolution of colloidal-size clay minerals entrapped in microporous silica gels. *Journal of Colloid and Interface Science*, *362*(2), 317–324. doi:10.1016/j.jcis.2011.07.031 PMID:21807370

Guha, S., & Maheshwari, S. C. (1964). In vitro production of embryos from anthers of Datura. *Nature*, *204*(4957), 497. doi:10.1038/204497a0

Gulrajani, M. L. (1999). Present status of natural dyes. In Book of papers of the convention on natural dyes. Department of Textile Technology, IIT Delhi.

Gulrajani, M. L. (2001). Present status of natural dyes. *Indian Journal of Fibre and Textile Research*, *26*, 191–201.

Gupta, A. K., Deva, D., Sharma, A., & Verma, N. (2010). Fe-grown carbon nanofibers for removal of arsenic (V) in waste water. *Industrial & Engineering Chemistry Research*, *49*(15), 7074–7084. doi:10.1021/ie100392q

Gupta, A. K., Dwivedi, S., Sinha, S., Tripathi, R. D., Rai, U. N., & Singh, S. N. (2007). Metal accumulation and growth performance of *Phaseolus vulgaris* grown in fly ash amended soil. *Bioresource Technology*, *98*(17), 3404–3407. doi:10.1016/j.biortech.2006.08.016 PMID:17451948

Gupta, D. K., Tripathi, R. D., Rai, U. N., Dwivedi, S., Mishra, S., Srivastava, S., & Inouhe, M. (2006). Changes in amino acid profile and and metal content in seeds of *Cicer arietinum* L. (Chick pea) grown under various fly ash amendments. *Chemosphere*, *65*(6), 939–945. doi:10.1016/j.chemosphere.2006.03.047 PMID:16682068

Gupta, H., & Gogate, P. R. (2016). Intensified removal of copper from waste water using activated watermelon based biosorbent in the presence of ultrasound. *Ultrasonics Sonochemistry*, *30*, 113–122. doi:10.1016/j.ultsonch.2015.11.016 PMID:26627740

Gupta, K., Bhattacharya, S., Chattopadhyay, D., Mukhopadhyay, A., Biswas, H., Dutta, J., & Ghosh, U. C. (2011). Ceria associated manganese oxide nanoparticles: Synthesis, characterization and arsenic (V) sorption behavior. *Chemical Engineering Journal*, *172*(1), 219–229. doi:10.1016/j.cej.2011.05.092

Gupta, V. K., Ali, I., Saini, V. K., Van Gerven, T., Van der Bruggen, B., & Vandecasteele, C. (2005). Removal of dyes from wastewater using bottom ash. *Industrial & Engineering Chemistry Research*, *44*(10), 3655–3664. doi:10.1021/ie0500220

Gupta, V. K., Ali, I., Saleh, T. A., Nayak, A., & Agarwal, S. (2012). Chemical treatment technologies for waste-water recycling—an overview. *RSC Advances*, *2*(16), 6380–6388. doi:10.1039/c2ra20340e

Gupta, V. K., Gupta, M., & Sharma, S. (2001). Process development for the removal of lead and chromium from aqueous solutions using red mud—an aluminium industry waste. *Water Research*, *35*(5), 1125–1134. doi:10.1016/S0043-1354(00)00389-4 PMID:11268832

Gupta, V. K., Rastogi, A., & Nayak, A. (2010). Biosorption of nickel onto treated alga (Oedogonium hatei): Application of isotherm and kinetic models. *Journal of Colloid and Interface Science*, *342*(2), 533–539. doi:10.1016/j.jcis.2009.10.074 PMID:20004906

Gupta, V. K., Tyagi, I., Sadegh, H., Shahryari-Ghoshekandi, R., Makhlouf, A. S. H., & Maazinejad, B. (2015). Nanoparticles as adsorbent; a positive approach for removal of noxious metal ions: A review. *Sci. Technol. Dev*, *34*(3), 195–214. doi:10.3923/std.2015.195.214

Gupta, V., Nayak, A., Bhushan, B., & Agarwal, S. (2015). A critical analysis on the efficiency of activated carbons from low-cost precursors for heavy metals remediation. *Critical Reviews in Environmental Science and Technology*, *45*(6), 613–668. doi:10.1080/10643389.2013.876526

Haberlandt, G. (1902). Culturversuchemit isolierten Pflanzenzellen. *Sitzungsber. Math. Naturrwiss. Kl. Kias. Akad. Wiss. Wien*, *111*, 69–92.

Habibi, B., & Mohammadyari, S. (2015). Facile synthesis of Pd nanoparticles on nano carbon supports and their application as an electrocatalyst for oxidation of ethanol in alkaline media: The effect of support. *International Journal of Hydrogen Energy*, *40*(34), 10833–10846. doi:10.1016/j.ijhydene.2015.07.021

Hai, F. I., Yamamoto, K., & Fukushi, K. (2006). Development of a submerged membrane fungi reactor for textile wastewater. *Desalination*, *192*(1-3), 315–322. doi:10.1016/j.desal.2005.06.050

Hakanson, L. (1980). An ecological risk index aquatic pollution control: A sedimentological approach. *Water Research*, *14*(8), 975–1001. doi:10.1016/0043-1354(80)90143-8

Halim, M., Conte, P., & Piccolo, A. (2002). Potential availability of heavy metals to phytoextraction from contaminated soils induced by exogenous humic substances. *Chemosphere*, *52*, 26–75. PMID:12729711

Halimoon, N., & Yin, R. G. S. (2010). Removal of Heavy Metals from Textile Wastewater using Zeolite. *Environment Asia*, *3*(SI), 124–130. Available online at www.tshe.org/EA

Hameed, A., Rasool, S., Azooz, M.M., Hossain, M.A., Ahanger, M.A, & Ahmad, P. (2016). Heavy metal stress: Plant responses and signalling. In *Plant metal interaction*. Academic Press.

Han, H. (1987). Application of pollen derived plants in crop improvement: Natesh, S. et al. Biotechnology in Agriculture. Oxford and IBH Publishing Co. Pvt. Ltd.

Hanger, B. C. (1979). The movement of calcium in plants. *Communications in Soil Science and Plant Analysis*, *10*(1-2), 171–193. doi:10.1080/00103627909366887

Harada, M. (1995). Minamata disease: Methylmercury poisoning in Japan caused by environmental pollution. *Critical Reviews in Toxicology*, *25*(1), 1–24. doi:10.3109/10408449509089885 PMID:7734058

Harnessing Biotechnology for the 21st Century. (n.d.). American Chemical Society.

Hashim, M. A., Mukhopadhyay, S., Sahu, J. N., & Sengupta, B. (2011). Remediation technologies for heavy metal contaminated groundwater. *Journal of Environmental Management*, *92*(10), 2355–2388. doi:10.1016/j.jenvman.2011.06.009 PMID:21708421

Hazotte, A. A., Peron, O., Abdelouas, A., Montavon, G., & Lebeau, T. (2016). Microbial mobilization of cesium from illite: The role of organic acids and siderophores. *Chemical Geology*, *428*, 8–14. doi:10.1016/j.chemgeo.2016.02.024

Heiligtag, F., A., Niederberge, M., (2013). *The fascinating world of nanoparticle research*. Academic Press.

Henry, J. R. (2000). An overview of the phytoremediation of lead and mercury. US Environmental Protection Agency, Office of Solid Waste and Emergency Response, Technology Innovation Office.

Herrmann, J. M. (1999). Heterogeneous photocatalysis: Fundamentals and applications to the removal of various types of aqueous pollutants. *Catalysis Today*, *53*(1), 115–129. doi:10.1016/S0920-5861(99)00107-8

HILSON, G. (2001). Putting theory into practice: How has the gold mining industry interpreted the concept of sustainable development? *Mineral Resources Engineering*, *10*(04), 397–413.

Hinsinger, P., Plassard, C., Tang, C. X., & Jaillard, B. (2003). Origins of root mediated pH changes in the rhizosphere and their responses to environmental constraints: A review. *Plant and Soil*, *248*(1/2), 43–59. doi:10.1023/A:1022371130939

Hong, S. G., Kim, B. C., Na, H. B., Lee, J., Youn, J., Chung, S. W., ... Kim, S. H. (2017). Single enzyme nanoparticles armored by a thin silicate network: Single enzyme caged nanoparticles. *Chemical Engineering Journal*, *322*, 510–515. doi:10.1016/j.cej.2017.04.022

Hong, S., Candelone, J.-P., Patterson, C. C., & Boutron, C. F. (1996). History of ancient copper smelting pollution during Roman and medieval times recorded in Greenland ice. *Science*, *272*(5259), 246–249. doi:10.1126/science.272.5259.246

Horsch, R. B., Fraley, R. T., Rogers, S. G., Sanders, P., Lloyd, A., & Hoffman, N. (1984). Inheritance of functional foreign gene in plants. *Science*, *223*(4635), 496–498. doi:10.1126/science.223.4635.496 PMID:17781445

Hossain, M. A., Ashrafuzzaman, M., Hossain, A. K. M. Z., Ismail, M. R., & Koyama, H. (2014). Role of accumulated calcium in alleviating aluminum injury in wheat plants. *The Scientific World Journal*. PMID:24701169

Hua, M., Zhang, S., Pan, B., Zhang, W., Lv, L., & Zhang, Q. (2012). Heavy metal removal from water/wastewater by nanosized metal oxides: A review. *Journal of Hazardous Materials*, *211*, 317–331. doi:10.1016/j.jhazmat.2011.10.016 PMID:22018872

Huang, C. P., Huang, C. P., & Morehart, A. L. (1990). The removal of Cu (II) from dilute aqueous solution by Saccharomyces cerevisiae. *Water Research*, *24*(4), 433–439. doi:10.1016/0043-1354(90)90225-U

Huang, C., & Huang, C. (1996). Application of Aspergillus oryze and Rhizopus oryzae for Cu (II) removal. *Water Research*, *30*(9), 1985–1990. doi:10.1016/0043-1354(96)00020-6

Huang, J., & Bachelard, E. P. (1993). Effects of aluminium on growth and cation uptake in seedlings of *Eucalyptus mannifera and Pinus radiate*. *Plant and Soil*, *149*(1), 121–127. doi:10.1007/BF00010769

Huang, J., Cao, Y., Liu, Z., Deng, Z., Tang, F., & Wang, W. (2012). Efficient removal of heavy metal ions from water system by titanate nanoflowers. *Chemical Engineering Journal*, *180*, 75–80. doi:10.1016/j.cej.2011.11.005

Hudlický, M. (1996). *Reductions in Organic Chemistry*. Washington, DC: American Chemical Society.

Hu, J., Chen, C., Zhu, X., & Wang, X. (2009). Removal of chromium from aqueous solution by using oxidized multiwalled carbon nanotubes. *Journal of Hazardous Materials*, *162*(2), 1542–1550. doi:10.1016/j.jhazmat.2008.06.058 PMID:18650001

Hu, S., Lau, K. W. K., & Wu, M. (2001). Cadmium sequestration in *Chlamydomonas* re- inhardtii. *Plant Science*, *161*(5), 987–996. doi:10.1016/S0168-9452(01)00501-5

Husseiny, M. I., El-Aziz, M. A., Badr, Y., & Mahmoud, M. A. (2007). Biosynthesis of gold nanoparticles using *Pseudomonas aeruginosa*. *Spectrochimica Acta. Part A: Molecular and Biomolecular Spectroscopy*, *67*(3-4), 3–4. doi:10.1016/j.saa.2006.09.028 PMID:17084659

Hu, W., Yin, J., Deng, B., & Hu, Z. (2015). Application of nano TiO 2 modified hollow fiber membranes in algal membrane bioreactors for high-density algae cultivation and wastewater polishing. *Bioresource Technology*, *193*, 135–141. doi:10.1016/j.biortech.2015.06.070 PMID:26125613

Hyung, H., Fortner, J. D., Hughes, J. B., & Kim, J. H. (2007). Natural organic matter stabilizes carbon nanotubes in the aqueous phase. *Environmental Science & Technology*, *41*(1), 179–184. doi:10.1021/es061817g PMID:17265945

Igwe, J. C., Ogunewe, D. N., & Abia, A. A. (2005). Competitive adsorption of Zn (II), Cd (II) and Pb (II) ions from aqueous and non-aqueous solution by maize cob and husk. *African Journal of Biotechnology*, *4*(10), 1113–1116.

Indriolo, E., Na, G., Ellis, D., Salt, D. E., & Banks, J. A. (2010). A vacuolar arsenite transporter necessary for arsenic tolerance in the arsenic hyperaccumulating fern *pteris vittta* is missing in flowering plants. *The Plant Cell*, *22*(6), 2045–2057. doi:10.1105/tpc.109.069773 PMID:20530755

Inoue, K.I. (2013). Heavy Metal Toxicity. *Journal of Clinical Toxicology*. doi: 10.4172/2161-0495.S3-007

Israr, M., Sahi, S. V., & Jain, J. (2006). Cadmium accumulation and antioxidative responses in the *Sesbania drummondii* callus. *Archives of Environmental Contamination and Toxicology*, *50*(1), 121–127. doi:10.1007/s00244-005-5029-x PMID:16247566

Jalali, M., & Aboulghazi, F. (2013). Sunflower stalk, an agricultural waste, as an adsorbent for the removal of lead and cadmium from aqueous solutions. *Journal of Material Cycles and Waste Management*, *15*(4), 548–555. doi:10.1007/s10163-012-0096-3

Jan, A.T., Azam, M., Siddiqui, K., Ali, A., Choi, I., & Rizwanul-Haq, Q.M. (2015). Heavy Metals and Human Health: Mechanistic Insight into Toxicity and Counter Defense System of Antioxidants. *International Journal of Molecular Sciences*, *16*, 29592–29630. doi:10.3390/ijms161226183

Jang, H. D., Kim, S. K., & Kim, S. J. (2001). Effect of particle size and phase composition of titanium dioxide nanoparticles on the photocatalytic properties. *Journal of Nanoparticle Research*, *3*(2/3), 141–147. doi:10.1023/A:1017948330363

Javaid, A., Bajwa, R., Shafique, U., & Anwar, J. (2011).Removal of heavy metals by adsorption on Pleurotus ostreatus. *Biomass and Bioenergy*, *35*(5), 1675-1682.

Javed, M. M., Ikram-ul-Haq, , & Shahbaz, F. (2007). Biosorption of mercury from industrial effluent by fungal consortia. *Bioremediation Journal*, *11*(3), 149–153. doi:10.1080/10889860701548705

Jiang, B., Tan, L., Ning, S., & Shi, S. (2016). A novel integration system of magnetically immobilized cells and a pair of graphite plate-stainless iron mesh electrodes for the bioremediation of coking wastewater. *Bioresource Technology*, *216*, 684–690. doi:10.1016/j.biortech.2016.06.009 PMID:27289060

Jiang, W., Pelaez, M., Dionysiou, D. D., Entezari, M. H., Tsoutsou, D., & O'Shea, K. (2013). Chromium (VI) removal by maghemite nanoparticles. *Chemical Engineering Journal, 222*, 527–533. doi:10.1016/j.cej.2013.02.049

Jindal, A.K. (2011). *Comparison of financial investment & return on the capital in the mining industries of the developing countries–case studies*. Academic Press.

Jin, X., Yang, X., Islam, E., Liu, D., & Mahmood, Q. (2008). Effects of cadmium on ultrastructure and antioxidative defense system in hyperaccumulator and nonhyperaccumulator ecotypes of *Sedum alfredii* Hance. *Journal of Hazardous Materials, 156*(1-3), 387–397. doi:10.1016/j.jhazmat.2007.12.064 PMID:18242844

Johnson, D. B. (1998). Biodiversity and ecology of acidophilic microorganisms. *FEMS Microbiology Ecology, 27*(4), 307–317. doi:10.1111/j.1574-6941.1998.tb00547.x

Johnson, D. B. (2014). Biomining—biotechnologies for extracting and recovering metals from ores and waste materials. *Current Opinion in Biotechnology, 30*, 24–31. doi:10.1016/j.copbio.2014.04.008 PMID:24794631

Johnson, D. B., Rolfe, S., Hallberg, K. B., & Iversen, E. (2001). Isolation and phylogenetic characterization of acidophilic microorganisms indigenous to acidic drainage waters at an abandoned Norwegian copper mine. *Environmental Microbiology, 3*(10), 630–637. doi:10.1046/j.1462-2920.2001.00234.x PMID:11722543

Jordanova, N., Petrovský, E., Kapicka, A., Jordanova, D., & Petrov, P. (2017). Application of magnetic methods for assessment of soil restoration in the vicinity of metallurgical copper-processing plant in Bulgaria. *Environmental Monitoring and Assessment, 189*(4), 158. doi:10.1007/s10661-017-5834-5 PMID:28285437

Joshi, M., Bansal, R., & Purwar, R. (2004). Colour removal from Textile Dyes. *Indian Journal of Fibre and Textile Research, 29*(6), 239–259.

Joutey, N. T., Bahafid, W., Sayel, H., & Ghachtouli, N. E. (2013). *Overview of bioremediation approaches*. Agricultural and Biological Sciences.

Kabbashi, N. A., Atieh, M. A., Al-Mamun, A., Mirghami, M. E., Alam, M. D. Z., & Yahya, N. (2009). Kinetic adsorption of application of carbon nanotubes for Pb (II) removal from aqueous solution. *Journal of Environmental Sciences (China), 21*(4), 539–544. doi:10.1016/S1001-0742(08)62305-0 PMID:19634432

Kaksonen, A. H., Lavonen, L., Kuusenaho, M. K., Kolli, A., Närhi, H. M., Vestola, E. A., ... Tuovinen, O. H. (2011). Bioleaching and recovery of metals from final slag waste of the copper smelting industry. *Minerals Engineering, 24*(11), 1113–1121. doi:10.1016/j.mineng.2011.02.011

Kandah, M. I., & Meunier, J. L. (2007). Removal of nickel ions from water by multi-walled carbon nanotubes. *Journal of Hazardous Materials, 146*(1), 283–288. doi:10.1016/j.jhazmat.2006.12.019 PMID:17196328

Kandeler, F., Kampichler, C., & Horak, O. (1996). Influence of heavy metals on the functional diversity of soil microbial communities. *Biology and Fertility of Soils*, *23*(3), 299–306. doi:10.1007/BF00335958

Kandile, N. G., Mohamed, H. M., & Mohamed, M. I. (2015). New heterocycle modified chitosan adsorbent for metal ions (II) removal from aqueous systems. *International Journal of Biological Macromolecules*, *72*, 110–116. doi:10.1016/j.ijbiomac.2014.07.042 PMID:25128098

Kanel, S., Greneche, J., & Choi, H. (2006). Arsenic (V) removal from groundwater using nano scale zero-valent iron as a colloidal reactive barrier material. *Environmental Science & Technology*, *40*(6), 2045–2050. doi:10.1021/es0520924 PMID:16570634

Kannan, N., & Sundaram, M. M. (2001). Kinetics and mechanism of removal of methylene blue by adsorption on various carbons—a comparative study. *Dyes and Pigments*, *51*(1), 25–40. doi:10.1016/S0143-7208(01)00056-0

Kapoor, A., & Viraraghvan, T. (1995). Fungal biosorption – An alternative treatment option for heavy metal bearing wastewater: A review. *Bioresource Technology*, *53*, 195–206.

Karami, A., & Shamsuddin, Z. H. (2010). Phytoremediation of heavy metals with several efficiency enhancer methods. *African Journal of Biotechnology*, *9*, 3689–3698.

Karn, B., Kuiken, T., & Otto. M., (2009). *Nanotechnology and in Situ Remediation: A Review of the Benefits and Potential Risks*. Academic Press.

Karn, B., Kuiken, T., & Otto, M. (2009). Nanotechnology and in situ remediation: A review of the benefits and potential risks. *Environmental Health Perspectives*, *117*(12), 1823–1831. doi:10.1289/ehp.0900793 PMID:20049198

Karnib, M., Kabbani, A., Holail, H., & Olama, Z. (2004). Heavy metals Removal using activated carbon, silica and silica activated carbon composite. *Energy Procedia*, *50*, 113–120. doi:10.1016/j.egypro.2014.06.014

Kasai, M., Sasaki, M., Yamamoto, Y., & Matsumoto, H. (1992). Aluminum stress increases Kp efflux and activities of ATP- and PPi dependent Hp pumps of tonoplast-enriched membrane vesicles from barley roots. *Plant & Cell Physiology*, *33*(7), 1035–1039.

Keharia, H., & Madamwar, D. (2004). Article. In A. Pandey (Ed.), Textile dye effluent in concise encyclopedia of Bioresearch Technology (pp. 167–175). New York: Food Product Press, The Haworth Preference Press Inc.

Kensa, M. (2011). Bioremediation - An Overview. *Jr. of Industrial Pollution Control*, *27*(2), 161–168.

Kerrich, R., Goldfarb, R., Groves, D., Garwin, S., & Jia, Y. (2000). The characteristics, origins, and geodynamic settings of supergiant gold metallogenic provinces. *Science in China. Series D, Earth Sciences*, *43*(1), 1–68. doi:10.1007/BF02911933

Khajeh, M., Laurent, S., & Dastafkan, K. (2013). Nanoadsorbents: Classification, preparation, and applications (with emphasis on aqueous media). *Chemical Reviews*, *113*(10), 7728–7768. doi:10.1021/cr400086v PMID:23869773

Khambhaty, Y., Mody, K., Basha, S., & Jha, B. (2009). Biosorption of inorganic mercury onto dead biomass of marine Aspergillus niger: Kinetic, equilibrium, and thermodynamic studies. *Environmental Engineering Science*, *26*(3), 531–539. doi:10.1089/ees.2008.0056

Khan, A. G., Bari, A., Chaudhry, T. M., & Qazilbash, A. A. (1997). Phytoremediation - a strategy to decontaminate heavy metal polluted soils and to conserve the biodiversity of Pakistan soils. In S. A. Mufti, C. A. Woods, & S. A. Hasan (Eds.), *Biodiversity of Pakistan* (pp. 133–144). Gainesville, FL: Pakistan Museum of Natural History, Islamabad & Florida Museum of Natural History.

Khan, M. R., & Khan, M. W. (1996). The effect of Fly-ash on plant growth and yield of Tomato. *Environmental Pollution*, *92*, 105–111. doi:10.1016/0269-7491(95)00098-4 PMID:15091390

Khan, S., Cao, Q., Zheng, Y., Huang, Y., & Zhu, Y. (2008). Health risks of heavy metals in contaminated soils and food crops irrigated with wastewater in Beijing, China. *Environmental Pollution*, *152*(3), 686–692. doi:10.1016/j.envpol.2007.06.056 PMID:17720286

Kirschling, T. L., & Tilton, R. D. (2010). Impact of nanoscale zero valent iron on geochemistry and microbial populations in trichloroethylene contaminated aquifer materials. *Environmental Science & Technology*, *44*(9), 3474–3480. doi:10.1021/es903744f PMID:20350000

Kiyono, M., & Pan-Hou, H. (2006). Genetic engineering of bacteria for environmental remediation of mercury. *Journal of Health Science*, *52*(3), 199–204. doi:10.1248/jhs.52.199

Kobya, M., Demirbas, E., Senturk, E., & Ince, M. (2005). Adsorption of heavy metal ions from aqueous solutions by activated carbon prepared from apricot stone. *Bioresource Technology*, *96*(13), 1518–1521. doi:10.1016/j.biortech.2004.12.005 PMID:15939281

Konishi, Y., Kubo, H., & Asai, S. (1992). Bioleaching of zinc sulfide concentrate by Thiobacillusferrooxidans. *Biotechnology and Bioengineering*, *39*(1), 66–74. doi:10.1002/bit.260390111 PMID:18600888

Koók, L., Rózsenberszki, T., Nemestóthy, N., Bélafi-Bakó, K., & Bakonyi, P. (2016). Bioelectrochemical treatment of municipal waste liquor in microbial fuel cells for energy valorization. *Journal of Cleaner Production*, *112*, 4406–4412. doi:10.1016/j.jclepro.2015.06.116

Koropatnick, J., & Leibbrandt, M. E. I. (1995) Effects of Metal on Gene Expression. In Handbook of Experimental Pharmacology, Toxicology of metals, Biochemical Aspects. Springer Verlag. doi:10.1007/978-3-642-79162-8_5

Kratochvil, D., & Volesky, B. (1998). Advances in the biosorption of heavy metals. *Trends in Biotechnology*, *16*(7), 291–300. doi:10.1016/S0167-7799(98)01218-9

Kreuter, J. (2006). *Nanoparticles—a historical perspective*. Academic Press.

Kuan, W. H., Hu, C. Y., & Chiang, M. C. (2009). Treatment of As (V) and As (III) by electrocoagulation using Al and Fe electrode. *Water Science and Technology*, *60*(5), 1341–1346. doi:10.2166/wst.2009.405 PMID:19717922

Kumar, S., & Gopinath, P. (2012). Nano-Bioremediation Applications of Nanotechnology for Bioremediation. *Remediation of Heavy Metals in the Environment*, 27-40.

Kumar, S., & Trivedi, P.K. (2016). Heavy metal stress signalling in plants. In *Plant metal interaction*. Academic Press.

Kumar, A., Bisht, B. S., Joshi, V. D., & Dhewa, T. (2011). Review on Bioremediation of Polluted Environment: A Management Tool. *International Journal of Environmental Sciences*, *1*(6), 1079–1093.

Kumar, B., Smita, K., Sánchez, E., Stael, C., & Cumbal, L. (2016). Andean Sacha inchi (Plukenetia volubilis L.) shell biomass as new biosorbents for Pb^{2+} and Cu^{2+} ions. *Ecological Engineering*, *93*, 152–158. doi:10.1016/j.ecoleng.2016.05.034

Kumari, A. (2007). Bioamelioration of fly ash by some fern species. *Ind. Fern J.*, *24*, 164–171.

Kumari, A., & Tripathi, K. P. (2008). Phytosociological studies on pteridophytic flora in tarai forest of north India. *Indian Journal of Forestry*, *30*(4), 2008.

Kumari, B., & Singh, D. P. (2016). A review on multifaceted application of nanoparticles in the field of bioremediation of petroleum hydrocarbons. *Ecological Engineering*, *97*, 98–105. doi:10.1016/j.ecoleng.2016.08.006

Kumari, , & Lal, , Pakade, & Chand. (2011). Bioaccumulation of heavy metals by *Pteris vittata* in the vicinity of fly ash. *International Journal of Phytoremediation*, *13*, 169–175. doi:10.1080/15226514.2010.525561

Kumar, K. S., Dahms, H. U., Won, E. J., Lee, J. S., & Shin, K. H. (2015). Microalgae – A promising tool for heavy metal remediation. *Ecotoxicology and Environmental Safety*, *113*, 329–352. doi:10.1016/j.ecoenv.2014.12.019 PMID:25528489

Kurian, M., & Nair, D. S. (2015). Heterogeneous Fenton behavior of nano nickel zinc ferrite catalysts in the degradation of 4-chlorophenol from water under neutral conditions. *Journal of Water Process Engineering*, *8*, e37–e49. doi:10.1016/j.jwpe.2014.10.011

Kurniati, E., Arfarita, N., Imai, T., Higuchi, T., Kanno, A., Yamamoto, K., & Sekine, M. (2014). Potential bioremediation of mercury-contaminated substrate using filamentous fungi isolated from forest soil. *Journal of Environmental Sciences (China)*, *26*(6), 1223–1231. doi:10.1016/S1001-0742(13)60592-6 PMID:25079829

Kurniawan, T. A., & Babel, S. (2003). A research study on Cr (VI) removal from contaminated wastewater using low-cost adsorbents and commercial activated carbon. *Second Int. Conf. on Energy Technology towards a Clean Environment (RCETE)*, 2, 1110-1117.

Kurniawan, T. A., Chan, G. Y. S., Lo, W. H., & Babel, S. (2006). Physico-chemical treatment techniques for wastewater laden with heavy metals. *Chemical Engineering Journal, 118*(1-2), 83–98. doi:10.1016/j.cej.2006.01.015

Kuyucak, N., & Volesky, B. (1988). Biosorbents for recovery of metals from industrial solutions. *Biotechnology Letters, 10*(2), 137–142. doi:10.1007/BF01024641

Kwankua, W., Sengsai, S., Muangphra, P., & Euawong, N. (2012). Screening for plants sensitive to heavy metals using cytotoxic and genotoxic biomarkers. *Witthayasan Kasetsat Witthayasat, 46*, 10–23.

Kyzas, G. Z., Terzopoulou, Z., Nikolaidis, V., Alexopoulou, E., & Bikiaris, D. N. (2015). Low-cost hemp biomaterials for nickel ions removal from aqueous solutions. *Journal of Molecular Liquids, 209*, 209–218. doi:10.1016/j.molliq.2015.05.060

Lafleur, J. P., Senkbeil, S., Jensen, T. G., & Kutter, J. P. (2012). Gold nanoparticle-based optical microfluidic sensors for analysis of environmental pollutants. *Lab on a Chip, 12*(22), 4651–4656. doi:10.1039/c2lc40543a PMID:22824920

Lasat, M. M. (2002). Phytoextraction of toxic metals. *Journal of Environmental Quality, 31*(1), 109–120. doi:10.2134/jeq2002.1090 PMID:11837415

Latha, N., & Gowri, M. (2014). Bio Synthesis and Characterisation of Fe_3O_4 Nanoparticles Using Caricaya Papaya Leaves Extract. *Synthesis, 12*, 17.

Leblanc, M., Petit, D., Deram, A., Robinson, B., & Brooks, R. R. (1999). The phytomining and environmental significance of hyperaccumulation of thallium by Iberisintermedia from southern France. *Economic Geology and the Bulletin of the Society of Economic Geologists, 94*(1), 109–113. doi:10.2113/gsecongeo.94.1.109

Lee, Y. C., & Yang, J. W. (2012). Self-assembled flower-like TiO_2 on exfoliated graphite oxide for heavy metal removal. *Journal of Industrial and Engineering Chemistry, 18*(3), 1178–1185. doi:10.1016/j.jiec.2012.01.005

Lefevre, E., Bossa, N., Wiesner, M., & Gunsch, C. (2016). A review of the environmental implications of in situ remediation by nanoscale zero valent iron (nZVI): Behavior, transport and impacts on microbial communities. *The Science of the Total Environment, 15*, 889–901. doi:10.1016/j.scitotenv.2016.02.003 PMID:26897610

Le, T. T., Nguyen, K.-H., Jeon, J.-R., Francis, A. J., & Chang, Y.-S. (2015). Nano/bio treatment of polychlorinated biphenyls with evaluation of comparative toxicity. *Journal of Hazardous Materials, 287*, 335–341. doi:10.1016/j.jhazmat.2015.02.001 PMID:25679799

Leung, C. M., & Jiao, J. J. (2006). Heavy metal and trace element distributions in groundwater in natural slopes and highly urbanized spaces in Mid-Levels area, Hong Kong. *Water Research, 40*(4), 753–767. doi:10.1016/j.watres.2005.12.016 PMID:16448684

Liang, P., Ding, Q., & Song, F. (2005). Application of multiwalled carbon nanotubes as solid phase extraction sorbent for preconcentration of trace copper in water samples. *Journal of Separation Science*, *28*(17), 2339–2343. doi:10.1002/jssc.200500154 PMID:16342800

Liao, H., Wan, H., Shaff, J., Wang, X., Yan, X., & Kochian, L. V. (2006). Phosphorus and aluminium interactions in soybean in relation to alumunium tolerance. Exudation of specific organic acids from different regions of the intact root system. *Plant Physiology*, *141*(2), 674–684. doi:10.1104/pp.105.076497 PMID:16648222

Li, H., & Jiang, G. (2009). Effects of waterborne nano-iron on medaka (oryziaslatipes): Antioxidant enzymatic activity, lipid peroxidation and histopathology. *Ecotoxicology and Environmental Safety*, *72*(3), 684–692. doi:10.1016/j.ecoenv.2008.09.027 PMID:19058851

Li, J., Guo, S., Zhai, Y., & Wang, E. (2009). Nafion–graphene nanocomposite film as enhanced sensing platform for ultrasensitive determination of cadmium. *Electrochemistry Communications*, *11*(5), 1085–1088. doi:10.1016/j.elecom.2009.03.025

Lim, A. P., & Aris, A. Z. (2014). A review on economically adsorbents on heavy metals removal in water and wastewater. *Reviews in Environmental Science and Biotechnology*, *13*(2), 163–181. doi:10.1007/s11157-013-9330-2

Lim, P. E., Mak, K. Y., Mohamed, N., & Noor, A. M. (2003). Removal and speciation of heavy metals along the treatment path of wastewater in subsurface-flow constructed wetlands. *Water Science and Technology*, *48*, 307–313. PMID:14621178

Lin, C. C., & Lin, H. L. (2005). Remediation of soil contaminated with the heavy metal (Cd^{2+}). *Journal of Hazardous Materials*, *122*(1-2), 7–15. doi:10.1016/j.jhazmat.2005.02.017 PMID:15943924

Lingamdinne, L. P., Koduru, J. R., Choi, Y. L., Chang, Y. Y., & Yang, J. K. (2016). Studies on removal of Pb (II) and Cr (III) using graphene oxide based inverse spinel nickel ferrite nano-composite as sorbent. *Hydrometallurgy*, *165*, 64–72. doi:10.1016/j.hydromet.2015.11.005

Lingamdinne, L. P., Koduru, J. R., Roh, H., Choi, Y. L., Chang, Y. Y., & Yang, J. K. (2016). Adsorption removal of Co (II) from waste-water using graphene oxide. *Hydrometallurgy*, *165*, 90–96. doi:10.1016/j.hydromet.2015.10.021

Lin, L., Wang, H., Luo, H., & Xu, P. (2015). Enhanced photocatalysis using side-glowing optical fibers coated with Fe-doped TiO 2 nanocomposite thin films. *Journal of Photochemistry and Photobiology A Chemistry*, *307*, 88–98. doi:10.1016/j.jphotochem.2015.04.010

Lin, S. T., Thirumavalavan, M., Jiang, T. Y., & Lee, J. F. (2014). Synthesis of ZnO/Zn nano photocatalyst using modified polysaccharides for photodegradation of dyes. *Carbohydrate Polymers*, *105*, 1–9. doi:10.1016/j.carbpol.2014.01.017 PMID:24708945

Li, Q., Liu, H., Alattar, M., Jiang, S., Han, J., Ma, Y., & Jiang, C. (2015). The preferential accumulation of heavy metals in different tissues following frequent respiratory exposure to PM2. 5 in rats. *Scientific Reports*, *5*. PMID:26582271

Little Pro. (n.d.). *How to Calculate German Water Hazard Class*. Retrieved March 15, 2017, from http://www.chemsafetypro.com/Topics/Other/How_to_Calculate_German_Water_Hazard_Class_WGK.html

Liu, C., Jin, R. N., Ouyang, X. K., & Wang, Y. G. (2017). Adsorption behavior of carboxylated cellulose nanocrystal—polyethyleneimine composite for removal of Cr (VI) ions. *Applied Surface Science*, *408*, 77–87. doi:10.1016/j.apsusc.2017.02.265

Liu, H., Feng, S., Zhang, N., Du, X., & Liu, Y. (2014). Removal of Cu (II) ions from aqueous solution by activated carbon impregnated with humic acid. *Frontiers of Environmental Science & Engineering*, *8*(3), 329–336. doi:10.1007/s11783-013-0553-9

Liu, M., Wang, Z., Zong, S., Chen, H., Zhu, D., Wu, L., & Cui, Y. (2014). SERS detection and removal of mercury (II)/silver (I) using oligonucleotide-functionalized core/shell magnetic silica sphere@ Au nanoparticles. *ACS Applied Materials & Interfaces*, *6*(10), 7371–7379. doi:10.1021/am5006282 PMID:24738775

Liu, Q., Zheng, T., Wang, P., Jiang, J., & Li, N. (2010). Adsorption isotherm, kinetic and mechanism studies of some substituted phenols on activated carbon fibers. *Chemical Engineering Journal*, *157*(2-3), 348–356. doi:10.1016/j.cej.2009.11.013

Liu, W. T. (2006). Nanoparticles and Their Biological and Environmental Applications. *Journal of Bioscience and Bioengineering*, *102*(1), 1–7. doi:10.1263/jbb.102.1 PMID:16952829

Liu, W., Yang, J., & Xiao, B. (2009). Review on treatment and utilization of bauxite residues in China. *International Journal of Mineral Processing*, *93*(3), 220–231. doi:10.1016/j.minpro.2009.08.005

Li, W. C., Law, F. Y., & Chan, Y. H. M. (2017). Biosorption studies on copper (II) and cadmium (II) using pretreated rice straw and rice husk. *Environmental Science and Pollution Research International*, *24*(10), 8903–8915. doi:10.1007/s11356-015-5081-7 PMID:26206132

Li, W., Zhang, Y., Tian, G., Xie, S., Xu, Q., Wang, L., ... Bu, Y. (2016). Fabrication of graphene-modified nano-sized red phosphorus for enhanced photocatalytic performance. *Journal of Molecular Catalysis A Chemical*, *423*, 356–364. doi:10.1016/j.molcata.2016.07.039

Li, Y. H., Wang, S., Luan, Z., Ding, J., Xu, C., & Wu, D. (2003). Adsorption of cadmium (II) from aqueous solution by surface oxidized carbon nanotubes. *Carbon*, *41*(5), 1057–1062. doi:10.1016/S0008-6223(02)00440-2

Li, Y. H., Wang, S., Wei, J., Zhang, X., Xu, C., Luan, Z., ... Wei, B. (2002). Lead adsorption on carbon nanotubes. *Chemical Physics Letters*, *357*(3), 263–266. doi:10.1016/S0009-2614(02)00502-X

Li, Z., Ma, Z., Kuijp, T. J. D., Yuan, Z., & Huang, L. (2014). A review of soil heavy metal pollution from mines in China: Pollution and health risk assessment. *The Science of the Total Environment*, *468-469*, 843–853. doi:10.1016/j.scitotenv.2013.08.090 PMID:24076505

Li, Z., & Xing, D. (2010). Mitochondrial pathway leading to programmed cell death induced by aluminum phyto-toxicity in *Arabidopsis*. *Plant Signaling & Behavior*, *5*(12), 1660–1662. doi:10.4161/psb.5.12.14014 PMID:21512328

Lloyd, J. R., & Lovley, D. R. (2001). Microbial detoxification of metals and radionuclides. *Current Opinion in Biotechnology*, *12*(3), 248–253. doi:10.1016/S0958-1669(00)00207-X PMID:11404102

Löbbert, G. (2002). Phthalocyanines. In Ullmann's Encyclopedia of Industrial Chemistry. Wiley-VCH Verlag GmbH & Co. KGaA.

Logroño, W., Pérez, M., Urquizo, G., Kadier, A., Echeverría, M., Recalde, C., & Rákhely, G. (2017). Single chamber microbial fuel cell (SCMFC) with a cathodic microalgal biofilm: A preliminary assessment of the generation of bioelectricity and biodegradation of real dye textile wastewater. *Chemosphere*, *176*, 378–388. doi:10.1016/j.chemosphere.2017.02.099 PMID:28278426

Lombi, E., Terall, K. L., Howarth, J. R., Zhao, F. J., Hawkesford, M. J., & McGrath, S. P. (2002). Influence of iron status on cadmium and zinc uptake by different ecotypes of the hyperaccumulator *Thlaspi caerulescens*. *Plant Physiology*, *128*(4), 1359–1367. doi:10.1104/pp.010731 PMID:11950984

Long, E. R., Field, L. J., & McDonald, D. D. (1998). Predicting toxicity in marine sediments with numerical sediment quality guidelines. *Environmental Toxicology and Chemistry*, *17*(4), 714–727. doi:10.1002/etc.5620170428

Long, E. R., & MacDonald, D. D. (1998). Recommended uses of empirically derived, sediment quality guidelines for marine and estuarine ecosystems. *Human and Ecological Risk Assessment*, *4*(5), 1019–1039. doi:10.1080/10807039891284956

Long, E. R., MacDonald, D. D., Severn, C. G., & Hong, C. B. (2000). Classifying probabilities of acute toxicity in marine sediments with empirically derived sediment quality guidelines. *Environmental Toxicology and Chemistry*, *19*(10), 2598–2601. doi:10.1002/etc.5620191028

Long, R. Q., & Yang, R. T. (2001). Carbon nanotubes as superior sorbent for dioxin removal. *Journal of the American Chemical Society*, *123*(9), 2058–2059. doi:10.1021/ja003830l PMID:11456830

Long, X. X., Yang, X. E., & Ni, W. Z. (2002). Current status and prospective on phytoremediation of heavy metal polluted soils. *Journal of Applied Ecology*, *13*, 757–762. PMID:12216411

Long, Y., Xiao, L., & Cao, Q. (2017). Co-polymerization of catechol and polyethylenimine on magnetic nanoparticles for efficient selective removal of anionic dyes from water. *Powder Technology*, *310*, 24–34. doi:10.1016/j.powtec.2017.01.013

Looney, B. (2001). Ultralow concentration mercury treatment using chemical reduction and air stripping (0013-936X). Academic Press.

López-Serrano, A., Olivas, R. M., Landaluze, J. S., & Cámara, C. (2014). Nanoparticles: A global vision. Characterization, separation, and quantification methods. Potential environmental and health impact. *Analytical Methods*, *6*(1), 38–56. doi:10.1039/C3AY40517F

Lori, V., Pietrini, F., Massacci, A., & Zacchini, M. (2015). *Morphophysiological responses, Heavy metal accumulation and phytoremoval ability in four Willow clones exposed to cadmium under hydroponics*. Phytoremediation Springer International Publishing.

Lovley, D. R. (1995). Bioremediation of organic and metal contaminants with dissimilatory metal reduction. *Journal of Industrial Microbiology & Biotechnology*, *14*(2), 85–93. doi:10.1007/BF01569889 PMID:7766214

Lu, C., Chiu, H., & Liu, C. (2006). Removal of zinc (II) from aqueous solution by purified carbon nanotubes: Kinetics and equilibrium studies. *Industrial & Engineering Chemistry Research*, *45*(8), 2850–2855. doi:10.1021/ie051206h

Lu, L., Tian, S. K., Yang, X. E., Li, T. Q., & He, Z. I. (2009). Cadmiumuptake and xylem loading are active processes in the hyperaccumulator *Sedum alfredii*. *Journal of Plant Physiology*, *166*(6), 579–587. doi:10.1016/j.jplph.2008.09.001 PMID:18937997

Lunge, S. S., Singh, S., & Sinha, A. (2014). Magnetic nanoparticle: Synthesis and environmental applications. *International Conference on Chemical, Civil and Environmental Engineering*, 7-9.

Luo, T., Cui, J., Hu, S., Huang, Y., & Jing, C. (2010). Arsenic removal and recovery from copper smelting wastewater using TiO_2. *Environmental Science & Technology*, *44*(23), 9094–9098. doi:10.1021/es1024355 PMID:21053910

Lytle, C. M., Lytle, F. W., Yang, N., Qian, J.-H., Hansen, D., Zayed, A., & Terry, N. (1998). Reduction of Cr (VI) to Cr (III) by wetland plants: Potential for in situ heavy metal detoxification. *Environmental Science & Technology*, *32*(20), 3087–3093. doi:10.1021/es980089x

Maestri, E., Marmiroli, M., Visioli, G., & Marmiroli, N. (2010). Metal tolerance and hyperaccumulation: Costs and trade-offs between traits and environment. *Environmental and Experimental Botany*, *68*(1), 1–13. doi:10.1016/j.envexpbot.2009.10.011

Ma, H., Wang, H., & Na, C. (2015). Microwave-assisted optimization of platinum-nickel nanoalloys for catalytic water treatment. *Applied Catalysis B: Environmental*, *163*, 198–204. doi:10.1016/j.apcatb.2014.07.062

Mahanta, N., & Valiyaveettil, S. (2013). Functionalized poly (vinyl alcohol) based nanofibers for the removal of arsenic from water. *RSC Advances*, *3*(8), 2776–2783. doi:10.1039/c2ra22768a

Mahbub, K. R., Bahar, M. M., Labbate, M., Krishnan, K., Andrews, S., Naidu, R., & Megharaj, M. (2017). Bioremediation of mercury: Not properly exploited in contaminated soils! *Applied Microbiology and Biotechnology*, *101*(3), 963–976.

Mahmood, T. (2010). Phytoextraction of heavy metals the process and scope for remediation of contaminated soils. *Soil & Environment*, *29*(2), 91–109.

Mahmud, H. N. M. E., Huq, A. K. O., & Yahya, R. B. (2017). The removal of heavy metal ions from wastewater/aqueous solution using polypyrrole-based adsorbents: A review. *RSC Advances*, *6*(18), 14778–14791. doi:10.1039/C5RA24358K

Ma, L. Q., Komar, K. M., Tu, C., Zhang, W., Cai, Y., & Kennelley, E. D. (2001a). A fern that hyperaccumulates arsenic. *Nature*, *409*(6820), 579. doi:10.1038/35054664 PMID:11214308

Ma, L. Q., Komar, K. M., Tu, C., Zhang, W., Cai, Y., & Kennelley, E. D. (2001b). A fern that hyperaccumulates arsenic-addendum. *Nature*, *411*(6836), 438. doi:10.1038/35078151 PMID:11373666

Malik, A. (2004). Metal bioremediation through growing cells. *Environment International*, *30*(2), 261–278. doi:10.1016/j.envint.2003.08.001 PMID:14749114

Manahan, S. E. (1997). Environmental Science and Technology. Boca Raton, FL: Lewis Publishers.

Mandal, B. K., Vankayala, R., & Uday Kumar, L. (2011). Speciation of chromium in soil and sludge in the surrounding tannery region, Ranipet, Tamil Nadu. *ISRN Toxicology*. PMID:23724287

Manzoor, S., Shah, M.H., Shaheen, N., Khalique, A., & Jaffar, M. (2006). Multivariate Analysis of Trace Metals in Textile Effluents in Relation to Soil and Groundwater. *Journal of Hazardous Materials*, *137*, 31-37. 10.1016/j.jhazmat.2006.01.077

Mapolelo, M., & Torto, N. (2004). Trace enrichment of metal ions in aquatic environments by Saccharomyces cerevisiae. *Talanta*, *64*(1), 39–47. doi:10.1016/j.talanta.2003.10.058 PMID:18969566

Mariano, E. D., & Keltjens, W. G. (2005). Long term effects of aluminum exposure on nutrient uptake by maize genotypes differing in aluminum resistance. *Journal of Plant Nutrition*, *28*(2), 323–333. doi:10.1081/PLN-200047625

Martínez, F., López-Muñoz, M. J., Aguado, J., Melero, J. A., Arsuaga, J., Sotto, A., & Cerro, L. (2013). Coupling membrane separation and photocatalytic oxidation processes for the degradation of pharmaceutical pollutants. *Water Research*, *47*(15), 5647–5658. doi:10.1016/j.watres.2013.06.045 PMID:23863375

Martínez-Juárez, V. M., Cárdenas-González, J. F., Torre-Bouscoulet, M. E., & Acosta-Rodríguez, I. (2012). Biosorption of mercury (II) from aqueous solutions onto fungal biomass. *Bioinorganic Chemistry and Applications*. PMID:23028382

Massot, P., & Chen, Z.-M. (2013). China and the global uranium market: Prospects for peaceful coexistence. *The Scientific World Journal*. PMID:23606818

Mathur, A., Viswamohan, K., Mohanty, K., Murthy, V., & Seshadrinath, S. (2000). Uranium extraction using biogenic ferric sulfate a case study on quartz chlorite ore from Jaduguda, Singhbhum Thrust Belt (STB), Bihar, India. *Minerals Engineering*, *13*(5), 575–579. doi:10.1016/S0892-6875(00)00037-6

Matlochová, A., Plachá, D., & Rapantová, N. (2013). The application of nanoscale materials in groundwater remediation. *Polish Journal of Environmental Studies*, *22*(5), 1401–1410.

Mattuschka, B., & Straube, G. (1993). Biosorption of metals by a waste biomass. *Journal of Chemical Technology and Biotechnology (Oxford, Oxfordshire)*, *58*(1), 57–63. doi:10.1002/jctb.280580108

Mauter, M. S., & Elimelech, M. (2008). Environmental applications of carbon-based nanomaterials. *Environmental Science & Technology*, *42*(16), 5843–5859. doi:10.1021/es8006904 PMID:18767635

Ma, X., Wang, Y., Gao, M., Xu, H., & Li, G. (2010). A novel strategy to prepare ZnO/PbS heterostructured functional nanocomposite utilizing the surface adsorption property of ZnO nanosheets. *Catalysis Today*, *158*(3), 459–463. doi:10.1016/j.cattod.2010.07.013

Ma, Y., Rajkumar, M., Luo, Y., & Freitas, H. (2011). Inoculation of endophytic bacteria on host and non-host plants-effects on plant growth and Ni uptake. *Journal of Hazardous Materials*, *195*, 230–237. doi:10.1016/j.jhazmat.2011.08.034 PMID:21872991

Mazumder, D. N. G., Ghosh, A., Majumdar, K. K., Ghosh, N., Saha, C., & Mazumder, R. N. G. (2010). Arsenic Contamination of Ground Water and its Health Impact on Population of District of Nadia, West Bengal, India. *Indian Journal of Community Medicine: Official Publication of Indian Association of Preventive & Social Medicine*, *35*(2), 331–338. doi:10.4103/0970-0218.66897 PMID:20922118

McCready, A., Parnell, J., & Castro, L. (2003). Crystalline placer gold from the Rio Neuquen, Argentina: Implications for the gold budget in placer gold formation. *Economic Geology and the Bulletin of the Society of Economic Geologists*, *98*(3), 623–633.

Md. Rizwan, Singh, M., Chanchal, K., Mitra, & Morve, R. (2014). Ecofriendly Application of Nanomaterials: Nanobioremediation. *Journal of Nanoparticles, 7*.

Meagher, R. B. (2000). Phytoremediation of toxic elemental and organic pollutants. *Current Opinion in Plant Biology*, *3*(2), 153–162. doi:10.1016/S1369-5266(99)00054-0 PMID:10712958

Meagher, R. B., Rugh, C. L., Kandasamy, M. K., Gragson, G., & Wang, N. J. (2000). Engineered phytoremediation of mercury pollution in soil and water using bacterial genes. In N. Terry & G. Banuelos (Eds.), *Phytoremediation of contaminated soil and water* (pp. 203–221). Lewis.

Meers, E., Ruttens, A., Hopgood, M., Lesage, E., & Tack, F. M. (2005). Potential of *Brassic rapa, Cannabis sativa, Helianthus annuus and Zea mays* for phytoextraction of heavy metals from calcareous dredged sediment derived soils. *Chemosphere*, *61*(4), 561–572. doi:10.1016/j.chemosphere.2005.02.026 PMID:16202810

Meers, E., Vandecasteele, B., Ruttens, A., Vangronsveld, J., & Tack, F. M. G. (2007). Potential of five willow species (*Salix spp.*) for phytoextraction of heavy metals. *Environmental and Experimental Botany*, *60*(1), 57–68. doi:10.1016/j.envexpbot.2006.06.008

Mehdawi, E. (2011). Selenium hyperaccumulators facilitate selenium-tolerant neighbors via phytoenrichment and reduced herbivory curt. *Current Biology*, *21*(17), 1440–1449. doi:10.1016/j.cub.2011.07.033 PMID:21856156

Mehndiratta, P., Jain, A., Srivastava, S., & Gupta, N. (2013). Environmental Pollution and Nanotechnology. *Environment and Pollution*, *2*(2), 49–58. doi:10.5539/ep.v2n2p49

Mehra, A., Farago, M. E., & Banerjee, D. K. (1998). Impact of fly ash from coal fired Station in Delhi, with particular reference to metal contamination. *Environmental Monitoring and Assessment, 50*(1), 15–35. doi:10.1023/A:1005860015123

Mehrag, A. A. (2002). Variation in arsenic accumulation –hyperaccumulation in ferns and their allies. *The New Phytologist, 157*(1), 25–31. doi:10.1046/j.1469-8137.2003.00541.x

Mehta, K., Kumar, V., Pandey, B., & Tamrakar, P. (2011). Bioleaching-an alternate uranium ore processing technology for India. *Energy Procedia, 7*, 158–162. doi:10.1016/j.egypro.2011.06.021

Mehta, S. K., Singh, A., & Gaur, J. P. (2002). Kinetics of adsorption and up take of Cu2 by Chlorella vulgaris: Influence of pH, temperature, culture age and cations. J.En- viron. *Sci. Health Part A, 37*(3), 399–414. doi:10.1081/ESE-120002837 PMID:11929076

Mendez, J. O., Melian, J. H., Arana, J., Rodriguez, J. D., Diaz, O. G., & Pena, J. P. (2015). Detoxification of waters contaminated with phenol, formaldehyde and phenol–formaldehyde mixtures using a combination of biological treatments and advanced oxidation techniques. *Applied Catalysis B: Environmental, 163*, 63–73. doi:10.1016/j.apcatb.2014.07.032

Michael, B. S., & Jerry, M. (2007). Advanced Organic Chemistry: Reactions, Mechanisms, and Structure (6th ed.). New York: John Wiley & Sons.

Mie, R., Samsudin, M. W., Din, L. B., Ahmad, A., Ibrahim, N., & Adnan, S. N. A. (2014). Synthesis of silver nanoparticles with antibacterial activity using the lichen Parmotrema praesorediosum. *International Journal of Nanomedicine, 9*, 121. PMID:24379670

Milojković, J., Pezo, L., Stojanović, M., Mihajlović, M., Lopičić, Z., Petrović, J., & Kragović, M. (2016). Selected heavy metal biosorption by compost of Myriophyllum spicatum—a chemometric approach. *Ecological Engineering, 93*, 112–119. doi:10.1016/j.ecoleng.2016.05.012

Mirbagherp, S. A., & Hosseini, S. (2004). Pilot plant investigation on petrochemical wastewater treatment for the removal of copper and chromium with the objective of reuse. *Desalination, 171*(1), 85–93. doi:10.1016/j.desal.2004.03.022

Mise, S., & Patil, T. N. (2015). Adsorption Studies of Chromium (VI) on Activated Carbon Derived from Mangifera indica. *Journal of The Institution of Engineers (India): Series A, 96*(3), 237-247.

Mishra, R. (1968). *Ecology Work Book*. Oxford and IBH Publishing Company Culcutta.

Mishra, V., Balomajumder, C., & Agarwal, V. K. (2010). Biosorption of Zn (II) onto the surface of non-living biomasses: A comparative study of adsorbent particle size and removal capacity of three different biomasses. *Water, Air, and Soil Pollution, 211*(1-4), 489–500. doi:10.1007/s11270-009-0317-0

Modak, J. M., & Natarajan, K. A. (1995). Biosorption of metals using nonliving biomass-A review. *Minerals & Metallurgical Processing, 12*(4), 189–196.

Moghadam, M. R., Nasirizadeh, N., Dashti, Z., & Babanezhad, E. (2013). Removal of Fe (II) from aqueous solution using pomegranate peel carbon: Equilibrium and kinetic studies. *International Journal of Industrial Chemistry*, *4*(1), 19. doi:10.1186/2228-5547-4-19

Moghimifar, V., Raisi, A., & Aroujalian, A. (2014). Surface modification of polyethersulfone ultrafiltration membranes by corona plasma-assisted coating TiO 2 nanoparticles. *Journal of Membrane Science*, *461*, 69–80. doi:10.1016/j.memsci.2014.02.012

Mohammadi, T., Mohebb, A., Sadrzadeh, M., & Razmi, A. (2005). Modeling of metal ion removal from wastewater by electrodialysis. *Separation and Purification Technology*, *41*(1), 73–82. doi:10.1016/j.seppur.2004.04.007

Mohammed, T. J., & Ibrahim, R. I. (2016). Remediation of Cu (II) from Well Water of Iraq by Using Cortex of Fruits and Agricultural Waste. *Arabian Journal for Science and Engineering*, *41*(2), 345–355. doi:10.1007/s13369-015-1848-x

Mohanpuria, P., Rana, N. K., & Yadav, S. K. (2008). Biosynthesis of nanoparticles: Technological concepts and future applications. *Journal of Nanoparticle Research*, *10*(3), 507–517. doi:10.1007/s11051-007-9275-x

Mohanty, B. C., Chandranouli, K. V., & Nayak, N. D. (1984). Natural dyeing processes of India. Calico Museum of Textiles.

Mohiuddin, K. M., Zakir, H. M., Otomo, K., Sharmin, S., & Shikazono, N. (2010). Geochemical distribution of trace metal pollutants in water and sediments of downstream of an urban river. *International Journal of Environmental Science and Technology*, *7*(1), 17–28. doi:10.1007/BF03326113

Mokhtar, H., Morad, N., & Fizri, F. F. A. (2011). Phytoaccumulation of copper from aqueous solutions using *Eichhornia Crassipes* and *Centella asiatica*. *International Journal of Environmental Sciences and Development*, *3*(1), 89–95.

Mondal, P., Majumdar, C. B., & Mohanty, B. (2006). Laboratory based approaches for arsenic remediation from contaminated water: Recent developments. *Journal of Hazardous Materials*, *B137*(1), 464–479. doi:10.1016/j.jhazmat.2006.02.023 PMID:16616812

Moneim, M. A. (2009). Staff suggestion scheme (3Ss) within the UAE context: Implementation and critical success factors. *Education, Business and Society*, *2*(2), 153–167. doi:10.1108/17537980910960717

Monier, M., Ayad, D. M., Wei, Y., & Sarhan, A. A. (2010). Preparation and characterization of magnetic chelating resin based on chitosan for adsorption of Cu (II), Co (II), and Ni (II) ions. *Reactive & Functional Polymers*, *70*(4), 257–266. doi:10.1016/j.reactfunctpolym.2010.01.002

Monteiro, C. M., Castro, P. M. L., & Malcata, F. X. (2012). Metal up take by microalgae: Underlying mechanisms and practical applications. *Biotechnology Progress*, *28*(2), 299–311. doi:10.1002/btpr.1504 PMID:22228490

Mooney, H. A., Lubchenco, J., Dirzo, R., & Sala, Q. E. (1995). *Biodiversity and Ecosystem functioning: Ecosystem analysis.* In V. H. Heywood & R. T. Watson (Eds.), *Global Biodiversity assessment* (pp. 326–345). UNEP.

Morais, S., Costa, F. G., & Pereira, M. L. (2012). Heavy Metals and Human Health. In *Environmental Health - Emerging Issues and Practice* (pp. 227-246). Academic Press. Available from: http://www.intechopen.com/books/environmental-health-emerging-issuesand-practice/heavy-metals-and-human-health

Moreno-Barbosa, J. J., López-Velandia, C., del Pilar Maldonado, A., Giraldo, L., & Moreno-Piraján, J. C. (2013). Removal of lead (II) and zinc (II) ions from aqueous solutions by adsorption onto activated carbon synthesized from watermelon shell and walnut shell. *Adsorption*, *19*(2-4), 675–685. doi:10.1007/s10450-013-9491-x

Moreno-Castilla, C., Alvarez-Merino, M. A., López-Ramón, M. V., & Rivera-Utrilla, J. (2004). Cadmium ion adsorption on different carbon adsorbents from aqueous solutions. Effect of surface chemistry, pore texture, ionic strength, and dissolved natural organic matter. *Langmuir*, *20*(19), 8142–8148. doi:10.1021/la049253m PMID:15350085

Moreno-Piraján, J. C., Garcia-Cuello, V. S., & Giraldo, L. (2011). The removal and kinetic study of Mn, Fe, Ni and Cu ions from wastewater onto activated carbon from coconut shells. *Adsorption*, *17*(3), 505–514. doi:10.1007/s10450-010-9311-5

Moustakas, M., Ouzounidou, G., & Lannoye, R. (1995). Aluminum effects on photosynthesis and elemental uptake in an aluminium-tolerant and non-tolerant wheat cultivar. *Journal of Plant Nutrition*, *18*(4), 669–683. doi:10.1080/01904169509364930

Mubarak, N. M., Sahu, J. N., Abdullah, E. C., & Jayakumar, N. S. (2014). Removal of heavy metals from wastewater using carbon nanotubes. *Separation and Purification Reviews*, *43*(4), 311–338. doi:10.1080/15422119.2013.821996

Mubayi, A., Chatterji, S., & Watal, G. (2017). Green Fabrication of Metal Nanoparticles. In Intelligent Nanomaterials (2nd ed.). Scrivener Publishing LLC.

Mugica-Alvarez, V., Cortes-Jimenez, V., Vaca-Mier, M., & Dommguez-Soria, V. (2015). Phytoremediation of mine tailings using *Lolium multiflorum. International Journal of Environmental Sciences and Development*, *6*(4), 246–251. doi:10.7763/IJESD.2015.V6.599

Mukhopadhyay, S., & Maiti, S. K. (2010). Phytoremediation of metal enrhiched mine wate: A review. *Global Environmental Research*, *4*, 135–150.

Muller, G. (1969). Index of Geo-accumulation in sediments of the Rhine River. *Geological Journal*, *2*, 109–118.

Muraleedharan, T. R., & Venkobachar, C. (1990). Mechanism of biosorption of copper (II) by Ganoderma lucidum. *Biotechnology and Bioengineering*, *35*(3), 320–325. doi:10.1002/bit.260350314 PMID:18592525

Muralidhara, H., Maggin, B., & Phipps, H. Jr. (1982). Conversion of tannery waste to useful products. *Resources and Conservation*, *8*(1), 43–59. doi:10.1016/0166-3097(82)90052-9

Muscatello, J., Belknap, A., & Janz, D. (2008). Accumulation of selenium in aquatic systems downstream of a uranium mining operation in northern Saskatchewan, Canada. *Environmental Pollution*, *156*(2), 387–393. doi:10.1016/j.envpol.2008.01.039 PMID:18346828

Mwiganga, M., & Kansiime, F. (2005). The impact of Mpererwe landfill in Kampala–Uganda, on the surrounding environment. *Physics and Chemistry of the Earth*, *30*(11-16), 744–750. doi:10.1016/j.pce.2005.08.016

Nagajyoti, P. C., Lee, K. D., & Sreekanth, T. V. M. (2010). Heavy metals, occurrence and toxicity for plants: A review. *Environmental Chemistry Letters*, *8*(3), 199–216. doi:10.1007/s10311-010-0297-8

Nageswara, L. R., & Prabhakara, G. (2011). Removal of heavy metals by biosorption–an overall review J. *Engg. Res. Sci*, *II*(IV), 17–22.

Nascimento, A. M. A., & Chartone-Souza, E. (2003). Operon *mer*: Bacterial resistance to mercury and potential for bioremediation of contaminated environments. *Genetics and Molecular Research*, *2*, 92–101. PMID:12917805

Nelson & Sommers. (1972). Total carbon, organic carbon, and organic matter. In Methods of soil analysis, part 2 Chemical and microbiological properties. ASA.

Nematian, M. A., & Kazemeini, F. (2013). Accumulation of Pb, Zn, Cu and Fe in plants and hyperaccumulator choice in Galali iron mine area. Iran. *International Journal of Agriculture and Crop Sciences*, *5*(4), 426–432.

Nemerow, N. L. (1978). *Industrial Water Pollution: Origins, Characteristics and Treatment*. Reading, MA: Addison-Wesley.

Neto, C., Guinea, F., Peres, N., Novoselov, K. S., & Geim, A. K. (2009). The electronic properties of grapheme. *Reviews of Modern Physics*, *81*(1), 109–162. doi:10.1103/RevModPhys.81.109

Ng, I. S., Wu, X., Yang, X., Xie, Y., Lu, Y., & Chen, C. (2013). Synergistic effect of Trichoderma reesei cellulases on agricultural tea waste for adsorption of heavy metal Cr (VI). *Bioresource Technology*, *145*, 297–301. doi:10.1016/j.biortech.2013.01.105 PMID:23419991

Nguyen, T.H.D., Vardhanabhuti, B., Lin, M., & Mustapha, A. (2017). *Food Control 2017. Antibacterial properties of selenium nanoparticles and their toxicity to Caco-2 cells*. Academic Press.

Nichol, B. E., Oliveira, L. A., Glass, A. D. M., & Saiddiqi, M. Y. (1993). The effect of aluminum on the influx of calcium, potassium, ammonium nitrate and phosphate in an aluminum sensitive cultivar of barley (*Hordeum vulgare L.*). *Plant Physiology*, *101*(4), 1263–1266. doi:10.1104/pp.101.4.1263 PMID:12231781

Nies, D. H. (1999). Microbial heavy-metal resistance. *Applied Microbiology and Biotechnology*, *51*(6), 730–750. doi:10.1007/s002530051457 PMID:10422221

Nisha, S.N., Aysha, O.S., & Rahaman, J.S.N. (2014). Lemon peels mediated synthesis of silver nanoparticles and its antidermatophytic activity. *SpectrochimicaActa Part A: Molecular and Biomolecular Spectroscopy, 124*, 194–198.

Nithya, R., & Ragunathan, R. (2010). *Decolorization of the dye Congo red by Aspergillus nigernanoparticle, Nation. Semi.* Anna Univer., Trichy.

Nithya, R., & Ragunathan, R. (2010). *Proc. On, Decolorization of the dye Congo red byAspergillus nigernanoparticle, Nation.* Semi. Anna Univer., Trichy.

Nordstrom, D. K., Alpers, C. N., Ptacek, C. J., & Blowes, D. W. (2000). Negative pH and extremely acidic mine waters from Iron Mountain, California. *Environmental Science & Technology*, *34*(2), 254–258. doi:10.1021/es990646v

Norris, P. R., & Kelly, D. P. (1988). *Biohydrometallurgy*. Chippenham, UK: A. Rowe.

Norvill, Z. N., Toledo-Cervantes, A., Blanco, S., Shilton, A., Guieysse, B., & Muñoz, R. (2017). Photodegradation and sorption govern tetracycline removal during wastewater treatment in algal ponds. *Bioresource Technology*, *232*, 35–43. doi:10.1016/j.biortech.2017.02.011 PMID:28214443

Nourbakhsh, M., Sag, Y., Ozer, D. A., Aksu, Z., Kutsal, T., & Çağlar, A. (1994). Comparative study of various biosorbents for removal of Cr (VI) ions from industrial waste waters. *Process Biochemistry*, *29*(1), 1–5. doi:10.1016/0032-9592(94)80052-9

Obenauf, P., & Skavroneck, S. (1997). *Mercury source sector asssessment for the Greater Milwaukee Area*. Retrieved from Organization, W. H. (1976). Environmental health criteria. 1. Mercury. *Environmental Health Criteria*, 1.

Okafor, E. Ch., & Opuene, K. (2007). Preliminary assessment of trace metals and polycyclic aromatic hydrocarbons in the sediments. *International Journal of Environmental Science and Technology*, *4*(2), 233–240. doi:10.1007/BF03326279

Okoye, A. I., Ejikeme, P. M., & Onukwuli, O. D. (2010). Lead removal from wastewater using fluted pumpkin seed shell activated carbon: Adsorption modeling and kinetics. *International Journal of Environmental Science and Technology*, *7*(4), 793–800. doi:10.1007/BF03326188

Olson, S. R., & Sommers, L. E. (1982). Phosphorus. In Methods of soil analysis, Part 2 Chemical and Microbiological properties. ASA.

Olson, G. J. (1994). Microbial oxidation of gold ores and gold bioleaching. *FEMS Microbiology Letters*, *119*(1-2), 1–6. doi:10.1111/j.1574-6968.1994.tb06858.x

Olson, G. J., Sakai, C. K., Parks, E., & Brinckman, F. (1990). Bioleaching of cobalt from smelter wastes byThiobacillusferrooxidans. *Journal of Industrial Microbiology & Biotechnology*, *6*(1), 49–52. doi:10.1007/BF01576176

Olson, J. W., Mehta, N. S., & Maier, R. J. (2001). Requirement of nickel metabolism protein HypA and HypB for full activity of both hydrogenase and urease in *Helicobacter pylori. Molecular Microbiology, 39*(1), 176–182. doi:10.1046/j.1365-2958.2001.02244.x PMID:11123699

Onundi, Y. B., Mamun, A. A., Al Khatib, M. F., & Ahmed, Y. M. (2010). Adsorption of copper, nickel and lead ions from synthetic semiconductor industrial wastewater by palm shell activated carbon. *International Journal of Environmental Science and Technology, 7*(4), 751–758. doi:10.1007/BF03326184

Onyango, M. S., Kojima, Y., Kumar, A., Kuchar, D., Kubota, M., & Matsuda, H. (2006). Uptake of fluoride by Al^{3+} pretreated low-silica synthetic zeolites: Adsorption equilibrium and rate studies. *Separation Science and Technology, 41*(4), 683–704. doi:10.1080/01496390500527019

Organization, W. H. (1996). *Trace elements in human nutrition and health.* World Health Organization.

Othmer, K. (2004). Amine by Reduction. *Encylopedia of Chemical Technology, 2*, 476–490.

Owamah, H. I. (2014). Biosorptive removal of Pb (II) and Cu (II) from wastewater using activated carbon from cassava peels. *Journal of Material Cycles and Waste Management, 16*(2), 347–358. doi:10.1007/s10163-013-0192-z

Owlad, M., Aroua, M. K., Daud, W. A. W., & Baroutian, S. (2009). Removal of hexavalent chromium-contaminated water and wastewater: A review. *Water, Air, and Soil Pollution, 200*(1-4), 59–77. doi:10.1007/s11270-008-9893-7

Padil, V. V. T., Wacławek, S., & Černík, M. (2016). Green Synthesis: Nanoparticles and Nanofibres Based on Tree Gums for Environmental Applications. *Ecological Chemistry and Engineering. S, 23*(4), 533–557. doi:10.1515/eces-2016-0038

Padmavathiamma, P. K., & Li, L. Y. (2007). Phytoremediation technology: Hyperaccumulation metals in plants. *Water, Air, and Soil Pollution, 184*(1-4), 105–126. doi:10.1007/s11270-007-9401-5

Pan, B., & Xing, B. (2008). Adsorption mechanisms of organic chemicals on carbon nanotubes. *Environmental Science & Technology, 42*(24), 9005–9013. doi:10.1021/es801777n PMID:19174865

Pandey, B., & Fulekar, M. H. (2012). Bioremediation technology: A new horizon for environmental clean-up. *Biology and Medicine, 4*(1), 51–59.

Pandey, S., Kumari, M., Singh, S., Bhattacharya, A., Mishra, S., Chauhan, P., & Mishra, A. (2015). *Bioremediation via Nanoparticles: An Innovative Microbial Approach.* Academic Press; doi:10.4018/978-1-4666-8682-3.ch019

Pandey, V. C., Abhilash, P. C., Upadhyay, R. N., & Tewari, D. D. (2009). Application of fly ash on the growth performance and translocation of toxic heavy metals within *Cajanus cajan* L.: Implication for safe utilization of fly ash for agricultural production. *Journal of Hazardous Materials, 166*(1), 255–259. doi:10.1016/j.jhazmat.2008.11.016 PMID:19111395

Pandey, V. C., & Singh, B. (2012). Rehabilitation of coal fly ash basins: Current need to use ecological engineering. *Ecological Engineering*, *49*, 190–192. doi:10.1016/j.ecoleng.2012.08.037

Pandey, V. C., Singh, J. S., Kumar, A., & Tewari, D. D. (2010). Accumulation of heavy metal by chickpea grown in fly ash treated soil: Effect on antioxidants. CLEAN – Soil, Air. *Water (Basel)*, *38*, 1116–1123.

Pandey, V. C., Singh, J. S., Singh, R. P., Singh, N., & Yunus, M. (2011). Arsenic hazards in coal FA and its fate in Indian scenario. *Resources, Conservation and Recycling*, *55*(9-10), 819–835. doi:10.1016/j.resconrec.2011.04.005

Pan, F., Zhong, X., Xia, D., Yin, X., Li, F., Zhao, D., & Liu, W. (2017). Nanoscale zero-valent iron/persulfate enhanced upflow anaerobic sludge blanket reactor for dye removal: Insight into microbial metabolism and microbial community. *Scientific Reports*, 7. PMID:28300176

Pan, Y. M., Jiang, R. X., Yang, J. L., Zheng, H., & Yin, E. Q. (2016). Removal of Pb (II) ions from aqueous solutions by litchi pericarp and its leachate. *Journal of Central South University*, *23*(7), 1626–1632. doi:10.1007/s11771-016-3217-7

Parnian, A., Chorom, M., Jaafarzadeh, N., & Dinarvand, M. (2016). Use of two aquatic macrophytes for the removal of heavy metals from synthetic medium. *Ecohydrology & Hydrobiology*, *16*(3), 194–200. doi:10.1016/j.ecohyd.2016.07.001

Patel, J., Zhang, Q., & Michael, R. (2010). Genetic engineering of *Caulobacter crescentus* for removal of cadmium from water. *Applied Biochemistry and Biotechnology*, *160*(1), 232–243. doi:10.1007/s12010-009-8540-0 PMID:19214794

Patnaik, P. (2004). *Dean's Analytical Chemistry Handbook*. New York: McGraw-Hill.

Patra, A., Pradhan, D., Kim, D., Ahn, J., & Yoon, H. (2011). Review on bioleaching of uranium from low-grade ore. *Journal of the Korean Institute of Resources Recycling*, *20*(2), 30–44. doi:10.7844/kirr.2011.20.2.030

Patterson, J. W., & Passino, R. (1987). *Metals separation and recovery*. Metal Speciation Separation and Recovery.

Pavithra, S., & Shanthakumar, S. (2017). Removal of COD, BOD and Color from Municipal Solid Waste Leachate using Silica and Iron nano particles-A Comparative Study. *Global NEST Journal*, 19.

Pereira, M. R., Arroyo, P. A., de Barros, M. A. S. D., Sanches, V. M., da Silva, E. A., Fonseca, I. M., & Lovera, R. G. (2006). Chromium adsorption in olive stone activated carbon. *Adsorption*, *12*(2), 155–162. doi:10.1007/s10450-006-0377-z

Perpetuo, E. A., Souza, C. B., & Nascimento, C. A. O. (2011). Engineering bacteria for bioremediation. In A. Carpi (Ed.), *Progress in Molecular and Environmental Bioengineering from Analysis and Modeling toTechnology Applications* (pp. 605–632). Rijeka, Croatia: In Tech Publishers.

Petersen, J., & Dixon, D. (2002). Thermophilic heap leaching of a chalcopyrite concentrate. *Minerals Engineering*, *15*(11), 777–785. doi:10.1016/S0892-6875(02)00092-4

Philip, T. P. (2007). Process control in metallurgical plants–from an xstrata perspective. *IFAC Proceedings Volumes*, *40*(11), 377-389.

Phukan, P., Phukan, R., & Phukan, S. N. (2015). Heavy metal uptake capacity of *Hydrilla verticillata*: A commonly available aquatic plant. *Int. Res. J. Environ. Sci.*, *4*(3), 35–40.

Pielou, E. C. (1966). The measurement of diversity in different types of biological collections. *Journal of Theoretical Biology*, *13*, 131–144. doi:10.1016/0022-5193(66)90013-0

Pierik, R.L.M. (1987). *In vitro Culture of Higher Plants*. Academic Press.

Piper, C. S. (1966). *Soil and Plant Analysis*. New York: Inter Science.

Pobozy, E., Knell, M., Kilian, K., Kataky, R., & Trojanowicz, M. (2003). Capillary electrophoresis speciation of chromium in leather tanning liquor. *Electrophoresis*, *24*(1213), 2259–2263. doi:10.1002/elps.200305461 PMID:12858399

Pollard, A. J., Stewart, H. L., & Roberson, C. B. (2009). Manganese hyperaccumulation in *Phytolacca americana L.* from the southeastern United States. *Northeastern Naturalist*, *16*(sp5), 155–162. doi:10.1656/045.016.0513

Potysz, A., Lens, P. N. L., van de Vossenberg, J., Rene, E. R., Grybos, M., Guibaud, G., ... van Hullebusch, E. D. (2016). Comparison of Cu, Zn and Fe bioleaching from Cu-metallurgical slags in the presence of *Pseudomonas fluorescens* and *Acidithiobacillus thiooxidans*. *Applied Geochemistry*, *68*, 39–52. doi:10.1016/j.apgeochem.2016.03.006

Pradeep, T. (2007). *Nano: The Essentials*. New Delhi: Tata-McGraw Hill.

Prasad, M. N. V., & Freitas, H. (2003). Metal hyperaccumulation in plants: Biodiversity prospecting for phytoremediation technology. *Electronic Journal of Biotechnology*, *6*(3), 275–321. doi:10.2225/vol6-issue3-fulltext-6

Prigione, V., Grosso, I., Tigini, V., Anastasi, A., & Varese, G. C. (2012). Fungal waste-biomasses as potential low-cost biosorbents for decolorization of textile wastewaters. *Water (Basel)*, *4*(4), 770–784. doi:10.3390/w4040770

Puranik, P., Modak, J., & Paknikar, K. (1999). A comparative study of the mass transfer kinetics of metal biosorption by microbial biomass. *Hydrometallurgy*, *52*(2), 189–197. doi:10.1016/S0304-386X(99)00017-1

Purcino, A. A. C., Alves, V. M. C., Parentoni, S. N., Belele, C. L., & Loguercio, L. L. (2003). Aluminum effects on nitrogen uptake and nitrogen assimilating enzymes in maize genotypes with contrasting tolerance to aluminum toxicity. *Journal of Plant Nutrition*, *26*(1), 31–61. doi:10.1081/PLN-120016496

Qiao, X. Q., Zheng, Z. Z., Zhang, L. F., Wang, J. H., Shi, G. X., & Xu, X. Y. (2015). Lead tolerance mechanism in sterilized seedlings of *Potamogeton crispus L.* Subcellular distribution, poluamine and proline. *Chemosphere*, *120*, 179–187. doi:10.1016/j.chemosphere.2014.06.055 PMID:25033323

Qin, T., Wu, Y., & Wang, H. (1993). Effect of cadmium, lead and their interactions on the physiological and biochemical characteristics of Brassica chinensis. *Acta Ecologica Sinica*, *14*(1), 46–50.

Qu, X., Alvarez, P. J. J., & Li, Q. (2013). Applications of Nanotechnology in Water and Wastewater Treatment. *Water Research*, *47*(12), 3931–3946. doi:10.1016/j.watres.2012.09.058 PMID:23571110

Qu, X., Alvarez, P. J. J., & Li, Q. (2013). Applications of nanotechnology in water and wastewater treatment. *Water Research*, *47*, 3931–3946.

R'ıos, C. A., Williams, C. D., & Roberts, C. L. (2008). Removal of heavy metals from acid mine drainage (AMD) using coal fly ash, natural clinker and synthetic zeolites. *Journal of Hazardous Materials*, *156*(1–3), 23–35. doi:10.1016/j.jhazmat.2007.11.123 PMID:18221835

Raaijmakers, J. M., Paulitz, T. C., Steinberg, C., Alabouvette, C., & Moënne-Loccoz, Y. (2009). The rhizosphere: A playground and battlefield for soilborne pathogens and beneficial microorganisms. *Plant and Soil*, *321*(1-2), 341–361. doi:10.1007/s11104-008-9568-6

Radwanek-Bak, B. (2014). Opportunities and barriers of critical minerals management in Poland. *International Multidisciplinary Scientific GeoConference: SGEM: Surveying Geology & mining Ecology Management*, *3*, 411.

Raghavan, R., Mohanan, P., & Swarnkar, S. (2000). Hydrometallurgical processing of lead-bearing materials for the recovery of lead and silver as lead concentrate and lead metal. *Hydrometallurgy*, *58*(2), 103–116. doi:10.1016/S0304-386X(00)00108-0

Rahbar, A., Farjadfard, S., Leili, M., Kafaei, R., Haghshenas, V., & Ramavandi, B. (2016). Experimental data of biomaterial derived from Malva sylvestris and charcoal tablet powder for Hg 2+ removal from aqueous solutions. *Data in Brief*, *8*, 132–135. doi:10.1016/j.dib.2016.05.032 PMID:27294181

Rai, U. N., Pandey, K., Sinha, S., Sinh, A., Saxena, R., & Gupta, D. K. (2004). Revegetating fly-ash landfills with Prosopis *julifora* L.: Impact of different amendments and *Rhizobium* inoculation. *Environment International*, *30*(3), 293–300. doi:10.1016/S0160-4120(03)00179-X PMID:14987858

Rajakumar, G., & Rahuman, A. A. (2011). Larvicidal activity of synthesized silver nanoparticles using Ecliptaprostrata leaf extract against fiariasis and malaria vectors. *Acta Tropica*, *118*(3), 196–203. doi:10.1016/j.actatropica.2011.03.003 PMID:21419749

Rajaram, T., & Das, A. (2008). Water pollution by industrial effluents in India: Discharge scenarios and case for participatory ecosystem specific local regulation. *Futures*, *40*(1), 56–69. doi:10.1016/j.futures.2007.06.002

Rajendran, P., & Gunasekaran, P. (2010). Nanotechnology for Bioremediation of Heavy Metals. *Environmental Bioremediation Technologies*, 211-221.

Ramamoorthy, V., Viswanathan, R., Raguchander, T., Prakasam, V., & Samiyappan, R. (2001). Induction of systemic resistance by plant growth promoting rhizobacteria in crop plants against pests and diseases. *Crop Protection (Guildford, Surrey)*, *20*(1), 1–11. doi:10.1016/S0261-2194(00)00056-9

Rawlings, D. E. (2005). Characteristics and adaptability of iron-and sulfur-oxidizing microorganisms used for the recovery of metals from minerals and their concentrates. *Microbial Cell Factories*, *4*(1), 13. doi:10.1186/1475-2859-4-13 PMID:15877814

Rawlings, D. E., Dew, D., & du Plessis, C. (2003). Biomineralization of metal-containing ores and concentrates. *Trends in Biotechnology*, *21*(1), 38–44. doi:10.1016/S0167-7799(02)00004-5 PMID:12480349

Recillas, S., Colón, J., Casals, E., González, E., Puntes, V., Sánchez, A., & Font, X. (2010). Chromium VI adsorption on cerium oxide nanoparticles and morphology changes during the process. *Journal of Hazardous Materials*, *184*(1), 425–431. doi:10.1016/j.jhazmat.2010.08.052 PMID:20832170

Reddy, K. R. (2010). Nanotechnology for site remediation: Dehalogenation of organic pollutants in soils and groundwater by nanoscale iron particles. *Proceedings of the 6th International Congress on Environmental Geotecnics*, *1*, 163-180.

Reddy, K. R., & Adams, J. A. (2010). *Sustainable Remediation of Contaminated Sites*. New York, NY: Momentum Press.

Reddy, K. R., Khodadoust, A. P., & Darko-Kagya, K. (2014). Transport and reactivity of lactate-modified nanoscale iron particles for remediation of DNT in subsurface soils. *Journal of Environmental Engineering*, *140*(12), 04014042. doi:10.1061/(ASCE)EE.1943-7870.0000870

Reeves, R. D., & Baker, A. J. M. (2000). Metal accumulating plants. In I. Raskin & B. D. Ensley (Eds.), *Phytoextraction of toxic metals-using plants to clean the environment* (pp. 193–229). New York, NY: Wiley.

Reinert, J., & Bajaj, P. P. S. (1977). *Applied and Fundamental Aspect of Plant Cell, Tissue and Organ Culture*. Berlin: Springer-Verlag. doi:10.1007/978-3-662-02279-5

Rengel, Z., & Elliott, D. C. (1992). Mechanism of aluminium inhibition of net ^{45}Ca2p Uptake by Amaranths Protoplasts. *Plant Physiology*, *98*(2), 632–638. doi:10.1104/pp.98.2.632 PMID:16668688

Rether, A., & Schuster, M. (2003). Selective separation and recovery of heavy metal ions using water-soluble N-benzoylthiourea modified PAMAM polymers. *Reactive & Functional Polymers*, *57*(1), 13–21. doi:10.1016/j.reactfunctpolym.2003.06.002

Ribeiro, M. A. Q., de Almeida, A. A. F., Mielke, M. S., Gomes, F. P., Pires, M. V., & Baligar, V. C. (2013). Aluminum effects on growth, photosynthesis, and mineral nutrition of cacao genotypes. *Journal of Plant Nutrition*, *36*(8), 1161–1179. doi:10.1080/01904167.2013.766889

Riekkola-Vanhanen, M. (2010). TalvivaaraSotkamo mine—bioleaching of a polymetallic nickel ore in subarctic climate. *New Biotechnology*, *10*(1), 7–14. PMID:20709631

Rivela, B., Moreira, M., Bornhardt, C., Méndez, R., & Feijoo, G. (2004). Life cycle assessment as a tool for the environmental improvement of the tannery industry in developing countries. *Environmental Science & Technology*, *38*(6), 1901–1909. doi:10.1021/es034316t PMID:15074705

Rizwan, M., Singh, M., Mitra, C. K., & Morve, R. K. (2014). Ecofriendly application of nanomaterials: Nanobioremediation. *Journal of Nanoparticles*, 1–7.

Rizwan, Md., Singh, M., Mitra, C.K., & Morve, R.K. (2014). Ecofriendly Application of Nanomaterials: Nanobioremediation. *Journal of Nanoparticles*. 10.1155/2014/431787

Robinson, R. (1957). The Life and Work of Sir William Henry Perkin. In *Proceedings of the Perkin Centennial 1856-1956*. AATCC.

Robinson, B., Fernandez, J. E., Madejon, P., Maranon, T., Murillo, J., Green, S., & Clothier, B. (2003). Phytoextraction: An assessment of biogeochemical and economic viability. *Plant and Soil*, *249*(1), 117–125. doi:10.1023/A:1022586524971

Robinson, T., McMullan, G., Marchant, R., & Nigam, P. (2001). Remediation of dyes in textile effluent: A critical review on current treatment technologies with a proposed alternative. *Bioresource Technology*, *77*(3), 247–255. doi:10.1016/S0960-8524(00)00080-8 PMID:11272011

Rodriguez, E., Azevedo, R., Fernandes, P., & Santos, C. (2011). Cr (VI) induces DNA damage, cell cycle arrest and polyploidization: A flow cytometric and comet assay study in *Pisum sativum*. *Chemical Research in Toxicology*, *24*(7), 1040–1047. doi:10.1021/tx2001465 PMID:21667992

Roy, A. K., Sharma, A., & Talukder, G. (1988). Some aspects of aluminum toxicity in plants. *Botanical Review*, *54*(2), 145–178. doi:10.1007/BF02858527

Rubio, J., & Kitchener, J. A. (1976). The mechanism of adsorption of poly (ethylene oxide) flocculant on silica. *Journal of Colloid and Interface Science*, *57*(1), 132–142. doi:10.1016/0021-9797(76)90182-X

Rueping, M., & Nachtsheim, B. J. (2010). A review of new developments in the Friedel–Crafts alkylation – From green chemistry to asymmetric catalysis. *Beilstein Journal of Organic Chemistry*, *6*(6). doi:10.3762/bjoc.6.6 PMID:20485588

Ruparelia, J. P., Duttagupta, S. P., Chatterjee, A. K., & Mukherji, S. O. U. M. Y. A. (2008). Potential of carbon nanomaterials for removal of heavy metals from water. *Desalination*, *232*(1), 145–156. doi:10.1016/j.desal.2007.08.023

Saba, G., Parizanganeh, A. H., Zamani, A., & Saba, J. (2015). Phytoremediation of heavy metals contaminated environments: Screening for native accumulator plants in Zanjan-Iran. *International Journal of Environmental of Research*, *9*(1), 309–316.

Sadegh, H., Ali, G. A., Gupta, V. K., Makhlouf, A. S. H., Shahryari-ghoshekandi, R., Nadagouda, M. N., & Megiel, E. (2017). The role of nanomaterials as effective adsorbents and their applications in wastewater treatment. *Journal of Nanostructure in Chemistry*, 1-14.

Saha, K. (2003). Review of arsenicosis in west Bengal, India—a clinical perspective. *Critical Reviews in Environmental Science and Technology*, *33*(2), 127–163. doi:10.1080/10643380390814514

Saha, R., Mukherjee, K., Saha, I., Ghosh, A., Ghosh, S. K., & Saha, B. (2013). Removal of hexavalent chromium from water by adsorption on mosambi (Citrus limetta) peel. *Research on Chemical Intermediates*, *39*(5), 2245–2257. doi:10.1007/s11164-012-0754-z

Sakaguchi, T., & Nakajima, A. (1991). Accumulation of heavy metals such as uranium and thorium by microorganisms. In R. W. Smith & M. Misra (Eds.), *Mineral Bioprocessing*. The Minerals, Metals and Materials Society.

Sakulchaicharoen, N., O'Carroll, D. M., & Herrera, J. E. (2010). Enhanced stability and dechlorination activity of pre-synthesis stabilized nanoscaleFePd particles. *Journal of Contaminant Hydrology*, *118*(3-4), 117–127. doi:10.1016/j.jconhyd.2010.09.004 PMID:20934234

Salla, V., Hardaway, C. J., & Sneddon, J. (2011). Preliminary investigation of *Spartina alterniflora* for phytoextraction of selected heavy metals in soil from Southwest Louisiana. *Microchemical Journal*, *97*(2), 207–212. doi:10.1016/j.microc.2010.09.005

Salt, D. E., Smith, R. D., & Raskin, I. (1998). Phytoremediation. *Annual Review of Plant Physiology and Plant Molecular Biology*, *49*(1), 643–668. doi:10.1146/annurev.arplant.49.1.643 PMID:15012249

Samanta, A. K., & Agarwal, P. (2009). Application of natural dyes on textiles. *Indian Journal of Fibre and Textile Research*, *34*(4), 384–399.

Samanta, A. K., & Konar, A. (2011). *Dyeing of Textiles with Natural Dyes*. Available from https://www.researchgate.net/publication/221919685_Dyeing_of_Textiles_with_Natural_Dyes

San, N. O., Celebioglu, A., Tümtaş, Y., Uyar, T., & Tekinay, T. (2014). Reusable bacteria immobilized electrospun nanofibrous webs for decolorization of methylene blue dye in wastewater treatment. *RSC Advances*, *4*(61), 32249–32255. doi:10.1039/C4RA04250F

Santhosh, C., Velmurugan, V., Jacob, G., Jeong, S. K., Grace, A. N., & Bhatnagar, A. (2016). Role of nanomaterials in water treatment applications: A review. *Chemical Engineering Journal*, *306*, 1116–1137. doi:10.1016/j.cej.2016.08.053

Santhy, K., & Selvapathy, P. (2004). Removal of heavy metals from wastewater by adsorption on coir pith activated carbon. *Separation Science and Technology*, *39*(14), 3331–3351. doi:10.1081/SS-200036561

Sarikaya, M., Tamerler, C., Jen, A. K., Schulten, K., & Baneyx, F. (2003). Molecular biomimetics: Nanotechnology through biology. *Nature Materials*, *2*(9), 577–585. doi:10.1038/nmat964 PMID:12951599

Sarioglu, O. F., San Keskin, N. O., Celebioglu, A., Tekinay, T., & Uyar, T. (2017). Bacteria encapsulated electrospun nanofibrous webs for remediation of methylene blue dye in water. *Colloids and Surfaces. B, Biointerfaces*, *152*, 245–251. doi:10.1016/j.colsurfb.2017.01.034 PMID:28119219

Sarioglu, O. F., Yasa, O., Celebioglu, A., Uyar, T., & Tekinay, T. (2013). Efficient ammonium removal from aquatic environments by Acinetobacter calcoaceticus STB1 immobilized on an electrospun cellulose acetate nanofibrous web. *Green Chemistry*, *15*(9), 2566–2572. doi:10.1039/c3gc40885j

Savage, N., & Diallo, M. S. (2005). Nanomaterials and water purification: Opportunities and challenges. *Journal of Nanoparticle Research*, *7*(4-5), 331–342. doi:10.1007/s11051-005-7523-5

Saxena, S., & Raja, A. S. M. (2014). Natural Dyes: Sources, Chemistry, Application and Sustainability Issues. In Roadmap to Sustainable Textiles and Clothing, Textile Science and Clothing Technology. Springer Science + Business Media.

Say, R., Yılmaz, N., & Denizli, A. (2003). Biosorption of cadmium, lead, mercury, and arsenic ions by the fungus Penicillium purpurogenum. *Separation Science and Technology*, *38*(9), 2039–2053. doi:10.1081/SS-120020133

Schippers, A., Hedrich, S., Vasters, J., Drobe, M., Sand, W., & Willscher, S. (2013). Biomining: metal recovery from ores with microorganisms. In *Geobiotechnology I* (pp. 1–47). Springer. doi:10.1007/10_2013_216

Schnell, H. A. (1997). *Bioleaching of copper.in: Biomining*. Springer. doi:10.1007/978-3-662-06111-4_2

Schoolinglin-Jordan, P.M., & Cheeing, K.M. (1999). Biosynthesis of heme in comprehensive natural products vol. 4 Aminoacid, Porphyrins and alkaloid chemistry. Elsevier Science Ltd.

Sehlin, H. M., & Lindström, E. B. (1992). Oxidation and reduction of arsenic by Sulfolobusacidocaldarius strain BC. *FEMS Microbiology Letters*, *93*(1), 87–92. doi:10.1111/j.1574-6968.1992.tb05045.x PMID:1381326

Seidel, D. (1976). Stored technology for possible use in uranium ore processing. *Uranium ore processing (Proceedings of the Mortgage Advisory Group Washington, DC, 1975)*, 79-89.

Şencan, A., Karaboyacı, M., & Kılıç, M. (2015). Determination of lead (II) sorption capacity of hazelnut shell and activated carbon obtained from hazelnut shell activated with ZnCl2. *Environmental Science and Pollution Research International*, *22*(5), 3238–3248. doi:10.1007/s11356-014-2974-9 PMID:24801293

Şengör, S. S., Singh, G., Dohnalkova, A., Spycher, N., Ginn, T. R., Peyton, B. M., & Sani, R. K. (2016). Impact of different environmental conditions on the aggregation of biogenic U (IV) nanoparticles synthesized by Desulfovibrio alaskensis G20. *Biometals*, *29*(6), 965–980. doi:10.1007/s10534-016-9969-6 PMID:27623995

Shafeeq, A., Muhammad, A., Sarfraz, W., Toqeer, A., Rashid, S., & Rafiq, M. (2012). Mercury removal techniques for industrial waste water. *World Academy of Science, Engineering and Technology*, *6*, 12–26.

Shameli, K., Ahmad, M. B., Zamanian, A., Sangpour, P., Shabanzadeh, P., Abdollahi, Y., & Zargar, M. (2012). Green biosynthesis of silver nanoparticles using Curcuma longa tuber powder. *International Journal of Nanomedicine*, *7*, 5603–5610. doi:10.2147/IJN.S36786 PMID:23341739

Shankar, S. S., Ahmad, A., Parsricha, R., & Sastry, M. (2003). Bioreduction of chloroaurate ions by geranium leaves and its endophytic fungus yields gold nanoparticles of different shapes. *Journal of Materials Chemistry*, *13*(7), 1822–1826. doi:10.1039/b303808b

Shannon, C. E., & Weaver, W. (1963). *The Mathematical theory of Communication*. Urbana, IL: University of Illinois Press.

Sharma, D. M., Sharma, C. P., & Tripathi, R. D. (2003). Phytotoxic lesions of chromium in maize. *Chemosphere*, *51*(1), 63–68. doi:10.1016/S0045-6535(01)00325-3 PMID:12586157

Sharma, N., Bhatnagar, P., Chatterjee, S., John, P. J., & Soni, I. P. (2017). Indian Journal of Pharmaceutical and Biological Research (IJPBR). *The Indian Journal of Pharmacy*, *5*(1), 17–24.

Shim, J. W., Park, S. J., & Ryu, S. K. (2001). Effect of modification with HNO_3 and NaOH on metal adsorption by pitch-based activated carbon fibers. *Carbon*, *39*(11), 1635–1642. doi:10.1016/S0008-6223(00)00290-6

Shirsath, D. S., & Shirivastava, V. S. (2015). Adsorptive removal of heavy metals by magnetic nanoadsorbent: An equilibrium and thermodynamic study. *Applied Nanoscience*, *5*(8), 927–935. doi:10.1007/s13204-014-0390-6

Shi, W., Becker, J., Bischoff, M., Turco, R. F., & Konopka, A. E. (2002). Association of Microbial Community Composition and activity with Lead, Chromium and Hydrocarbon contamination. *Applied and Environmental Microbiology*, *68*(8), 3859–3866. doi:10.1128/AEM.68.8.3859-3866.2002 PMID:12147482

Shukla, S. R. (2006). Heavy metals. *Colourage*, *53*(8), 74–77.

Silver, S., & Phung, L. T. (1996). Bacterial heavy metal resistance: New surprises. *Annual Review of Microbiology*, *50*(1), 753–789. doi:10.1146/annurev.micro.50.1.753 PMID:8905098

Simpson, E. M. (1949). Measurement of diversity. *Nature*, *163*(4148), 688. doi:10.1038/163688a0

Singh, N., Srivastava, P.K., Tripathi, R.D., Srivastava, S., & Vaish, A. (2014). Microbial in-situ mitigation of arsenic contamination in plants and soils. *In-Situ Remediation of Arsenic-Contaminated Sites*, 115.

Singh, J. S., Pandey, V. C., & Singh, D. P. (2011). Coal fly ash and farmyard manure amendments in dry-land paddy agriculture field: Effect on N-dynamics and paddy productivity. *Applied Soil Ecology*, *47*(2), 133–140. doi:10.1016/j.apsoil.2010.11.011

Singh, N., Ma, L. Q., Srivastava, M., & Rathinasabapathi, B. (2006). Metabolic adaptations to arsenic-induced oxidative stress in *P. vittata* L. and *Pteris ensiformis* L. *Plant Science*, *170*(2), 274–282. doi:10.1016/j.plantsci.2005.08.013

Singh, O. V., Labana, S., Pandey, G., Budhiraja, R., & Jain, R. K. (2003). Phytoremediation: An overview of metallic ion decontamination from soil. *Applied Microbiology and Biotechnology*, *61*(5-6), 405–412. doi:10.1007/s00253-003-1244-4 PMID:12764555

Singh, P., Kim, Y. J., Zhang, D., & Yang, D.-C. (2016). Biological Synthesis of Nanoparticles from Plants and Microorganisms. *Trends in Biotechnology*, *34*(7), 588–599. doi:10.1016/j.tibtech.2016.02.006 PMID:26944794

Singh, R., Manickam, N., Mudiam, M. K. R., Murthy, R. C., & Misra, V. (2013). An integrated (nano-bio) technique for degradation of -HCH contaminated soil. *Journal of Hazardous Materials*, *35*, 258–259. PMID:23692681

Singh, S. A., & Shukla, S. R. (2016). Adsorptive removal of cobalt ions on raw and alkali-treated lemon peels. *International Journal of Environmental Science and Technology*, *13*(1), 165–178. doi:10.1007/s13762-015-0801-6

Smith, R. W., & Misra, M. (1991). *Mineral Bioprocessing*. The Minerals, Metals and Materials Society.

Sobhanardakani, S., Parvizimosaed, H., & Olyaie, E. (2013). Heavy metals removal from wastewaters using organic solid waste—rice husk. *Environmental Science and Pollution Research International*, *20*(8), 5265–5271. doi:10.1007/s11356-013-1516-1 PMID:23381799

Soderland, P., Lovekar, S., Weiner, D. E., Brooks, D. R., & Kaufman, J. S. (2010). Chronic kidney disease associated with environmental toxins and exposures. *Advances in Chronic Kidney Disease*, *17*(3), 254–264. doi:10.1053/j.ackd.2010.03.011 PMID:20439094

Song, S. T., Saman, N., Johari, K., & Mat, H. (2014). Surface chemistry modifications of rice husk toward enhancement of Hg (II) adsorption from aqueous solution. *Clean Technologies and Environmental Policy*, *16*(8), 1747–1755. doi:10.1007/s10098-014-0803-y

Sounthararajah, D. P., Loganathan, P., Kandasamy, J., & Vigneswaran, S. (2015). Adsorptive removal of heavy metals from water using sodium titanate nanofibres loaded onto GAC in fixed-bed columns. *Journal of Hazardous Materials*, *287*, 306–316. doi:10.1016/j.jhazmat.2015.01.067 PMID:25668299

Srivastava, S., Agrawal, S. B., & Mondal, M. K. (2016). Characterization, isotherm and kinetic study of Phaseolus vulgaris husk as an innovative adsorbent for Cr (VI) removal. *Korean Journal of Chemical Engineering*, *33*(2), 567–575. doi:10.1007/s11814-015-0165-0

Srivastava, S., & Thakur, I. S. (2006). Evaluation of bioremediation and detoxification potentiality of Aspergillus niger for removal of hexavalent chromium in soil microcosm. *Soil Biology & Biochemistry*, *38*(7), 1904–1911. doi:10.1016/j.soilbio.2005.12.016

Stavreva Veselinovska, S. (2011). Microorganisms-indicators of the level of soil pollution with lead. *1st National Agriculture Congress and Exposition on behalf of Ali Numan Kıraç with International Participation Proceedings*, 279-287.

Stillman, M. J., & Presta, A. (2000). Characterizing Metal Ion Interaction with Biological Molecules—the Spectroscopy of Metallothionine. *Mol. Biol. Toxicol. Metals*, *1*, 276–290.

Street, H. S. (Ed.). (1973). *Plant Tissue and Cell Culture*. Oxford, UK: Blackwell Scientific Publications.

Sublet, R., Simonnot, M. O., Boireau, A., & Sardin, M. (2003). A kinetic study of 3-chlorophenol enhanced hydroxyl radical generation during ozonation. *Water Research*, *37*, 4904–4912. doi:10.1016/j.watres.2003.08.010 PMID:14604636

Sudha, R., & Srinivasan, K. (2015). Nickel (II) removal using modified Citrus limettioides peel. *International Journal of Environmental Science and Technology*, *12*(12), 3993–4004. doi:10.1007/s13762-015-0780-7

Sudha, R., Srinivasan, K., & Premkumar, P. (2016). Kinetic, mechanism and equilibrium studies on removal of Pb (II) using Citrus limettioides peel and seed carbon. *Research on Chemical Intermediates*, *42*(3), 1677–1697. doi:10.1007/s11164-015-2111-5

Sukla, L. B., Kar, R. N., & Panchanadikar, V. (1992). Leaching of copper converter slag withAspergillusniger culture filtrate. *Biometals*, *5*(3), 169–172. doi:10.1007/BF01061324

Sukla, L., Panchanadikar, V., & Kar, R. (1993). Microbial leaching of lateritic nickel ore. *World Journal of Microbiology & Biotechnology*, *9*(2), 255–257. doi:10.1007/BF00327850 PMID:24419960

Svecova, L., Spanelova, M., Kubal, M., & Guibal, E. (2006). Cadmium, lead and mercury biosorption on waste fungal biomass issued from fermentation industry. I. Equilibrium studies. *Separation and Purification Technology*, *52*(1), 142–153. doi:10.1016/j.seppur.2006.03.024

Swamy, N. K., Singh, P., & Sarathy, I. P. (2011). Precipitation of phenols from paper industry wastewater using ferric chloride. *Rasayan Journal of Chemistry*, *4*, 452–456.

Szynkowska, M. I., Czerski, K., Paryjczak, T., Rybicki, E., & Wlochowicz, A. (2006). Testing Textiles Using the LA-ICP-MS-TOF Method. *Fibres & Textiles in Eastern Europe*, *14*, 87–90.

Taherian, F., Marcon, V., van der Vegt, N. F., & Leroy, F. (2013). What is the contact angle of water on graphene? *Langmuir*, *29*(5), 1457–1465. doi:10.1021/la304645w PMID:23320893

Tambe Patil, B. B. (2015). Wastewater Treatment Using Nanoparticles. *J. Advances in Chemical Engineering*, *5*, 131.

Taylor, M.P., Van Zyl, L., Tuffin, M., Cowan, D. (2012). Extremophiles and biotechnology: how far have we come. *Extremophiles: microbiolbiotechnol*, *1*, 1-24.

Teitzel, G. M., & Parsek, M. R. (2003). Heavy metal resistance of biofilm and planktonic *Pseudomonas aeruginosa. Applied and Environmental Microbiology*, *69*(4), 2313–2320. doi:10.1128/AEM.69.4.2313-2320.2003 PMID:12676715

Thenmozhi, R., & Santhi, T. (2015). Kinetics and equilibrium studies of the adsorption of Cu (II) from aqueous solution using Acacia nilotica seed pods on $ZnCl_2$ activation. *Research on Chemical Intermediates*, *41*(3), 1327–1341. doi:10.1007/s11164-013-1276-z

Thome, A., Reddy, K. R., Reginatto, C., & Cecchin, I. (2015). Review of nanotechnology forsoil and groundwater remediation: Brazilian perspectives. *Water, Air, and Soil Pollution*, *226*(4), 1–20. doi:10.1007/s11270-014-2243-z

Thorpe, T. A. (1981). *Plant Tissue Culture: Methods and Applications in Agriculture*. New York: Academic Press Inc.

Tinge, J. T., Mencke, K., & Drinkenburg, A. A. H. (1987). The absorption of propane and ethene in slurries of activated carbon in water—I. *Chemical Engineering Science*, *42*(8), 1899–1907. doi:10.1016/0009-2509(87)80136-7

Tofighy, M. A., & Mohammadi, T. (2011). Adsorption of divalent heavy metal ions from water using carbon nanotube sheets. *Journal of Hazardous Materials*, *185*(1), 140–147. doi:10.1016/j.jhazmat.2010.09.008 PMID:20926186

Tomlinson, D. L., Wilson, J. G., Harris, C. R., & Jeffrey, D. W. (1980). Problem in the assessment of heavy metals levels in estuaries and the formation of a pollution index. *Helgoländer Meeresuntersuchungen*, *33*(1-4), 566–575. doi:10.1007/BF02414780

Tong, X., Yuan, L., Luo, L., Yin, X., 2014. Characterization of a Selenium-Tolerant Rhizosphere Strain from a Novel Se-Hyperaccumulating Plant *Cardamine hupingshanesis*. *Scitific World J*, 1-8.

Tratnyek & Johnson. (2006). Nanotechnologies for environmental cleanup. Nanatoday, 1(2), 44-48.

Tripathi, K. P. (2001). *Ecology of a rehabilitated forest on sodic wasteland* (Ph.D Thesis). Lucknow University, Lucknow, India.

Tripathy, S. S., Bersillon, J. L., & Gopal, K. (2006). Removal of fluoride from drinking water by adsorption onto alum-impregnated activated alumina. *Separation and Purification Technology*, *50*(3), 310–317. doi:10.1016/j.seppur.2005.11.036

Tsezos, M. (1983). The role of chitin in uranium adsorption by R. arrhizus. *Biotechnology and Bioengineering*, *25*(8), 2025–2040. doi:10.1002/bit.260250812 PMID:18551548

Tsezos, M., McCready, R., & Bell, J. (1989). The continuous recovery of uranium from biologically leached solutions using immobilized biomass. *Biotechnology and Bioengineering*, *34*(1), 10–17. doi:10.1002/bit.260340103 PMID:18588046

Tsezos, M., & Volesky, B. (1981). Biosorption of uranium and thorium. *Biotechnology and Bioengineering*, *23*(3), 583–604. doi:10.1002/bit.260230309

Tsezos, M., & Volesky, B. (1982). The mechanism of uranium biosorption by Rhizopus arrhizus. *Biotechnology and Bioengineering*, *24*(2), 385–401. doi:10.1002/bit.260240211 PMID:18546310

Tuncuk, A., Stazi, V., Akcil, A., Yazici, E. Y., & Deveci, H. (2012). Aqueous metal recovery techniques from e-scrap: Hydrometallurgy in recycling. *Minerals Engineering*, *25*(1), 28–37. doi:10.1016/j.mineng.2011.09.019

Uçar, S., Erdem, M., Tay, T., & Karagöz, S. (2015). Removal of lead (II) and nickel (II) ions from aqueous solution using activated carbon prepared from rapeseed oil cake by Na2CO3 activation. *Clean Technologies and Environmental Policy*, *17*(3), 747–756. doi:10.1007/s10098-014-0830-8

Ueno, D., Koyama, E., Yamaji, N., & Ma, J. F. (2011). Physiological, genetic, and molecular characterization of a high-Cd-accumulating rice cultivar, Jarjan. *Journal of Experimental Botany*, *62*(7), 2265–2272. doi:10.1093/jxb/erq383 PMID:21127026

Ullah, A., Heng, S., Munis, M. F. H., Fahad, S., & Yang, X. (2015). Phytoremediation of heavy metals assisted by plant growth promoting (PGP) bacteria: A review. *Environmental and Experimental Botany*, *117*, 28–40. doi:10.1016/j.envexpbot.2015.05.001

Underhill, D. (1992). In situ leach uranium mining in the United States of America: past, present and future. *Uranium in situ leaching*.

UNESCAP. (1999). Integrating environmental consideration into economic policy making process: background readings: Vol. 1. Institional arrangement and mechanism at national level (ST/ESCAP/1944). UNESCAP.

UNESCAP. (2000*). Integrating environmental consideration into economic policy making: Institutional issues* (ST/ESCAP/ 1990). UNESCAP.

Urbansky, E. T., & Schock, M. (1999). Issues in managing the risks associated with perchlorate in drinking water. *Journal of Environmental Management*, *56*(2), 79–95. doi:10.1006/jema.1999.0274

Valenzuela, L., Chi, A., Beard, S., Orell, A., Guiliani, N., Shabanowitz, J., ... Jerez, C. A. (2006). Genomics, metagenomics and proteomics in biomining microorganisms. *Biotechnology Advances*, *24*(2), 197–211. doi:10.1016/j.biotechadv.2005.09.004 PMID:16288845

Van Dillewijn, P., Caballero, A., Paz, J. A., González-Pérez, M. M., Oliva, J. M., & Ramos, J. L. (2007). Bioremediation of 2, 4, 6-trinitrotoluene under field conditions. *Environmental Science & Technology*, *41*(4), 1378–1383. doi:10.1021/es062165z PMID:17593745

Vassileva, P., Tzvetkova, P., Lakov, L., & Peshev, O. (2008). Thiouracil modified activated carbon as a sorbent for some precious and heavy metal ions. *Journal of Porous Materials*, *15*(5), 593–599. doi:10.1007/s10934-007-9138-y

Veglio, F., & Beolchini, F. (1997). Removal of metals by biosorption: A review. *Hydrometallurgy*, *44*(3), 301–316. doi:10.1016/S0304-386X(96)00059-X

Veglio, F., Beolchini, F., & Gasbarro, A. (1997). Biosorption of toxic heavy metals: An equilibrium study using free cells of *Arthrobacter* sp. *Process Biochemistry (Barking, London, England)*, *32*(2), 99–105. doi:10.1016/S0032-9592(96)00047-7

Venkatesan, G., Senthilnathan, U., & Rajam, S. (2014). Cadmium removal from aqueous solutions using hybrid eucalyptus wood based activated carbon: Adsorption batch studies. *Clean Technologies and Environmental Policy*, *16*(1), 195–200. doi:10.1007/s10098-013-0628-0

Venkatraman, K. (1952-1972). *The Chemistry of Synthetic Dyes* (Vols. 1-8). Academic Press Inc.

Venkobachar, C. (1990). Metal removal by waste biomass to upgrade wastewater treatment plants. *Water Science and Technology*, *22*, 319–320.

Vidali, M. (2001). Bioremediation. An overview. *Pure and Applied Chemistry*, *73*(7), 1163–1172. doi:10.1351/pac200173071163

Vijayaraghavan, K., & Balasubramanian, R. (2015). Is biosorption suitable for decontamination of metal-bearing wastewaters? A critical review on the state-of-the-art of biosorption processes and future directions. *Journal of Environmental Management*, *160*, 283–296. doi:10.1016/j.jenvman.2015.06.030 PMID:26143501

Vilks, P., Cramer, J., Bachinski, D., Doern, D., & Miller, H. (1993). Studies of colloids and suspended particles, Cigar Lake uranium deposit, Saskatchewan, Canada. *Applied Geochemistry*, *8*(6), 605–616. doi:10.1016/0883-2927(93)90016-A

Vishwanathan, M. (2011). *Principles of Organic Chemistry*. Kollam: Jai Sai Publications.

Volesky, B. (1992). Removal of heavy metals by biosorption. In M. R. Ladisch & A. Bose (Eds.), *Harnessing Biotechnology for the 21st Century*. Washington, DC: American Chemical Society.

Volesky, B., & Holan, Z. (1995). Biosorption of heavy metals. *Biotechnology Progress*, *11*(3), 235–250. doi:10.1021/bp00033a001 PMID:7619394

Vu, H. C., Dwivedi, A. D., Le, T. T., Seo, S. H., Kim, E. J., & Chang, Y. S. (2017). Magnetite graphene oxide encapsulated in alginate beads for enhanced adsorption of Cr (VI) and As (V) from aqueous solutions: Role of crosslinking metal cations in pH control. *Chemical Engineering Journal*, *307*, 220–229. doi:10.1016/j.cej.2016.08.058

Wagner-Döbler, I., Von Canstein, H., Li, Y., Timmis, K. N., & Deckwer, W.-D. (2000). Removal of mercury from chemical wastewater by microoganisms in technical scale. *Environmental Science & Technology*, *34*(21), 4628–4634. doi:10.1021/es0000652

Walkely, Y. A., & Black, I. A. (1934). An examination of the Degtjareff method for determining soil organic matter and a proposed modification of the chromic acid titration method. *Soil Science, 37*(1), 29–38. doi:10.1097/00010694-193401000-00003

Walsh, F. C., & Reade, G. W. (1994). Electrochemical techniques for the treatment of dilute metal-ion solutions. *Studies in Environmental Science, 59*, 3-44.

Wang, L. K., Vaccari, D. A., Li, Y., & Shammas, N. K. (2004). Chemical precipitation. In Physicochemical Treatment Processes (vol. 3, pp. 141-198). Humana Press.

Wang, H., Yuan, X., Wu, Y., Zeng, G., Chen, X., Leng, L., ... Li, H. (2015). Facile synthesis of amino-functionalized titanium metal-organic frameworks and their superior visible-light photocatalytic activity for Cr (VI) reduction. *Journal of Hazardous Materials, 286*, 187–194. doi:10.1016/j.jhazmat.2014.11.039 PMID:25585267

Wang, J., & Chen, C. (2006). Biosorption of heavy metals by Saccharomyces cerevisiae: A review. *Biotechnology Advances, 24*(5), 427–451. doi:10.1016/j.biotechadv.2006.03.001 PMID:16737792

Wang, J., & Chen, C. (2009). Biosorbents for heavy metals removal and their future. *Biotechnology Advances, 27*(2), 195–226. doi:10.1016/j.biotechadv.2008.11.002 PMID:19103274

Wang, J., Shen, S., Kang, J., Li, H., & Guo, Z. (2010). Effect of ore solid concentration on the bioleaching of phosphorus from high-phosphorus iron ores using indigenous sulfur-oxidizing bacteria from municipal wastewater. *Process Biochemistry, 45*(10), 1624–1631. doi:10.1016/j.procbio.2010.06.013

Wang, Q. Q., Wang, L. H., Han, R. B., Yang, L. Y., Zhou, Q., & Huang, X. H. (2015). Effects of bisphenol A on antioxidant system in soybean seedling roots. *Environmental Toxicology and Chemistry, 34*(5), 1127–1133. doi:10.1002/etc.2904 PMID:25651304

Wang, S. G., Gong, W. X., Liu, X. W., Yao, Y. W., Gao, B. Y., & Yue, Q. Y. (2007). Removal of lead (II) from aqueous solution by adsorption onto manganese oxide-coated carbon nanotubes. *Separation and Purification Technology, 58*(1), 17–23. doi:10.1016/j.seppur.2007.07.006

Wang, S., & Peng, Y. (2010). Natural zeolites as effective adsorbents in water and wastewater treatment. *Chemical Engineering Journal, 156*(1), 11–24. doi:10.1016/j.cej.2009.10.029

Wang, X., Cai, W., Lin, Y., Wang, G., & Liang, C. (2010). Mass production of micro/nanostructured porous ZnO plates and their strong structurally enhanced and selective adsorption performance for environmental remediation. *Journal of Materials Chemistry, 20*(39), 8582–8590. doi:10.1039/c0jm01024c

Wang, X., Wang, T., Zheng, X., Shen, Y., & Lu, X. (2017). Isotherms, thermodynamic and mechanism studies of removal of low concentration uranium (VI) by Aspergillus niger. *Water Science and Technology, 75*(12), 2727–2736. doi:10.2166/wst.2017.055 PMID:28659512

Wang, Y. S., & Yang, Z. M. (2005). Nitric oxide reduces aluminum toxicity by preventing oxidative stress in the roots of *Cassia tora L. Plant & Cell Physiology, 46*(12), 1915–1923. doi:10.1093/pcp/pci202 PMID:16179356

Wang, Y., Wang, F., Wan, T., Cheng, S., Xu, G., Cao, R., & Gao, M. (2013). Enhanced adsorption of Pb (II) ions from aqueous solution by persimmon tannin-activated carbon composites. *Journal of Wuhan University of Technology-Mater. Science Editor*, *28*(4), 650–657.

Wang, Z., Feng, Y., Hao, X., Huang, W., & Feng, X. (2014). A novel potential-responsive ion exchange film system for heavy metal removal. *Journal of Materials Chemistry. A, Materials for Energy and Sustainability*, *2*(26), 10263–10272. doi:10.1039/C4TA00782D

Wase, J., & Foster, C. (1997). *Biosorbent for metal Ions*. London: Taylor and Francis, Ltd.

Watling, H. (2008). The bioleaching of nickel-copper sulfides. *Hydrometallurgy*, *91*(1), 70–88. doi:10.1016/j.hydromet.2007.11.012

Weast, R. C. (1984). *CRC handbook of chemistry and physics* (64th ed.). Boca Raton, FL: CRC Press, Inc.

Wei, W., Liu, X., Sun, P., Wang, X., Zhu, H., Hong, M., ... Zhao, J. (2014). Simple whole-cell biodetection and bioremediation of heavy metals based on an engineered lead-specific operon. *Environmental Science & Technology*, *48*(6), 3363–3371. doi:10.1021/es4046567 PMID:24564581

Weng, C. H., Lin, Y. T., Hong, D. Y., Sharma, Y. C., Chen, S. C., & Tripathi, K. (2014). Effective removal of copper ions from aqueous solution using base treated black tea waste. *Ecological Engineering*, *67*, 127–133. doi:10.1016/j.ecoleng.2014.03.053

Wheeler, D. M., & Follett, J. M. (1991). Effect of aluminium on onions, *Asparagus* and *Squash*. *Journal of Plant Nutrition*, *14*(9), 897–912. doi:10.1080/01904169109364251

White, C., & Gadd, G. M. (1990). Biosorption of radionuclides by fungal biomass. *Journal of Chemical Technology and Biotechnology (Oxford, Oxfordshire)*, *46*, 331–343. PMID:1366965

WHO. (2006). *Guidelines for drinking-water quality: recommendations*. World Health Organization.

WHO. (2015). *Drinking-water: Fact sheet No. 391*. Retrieved from http://www.who.int/mediacentre/factsheets/fs391/en/

WHO's drinking water standards. (1993). Retrieved from http://www.lenntech.com/applications/drinking/standards/who-s-drinking-water-standards.htm

Williams, J. W., & Silver, S. (1984). Bacterial resistance and detoxification of heavy metals. *Enzyme and Microbial Technology*, *6*(12), 530–537. doi:10.1016/0141-0229(84)90081-4

World Health Organization (WHO). (2015). *Progress on sanitation and drinking-water – 2015 update*. Geneva, Switzerland: World Health Organization and UNICEF.

World Health Organization. (1996). *Trace Elements in Human Nutrition and Health*. WHO.

Wu, C. H. (2007). Studies of the equilibrium and thermodynamics of the adsorption of Cu^{2+} onto as-produced and modified carbon nanotubes. *Journal of Colloid and Interface Science*, *311*(2), 338–346. doi:10.1016/j.jcis.2007.02.077 PMID:17442333

Wu, Y., Yilihan, P., Cao, J., & Jin, Y. (2013). Competitive adsorption of Cr (VI) and Ni (II) onto coconut shell activated carbon in single and binary systems. *Water, Air, and Soil Pollution, 224*(9), 1662. doi:10.1007/s11270-013-1662-6

Xia, B., Shen, S., & Xue, F. (2013). Phytoextraction of heavy metals from highly contaminated soils using *Sauropus androgynus*. Soil Sed. *Contamin., 22*(6), 631–640.

Xiao, J., Chu, S., Tian, G., Thring, R. W., & Cui, L. (2016). An Eco-tank system containing microbes and different aquatic plant species for the bioremediation of N, N-dimethylformamide polluted river waters. *Journal of Hazardous Materials, 320*, 564–570. doi:10.1016/j.jhazmat.2016.07.037 PMID:27501878

Xing, Y., Chen, X., & Wang, D. (2007). Electrically regenerated ion exchange for removal and recovery of Cr (VI) from wastewater. *Environmental Science & Technology, 41*(4), 1439–1443. doi:10.1021/es061499l PMID:17593754

Xiu, Z., & Alvarez, P. J. J. (2010). Effects of nano-scale zero-valent iron particles on a mixed culture dechlorinating trichloroethylene. *Bioresource Technology, 101*(4), 1141–1146. doi:10.1016/j.biortech.2009.09.057 PMID:19819128

Xi, Y., Mallavarapu, M., & Naidu, R. (2010). Reduction and adsorption of Pb 2 þ in aqueous solution by nano zero-valent iron: A SEM, TEM and XPS study. *Materials Research Bulletin, 45*(10), 1361–1367. doi:10.1016/j.materresbull.2010.06.046

Yadav, K. K., Singh, J. K., Gupta, N., & Kumar, V. (2017). *A Review of Nanobioremediation Technologies for Environmental Cleanup: A Novel Biological Approach*. Academic Press.

Yadav, K. K., Singh, J. K., Gupta, N., & Kumar, V. (2017). A Review of Nanobioremediation Technologies for Environmental Cleanup: A Novel Biological Approach. *Journal of Materials and Environmental Sciences, 8*(2), 740–757.

Yang, B., Geng, P., & Chen, G. (2015). One-dimensional structured IrO 2 nanorods modified membrane for electrochemical anti-fouling infiltration of oily wastewater. *Separation and Purification Technology, 156*, 931–941. doi:10.1016/j.seppur.2015.10.040

Yan, J., Han, L., Gao, W., Xue, S., & Chen, M. (2015). Biochar supported nanoscale zerovalent iron composite used as persulfate activator for removing trichloroethylene. *Bioresource Technology, 175*, 269–274. doi:10.1016/j.biortech.2014.10.103 PMID:25459832

YanJun, W., JingHui, Z., YiFei, T., & XinHua, X. (2009). Chromium (VI) reduction in aqueous solutions by Fe_3O_4-stabilized Fe0 nanoparticles. *Journal of Hazardous Materials, 172*(2/3), 1640–1645. PMID:19740609

Yan, W., Lien, H.-L., Koel, B. E., & Zhang, W.-X. (2013). Iron nanoparticles for environmental clean-up: Recent developments and future outlook. *Environmental Science. Processes & Impacts, 15*(1), 63–77. doi:10.1039/C2EM30691C PMID:24592428

Yao, Z., Li, J., Xie, H., & Yu, C. (2012). Review on remediation technologies soil contaminated by heavy metals. *Procedia Environmental Sciences, 16*, 722–729. doi:10.1016/j.proenv.2012.10.099

Yayintaş, Ö. T., Yilmaz, S., Türkoĝlu, M., Şolakoĝlu, F. A., & Şakir, F. (2007). Seasonal variation of some heavy metal pollution with environmental and microbiological parameters in sub-basin of Kocabas Stream (Biga, Canakkale, Turkey) by ICP-AES. *Environmental Monitoring and Assessment*, *134*(1), 321–331. doi:10.1007/s10661-007-9623-4 PMID:17345011

Yin, C. Y., Aroua, M. K., & Daud, W. M. A. W. (2008). Enhanced adsorption of metal ions onto polyethyleneimine-impregnated palm shell activated carbon: Equilibrium studies. *Water, Air, and Soil Pollution*, *192*(1-4), 337–348. doi:10.1007/s11270-008-9660-9

Yoonessi, M., & Gaier, J. R. (2010). Highly Conductive Multifunctional Graphene Polycarbonate Nanocomposites. *ACS Nano*, *4*(12), 7211–7220. doi:10.1021/nn1019626 PMID:21082818

Zafar, S., Aqil, F., & Ahmad, I. (2007). Metal tolerance and biosorption potential of filamentous fungi isolated from metal contaminated agricultural soil. *Bioresource Technology*, *98*(13), 2557–2561. doi:10.1016/j.biortech.2006.09.051 PMID:17113284

Zakhama, S., Dhaouadi, H., & M'henni, F. (2011). Nonlinear modelisation of heavy metal removal from aqueous solution using Ulva lactuca algae. *Bioresource Technology*, *102*(2), 786–796. doi:10.1016/j.biortech.2010.08.107 PMID:20855200

Zare-Dorabei, R., Ferdowsi, S. M., Barzin, A., & Tadjarodi, A. (2016). Highly efficient simultaneous ultrasonic-assisted adsorption of Pb (II), Cd (II), Ni (II) and Cu (II) ions from aqueous solutions by graphene oxide modified with 2, 2′-dipyridylamine: Central composite design optimization. *Ultrasonics Sonochemistry*, *32*, 265–276. doi:10.1016/j.ultsonch.2016.03.020 PMID:27150770

Zare, K., Najafi, F., Sadegh, H., & Ghoshekandi, R. (2013). Studies of ab initio and Monte Carlo simulation on interaction of fluorouracil anticancer drug with carbon nanotube. *Journal of Nanostructure in Chemistry*, *3*(1), 71. doi:10.1186/2193-8865-3-71

Zarghami, Z., Akbari, A., Latifi, A. M., & Amani, M. A. (2016). Design of a new integrated chitosan-PAMAM dendrimer biosorbent for heavy metals removing and study of its adsorption kinetics and thermodynamics. *Bioresource Technology*, *205*, 230–238. doi:10.1016/j.biortech.2016.01.052 PMID:26836608

Zeiner, M., Reziæ, I., & Steffan, I. (2006). TLC determination of allergenic and carcinogenic dyes. *Toxicology Letters*, *164*, S184. doi:10.1016/j.toxlet.2006.07.040

Zeiner, M., Reziæ, I., & Steffan, I. (2007). Analytical Methods for the Determination of Heavy Metals in the Textile Industry. *Kem. Ind.*, *56*(11), 587–595.

Zhang, S., Cheng, F., Tao, Z., Gao, F., & Chen, J. (2006). Removal of nickel ions from wastewater by Mg $(OH)_2$/MgO nanostructures embedded in Al_2O_3 membranes. *Journal of Alloys and Compounds*, *426*(1), 281–285. doi:10.1016/j.jallcom.2006.01.095

Zhang, S., Zeng, M., Xu, W., Li, J., Li, J., Xu, J., & Wang, X. (2013). Polyaniline nanorods dotted on graphene oxide nanosheets as a novel super adsorbent for Cr (vi). *Dalton Transactions (Cambridge, England)*, *42*(22), 7854. doi:10.1039/c3dt50149c

Zhang, Y. F., He, L. Y., Chen, Z. J., Wang, Q. Y., Qian, M., & Sheng, X. F. (2011). Characterization of ACC deaminase-producing endophytic bacteria isolated from coppertolerant plants and their potential in promoting the growth and copper accumulation of *Brassica napus. Chemosphere, 83*(1), 57–62. doi:10.1016/j.chemosphere.2011.01.041 PMID:21315404

Zhao, G., Li, J., Ren, X., Chen, C., & Wang, X. (2011). Few-layered graphene oxide nanosheets as superior sorbents for heavy metal ion pollution management. *Environmental Science & Technology, 45*(24), 10454–10462. doi:10.1021/es203439v PMID:22070750

Zhao, X. W., Zhou, M. H., Li, Q. B., Lu, Y. H., He, N., Sun, D. H., & Deng, X. (2005). Simultaneous mercury bioaccumulation and cell propagation by genetically engineered Escherichia coli. *Process Biochemistry, 40*(5), 1611–1616. doi:10.1016/j.procbio.2004.06.014

Zhitkovich, A. (2005). Importance of chromium– DNA adducts in mutagenicity and toxicity of chromium (VI). *Chemical Research in Toxicology, 18*(1), 3–11. doi:10.1021/tx049774+ PMID:15651842

Zhou, J. L., & Kiff, R. J. (1991). The uptake of copper from aqueous solution by immobilized fungal biomass. *Journal of Chemical Technology and Biotechnology (Oxford, Oxfordshire), 52*(3), 317–330. doi:10.1002/jctb.280520305

Zhou, J., Xia, B., Treves, D. S., Wu, L. Y., Marsh, T. L., O'Neill, R. V., ... Tiedje, J. M. (2002). Spatial and resource factors influencing high microbial diversity in soil. *Applied and Environmental Microbiology, 68*(1), 326–334. doi:10.1128/AEM.68.1.326-334.2002 PMID:11772642

Zhuang, X., Wan, Y., Feng, C., Shen, Y., & Zhao, D. (2009). Highly efficient adsorption of bulky dye molecules in wastewater on ordered mesoporous carbons. *Chemistry of Materials, 21*(4), 706–716. doi:10.1021/cm8028577

Zhu, L. N., Jun, M., & Yang, S. D. (2007). Removal of phenol by activated alumina bed in pulsed high-voltage electric field. *Journal of Environmental Sciences (China), 19*(4), 409–415. doi:10.1016/S1001-0742(07)60068-0 PMID:17915702

Zollinger, H. (1991). *Color Chemistry* (2nd ed.). Weinheim: VCH.

Zongo, I., Leclerc, J. P., Maïga, H. A., Wéthé, J., & Lapicque, F. (2009). Removal of hexavalent chromium from industrial wastewater by electrocoagulation: A comprehensive comparison of aluminium and iron electrodes. *Separation and Purification Technology, 66*(1), 159–166. doi:10.1016/j.seppur.2008.11.012

Zsoldos, F., Vashegyi, A., Bona, L., Pecsvaradi, A., & Szegletes, Z. (2000). Growth of and potassium transport in winter wheat and durum wheat as affects by various aluminium exposure times. *Journal of Plant Nutrition, 23*(7), 913–926. doi:10.1080/01904160009382070

About the Contributors

Ashok K. Rathoure (Independent Researcher) holds Doctorate degree in Bio-remediation. Currently he shares his knowledge and experience in the field of Environment Impact Assessment for M/s Eco Group of Companies, Surat India as GM-Consultancy. Previously he was associated with Vardan Environet Gurgaon, En-vision Group Surat (En-vision Environmental Services and En-vision Enviro Engineers Pvt. Ltd.) for EIA studies; Himachal Institute of Life Sciences Paonta and Beehive College of Ad. Studies Dehradun for teaching to Biotechnology, Microbiology, Biochemistry and other biosciences subjects. His area of research is environmental biotechnology and publication includes 70 full length research papers in international and national journals of repute, 24 Course books from reputed publishers in India, 8 Research books and 10 Book chapters. He is also associated as Editor-in-Chief for Octa Journal of Environmental Research, Managing Editor for Octa Journal of Biosciences and Executive Editor for Scientific India Magazine. He has traveled extensively in the major cities of Northern and Western India for his education and as a teacher, trainer and taught to widen the horizon of knowledge and to sharpen his intellect.

* * *

Vidushi Abrol is serving as Project Fellow and pursuing Ph.D. in Microbial Biotechnology Division at CSIR-Indian Institute of Integrative Medicine, Jammu. She received her B.Tech. and M.Tech. in Biotechnology from Lovely Professional University, Jalandhar, India in 2011 and 2013 respectively. In 2013-2014, she has served as Assistant Professor in C.T. Group of Institutes, Jalandhar, India. Her current research interests are isolation of secondary metabolites from important fungi, and their bioactive potential.

Bipin J. Agrawal, Professor with the The M.S. University of Baroda. Done Doctorate in the field of minimization of effluent load during the application of Reactive dyes. Have 76 publications and have presented 59 Research Papers at various Conferences and Seminars. Written 6 chapters for 4 books. Appointed the Chief Editor of one International Journal and Editorial Board Member of 56 other National/International Journals. Conducted Technical Sessions at several Conferences as Chairperson/President and included in the Advisory Committee by various National/International Conference Organizers. Associated with AICTE as an Expert Committee Member for Accreditation of various Textile Institutes all over India. Completed three Research Projects financed by UGC, AICTE and The M. S. University of Baroda respectively and at present undergoing a project under DST-PURSE scheme. Awarded Fellowship by six renowned Associations. Also awarded "Best International Teacher" award, "Indo Global Best Professor Award", "Guest of Honour", "Shiksha Rattan Puraskar", "Glory of India Gold Medal", "Best Educator Award 2014" and "Global teacher Role Model Award 2015" by various associations/organisations.

J. Anuradha (Assistant Professor) has received her doctoral degree in Pollution control and Environmental Technology from Pondicherry (Central) University. Her proficiency in field of Environmental Technology and Nanobiotechnology has been peer-recognized in the form of several international and national publications. She has thus far authored three patent abstracts; Twenty four research papers in high impact-factor international and national journals primarily in UGC approved Journals; Twenty eight of her papers have been taken part in international and nation scientific meet discussions. To add as an exemplary researcher, her research work has been awarded with nine best paper's in scientific discussions, and does presided as a 'Resource person and Subject specialist' in (ANVESION - West Zone) annual students science meet (2017) in India. She has been 'nominated' for the 'Dr APJ Abdul Kalam Innovation award' in Indian Academy of Biomedical Sciences (2017) and had delivered 'Resource person' lecture on 'Facile synthesis of gold nanoparticles and its application on cancer diagnosis and treatment'.

Aniruddha Chabukswar is currently working as a Professor in Pharmaceutical Chemistry at MAEER's Maharashtra Institute of Pharmacy, Pune 411 038. He has more than 17 years of teaching experience. He has done his B. Pharm from Pune University, M. Pharm from Mumbai University and completed his Ph. D. from Government college of Pharmacy, Karad. He has 04 books to his credit and contributed chapters for International book. He has published various research articles in reputed National and International journals. He has presented his research work at various seminar and conferences. His area of research includes drug design, synthesis of compounds and analytical method development.

Manisha Choudhary working as Assistant Professor since June 2014. Her research expertise in metal ion neurotoxicity. She is exploring neurochemical and behavioral changes during neurotoxicity. She has received Best PhD Scholar Award during her PhD and she has also received prestigious fellowship Award from ELSEVIER London to attend 12th Biennial conference of International Society of Developmental Neuroscience 2012. Dr. Choudhary received "Certificate of Merits" by the Nims University Jaipur during Women's Day 2016 for the outstanding contribution in the field of research and teaching. She has guided many graduate and postgraduate projects. She has published 25 research publications in national and international repute journal. She has attended more than 15 National and International conferences. She is the member of various International and National Societies (IAN, ISN, ACBI, SFRR, SOPI, and InSLAR).

Rajiv Dutta earned M.Tech. in Biotech. and Engg. IIT, Kharagpur and Ph.D. in Bioelectronics from BITS, Pilani. Prof. Dutta is Professor in Bioengineering at School of Engg. & Tech., Sharda University, India. He was also Faculty at the Department of Botany at Oklahoma State University, Stillwater, OK and Department of Biological Sciences at Purdue University, West Lafayette, IN, and carried out advanced research in the area of Ion channels electrophysiology and Membrane Nanopore. He is Fellow of American Association of Science & Technology (FAAST). He is Hon. Scientific Advisor to Serra BioLife Ltd., Melbourne, Australia. He had received various awards including JCI Outstanding Ten Young Person of India and ISBEM Dr. Ramesh Gulrajani Memorial Award 2006 and Outstanding Researcher Award-2011 at Ion Channel Retreat, Vancouver, Canada. He is Editorial Board Member of several journals including American Journal of Plant Physiology, American Journal of Science & Technology, International Journal of Engineering etc. He has published 44 publications in highly cited journals and three patents to his credits. Prof. Dutta visited USA, UK, Germany, France, Japan, Italy, and China.

Srijan Goswami is the founder Director of Indian School of Complementary Therapy & Allied Sciences (ISCTAS), a non-profit, non-government organization for promoting quality Education, Training, Treatment and social awareness about the various complementary medical sciences. He is currently associated as the Honorary Professor, Department of Biochemistry, Shree Krishna Biochemic College, India. And also associated as Assistant Professor, Department of Biotechnology, Institute of Genetic Engineering, India. He received his training in Medical Microbiology and Biochemistry from R.G.Kar Medical College and Calcutta National Medical College, Kolkata, India respectively. He is currently pursuing his post graduate medical training in the field of Biochemic System of Medicine from Indian Council of Biochemic Medicine and Yoga Therapy, India. He completed M.Sc in Medical

Biotechnology from The West Bengal University of Health Sciences, India in the year 2012. His research area of interest is medical informatics and medical sciences, biochemistry and complementary medical sciences. He has teaching and clinical research experience of around four years.

Swati Jagdale, M.Pharm, Ph.D, M.B.A., studied at Savitribai Phule Pune University. She is presently working as Professor and Head, Department of Pharmaceutics at MAEER's Maharashtra Institute of Pharmacy, MIT Campus, Kothrud, Pune, (MS), India. She has more than 16 years of teaching experience. She has published more than 80 National and International research articles in reputed journals. She has 04 books and three chapters to her credit. She has worked as a member of Board of Studies for Faculty of Pharmaceutics in Savitribai Phule Pune University. She has been awarded as Ideal Teacher award by MAEER's MIT and DR. G.R. Ekbote Women Pharmacist Award by MAHA-APTICON. She has guided more than 50 post graduate students and 01 Ph.D student. Presently 06 Ph.D and 06 M. Pharm students are working under her guidance. She is life member of Association of Pharmaceutical Teachers of India and Indian Pharmaceutical Association.

Sundeep Jaglan is working as Scientist in Microbial Biotechnology Division at CSIR-Indian Institute of Integrative Medicine, Jammu, India. Dr. Jaglan graduated in Biotechnology and obtained his Ph.D. from Guru Jambheshwar University of Science & Technology, Hisar, India. He has to his credit several international publications and patents. Dr. Jaglan is member of various professional bodies and reviewer for various prestigious journals. His research interests include microbial biotechnology, natural product chemistry and nanotechnology.

Shankar Khade is presently working as a PhD scholar on Enzyme Engineering at School of Biochemical Engineering, Indian Institute of Technology (BHU), Varanasi. He had achieved M. Tech. degree in the field of Biotechnology at NIT, Rourkela. He has published number of research articles in the reputed journals and also published book chapter on "Genetically modified microorganisms for the bioremediation of oil spills in marine environment." In addition to this, he has presented the research work at International conferences in India and Singapore and awarded the prizes for best poster presentations in two conferences. He has cleared national level competitive examinations like ASRB-NET, GATE Biotechnology (2011 & 2013).

Vinod Kumar has done Ph.D. in Chemistry from H.N. B. Garhwal University Srinagar Uttarakhand, INDIA. He has done MSc Biochemistry from same university. His research interest in Algal Biofuel, Biofuel, Biochemistry. He has more than 6 years teaching experience, Currently he is working as assistant professor in the Dept. of Chemistry at Uttaranchal University. He has published more than 26 research papers.

Sharada Mallubhotla, Ph. D., is an Associate Professor at Department of Biotechnology, Shri Mata Vaishno Devi University, Katra, J&K India with 21 years research experience in Plant Biotechnology. Has published 25 research articles in National and International journals. Has authored two books and is actively engaged in research on bioactive secondary Metabolites from Medicinal & Aromatic Plants, besides working on aspects for a safer and greener environment!

Sudhanshu Mishra is working as Lecturer of Microbiology in Department of Biotechnology, Nims University, Jaipur, Rajasthan, India. He has completed his PhD in Microbiology from Nims University, Jaipur, Rajasthan, India. He successfully completed a project work entitled a "Laccase Enzyme Production from Isolated Fungus and its Application in Pulp & Paper Industry" from Central Pulp and Paper Research Institute, Saharanpur, UP. He assisted several ongoing research projects in the department. He presented his research work as oral as well as poster presentation in various National & International conferences. His research area is to potentiate Anticataract, antimicrobial and antioxidants property of herbal medicines. He has published more than 10 research articles in National & International Journals.

Debraj Modak is currently associated as Assistant Professor in the Department of Biotechnology at Institute of Genetic Engineering, India. His area of specialization is Plant Tissue Culture techniques and Plant Biotechnology. He has 12 years of experience in tissue culture techniques.

Munmun Modak is currently associated as a lecturer in the Department of Microbiology, Institute of Genetic Engineering, India. She has teaching experience of about 7 years. Her area of specialization is Microbiology.

Manisha Nanda has done Ph.D. in Biotechnology from H.N. B. Garhwal University Srinagar Uttarakhand, India. She has done MSc Biotechnology from same university. His research interest in Bioremediation and Biofuel. She has more than 5 years teaching experience, Currently he is working as assistant professor in the Dept. of Biotechnology, Dolphin (PG) Institute of Biomedical and Natural Sciences. She has published more than 10 research papers.

Pragati Sahai is M.Sc. M.Phil, and holds 10 years of research and teaching experience with brands like Amity University, in department of Biotechnology, where she taught and research guided M.Tech,M.Sc., B.Tech and B.Sc. and Int.M.Tech students. She was also given certificate for her exemplary teaching pedagogy by Dale Carnegie foundation as HITS qualified Academician. She has submitted two projects in field of Bioremediation catering to environmental concerns in MoEF and DST, India. Recently she has published a chapter "Role of Carriers in Bioremediation" in book "Probiotics and Plant Heath", Springer. she is an active researcher working under Prof. Rajiv Dutta, a renowned researcher in field of Nanotechnology and Biochem. Engg in dept. of Biotech. at Sharda University. They both have published papers in international and National conferences in Bioelectromagnetism and Biophysics. Their recent areas of development are in fields of Nanomedicine and Nanobioremediation.

R. Sanjeevi received his doctoral degree in Environmental Science & Engineering from Pondicherry (Central) University and was awarded with Post Doctoral Fellowship form CSIR, New Delhi at selfsame institute. He has been peer-recognized with Senior Research Fellowship, awarded form CSIR, New Delhi during his doctoral research. His proficiency in field of Bioprocess technology has been peer-recognized in the form of several international and national publications. He has published papers in high impact-factor international journals primarily in Bioresource Technology (a paper referred with Scopus citation of 227) and Applied Biochemistry & Biotechnology. To add as an exemplary researcher, he has been awarded for three best paper's in scientific discussions, and does presided as a 'Resource person and Subject specialist' in (ANVESION - West Zone) annual students science meet (2017) in India.

Jyoti Srivastava has completed M.Sc. Biotechnology from Utkal University, Odisha. She has 3 years of teaching academic experience. Presently she is working on CoQ10 production and scale up from microbial source at IIT (BHU), Varanasi.

Pradeep Srivastava had completed PhD degree from School of Biochemical Engineering, IIT (BHU). He has expertise in the area of Instrumentation and process control, downstream processing, tissue engineering, biochemical engineering. He has wide experience in the industries, academic as well as entrepreneurship field and has published number of research papers as well as books.

Suresh Srivastava has expertise in the area of Enzyme Engineering, Food processing, Biochemical Engineering and Waste water treatment management. He has number of publications in the area of bioremediation of azo dyes. He has more than 30 years of academic and research experience and guided number of PhD students.

Abhishek Tripathi has expertise in the field of Biochemical Engineering, Enzyme Engineering, Food processing, Biotechnology, waste water management, etc.

Sandeep Tripathi is the Associate Professor of Biochemistry and Director, Institute of Public Health and Nutrition. His research degree was based on "Metal ion neurotoxicity" which was awarded as "Best PhD thesis" among the batch 2010 in the King George's Medical University, Lucknow, India. He has published 68 research articles and many book chapters. He has awarded Mrs. Abida Mehdi Award for Neuroscience Research by Indian Academy of Biomedical Sciences and Young Scientist Award by centre for education growth and research. He has established "National referral centre for fluoride poisoning (NRCFPI) in India", it was first fluoride referral centre in the country and continuously leading by Dr. Tripathi as Head, Research and Development and he is the recipient of Niloufer Chinoy Award- 2013 presented by the International Society for fluride Research. Dr. Tripathi is the member of editorial board of many journals. He is the member of various International and National Societies (IBRO, ISNR, SOT, IAN, ACBI, SFRR, SOPI, SPGHTN and InSLAR). He has attended more than 50 conferences and workshops as speaker/ chairperson / moderator or as delegate.

Index

O

P

R

S

T

U

W

Purchase Print, E-Book, or Print + E-Book

IGI Global books can now be purchased from three unique pricing formats: Print Only, E-Book Only, or Print + E-Book. Shipping fees apply.

www.igi-global.com

Recommended Reference Books

ISBN: 978-1-4666-9553-5
© 2016; 410 pp.
List Price: $210

ISBN: 978-1-5225-0302-6
© 2016; 362 pp.
List Price: $190

ISBN: 978-1-4666-9845-1
© 2016; 1,997 pp.
List Price: $1,845

ISBN: 978-1-5225-0803-8
© 2017; 1,647 pp.
List Price: $1,850

ISBN: 978-1-4666-9723-2
© 2016; 438 pp.
List Price: $240

ISBN: 978-1-4666-8814-8
© 2016; 711 pp.
List Price: $325

Looking for free content, product updates, news, and special offers?

Join IGI Global's mailing list today and start enjoying exclusive perks sent only to IGI Global members. Add your name to the list at **www.igi-global.com/newsletters.**

Publishing Information Science and Technology Research Since 1988

IGI Global
DISSEMINATOR OF KNOWLEDGE

www.igi-global.com Sign up at www.igi-global.com/newsletters facebook.com/igiglobal twitter.com/igiglobal

Stay Current on the Latest Emerging Research Developments

Become an IGI Global Reviewer for Authored Book Projects

The overall success of an authored book project is dependent on quality and timely reviews.

In this competitive age of scholarly publishing, constructive and timely feedback significantly decreases the turnaround time of manuscripts from submission to acceptance, allowing the publication and discovery of progressive research at a much more expeditious rate. Several IGI Global authored book projects are currently seeking highly qualified experts in the field to fill vacancies on their respective editorial review boards:

Applications may be sent to:
development@igi-global.com

Applicants must have a doctorate (or an equivalent degree) as well as publishing and reviewing experience. Reviewers are asked to write reviews in a timely, collegial, and constructive manner. All reviewers will begin their role on an ad-hoc basis for a period of one year, and upon successful completion of this term can be considered for full editorial review board status, with the potential for a subsequent promotion to Associate Editor.

If you have a colleague that may be interested in this opportunity, we encourage you to share this information with them.

www.igi-global.com

InfoSci®-Books

A Database for Information Science and Technology Research

Maximize Your Library's Book Collection!

Invest in IGI Global's InfoSci®-Books database and gain access to hundreds of reference books at a fraction of their individual list price.

The InfoSci®-Books database offers unlimited simultaneous users the ability to precisely return search results through more than 80,000 full-text chapters from nearly 3,900 reference books in the following academic research areas:

Business & Management Information Science & Technology • Computer Science & Information Technology
Educational Science & Technology • Engineering Science & Technology • Environmental Science & Technology
Government Science & Technology • Library Information Science & Technology • Media & Communication Science & Technology
Medical, Healthcare & Life Science & Technology • Security & Forensic Science & Technology • Social Sciences & Online Behavior

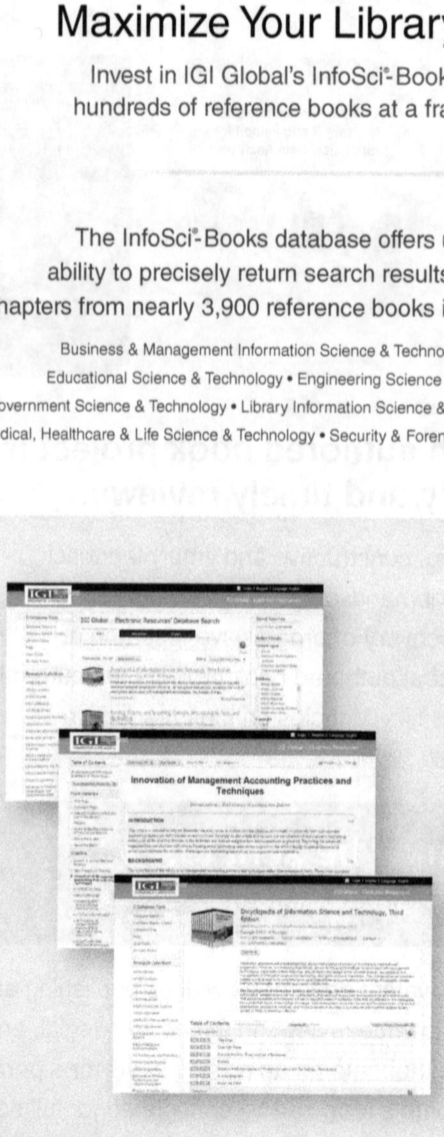

Peer-Reviewed Content:

- Cutting-edge research
- No embargoes
- Scholarly and professional
- Interdisciplinary

Award-Winning Platform:

- Unlimited simultaneous users
- Full-text in XML and PDF
- Advanced search engine
- No DRM

Librarian-Friendly:

- Free MARC records
- Discovery services
- COUNTER4/SUSHI compliant
- Training available

To find out more or request a free trial, visit:
www.igi-global.com/eresources

IGI Global
DISSEMINATOR OF KNOWLEDGE
www.igi-global.com

IGI Global
DISSEMINATOR OF KNOWLEDGE
www.igi-global.com

IGI Global Proudly Partners with

eContent Pro International

Expert Copy Editing

eContent Pro International copy editors, with over 70 years of combined experience, will provide complete and comprehensive care for your document by resolving all issues with spelling, punctuation, grammar, terminology, jargon, semantics, syntax, consistency, flow, and more. In addition, they will format your document to the style you specify (APA, Chicago, etc.). All edits will be performed using Microsoft Word's Track Changes feature, which allows for fast and simple review and management of edits.

Additional Services

eContent Pro International also offers fast and affordable proofreading to enhance the readability of your document, professional translation in over 100 languages, and market localization services to help businesses and organizations localize their content and grow into new markets around the globe.

IGI Global Authors Save 25% on eContent Pro International's Services!

Scan the QR Code to Receive Your 25% Discount

The 25% discount is applied directly to your eContent Pro International shopping cart when placing an order through IGI Global's referral link. Use the QR code to access this referral link. eContent Pro International has the right to end or modify any promotion at any time.

Email: customerservice@econtentpro.com

econtentpro.com

Information Resources Management Association

Advancing the Concepts & Practices of Information Resources Management in Modern Organizations

Become an IRMA Member

Members of the **Information Resources Management Association (IRMA)** understand the importance of community within their field of study. The Information Resources Management Association is an ideal venue through which professionals, students, and academicians can convene and share the latest industry innovations and scholarly research that is changing the field of information science and technology. Become a member today and enjoy the benefits of membership as well as the opportunity to collaborate and network with fellow experts in the field.

IRMA Membership Benefits:

- **One FREE Journal Subscription**
- **30% Off Additional Journal Subscriptions**
- **20% Off Book Purchases**
- Updates on the latest events and research on Information Resources Management through the IRMA-L listserv.
- Updates on new open access and downloadable content added to Research IRM.
- A copy of the Information Technology Management Newsletter twice a year.
- A certificate of membership.

IRMA Membership $195

Scan code or visit **irma-international.org** and begin by selecting your free journal subscription.

Membership is good for one full year.

www.irma-international.org

IGI Global
DISSEMINATOR OF KNOWLEDGE
www.igi-global.com

Available to Order Now

Order through www.igi-global.com with **Free Standard Shipping**.*

The Premier Reference for Information Science & Information Technology

100% Original Content

Contains 705 new, peer-reviewed articles with color figures covering over 80 categories in 11 subject areas

Diverse Contributions

More than 1,100 experts from 74 unique countries contributed their specialized knowledge

Easy Navigation

Includes two tables of content and a comprehensive index in each volume for the user's convenience

Highly-Cited

Embraces a complete list of references and additional reading sections to allow for further research

Included in:
InfoSci®-Books

Encyclopedia of Information Science and Technology Fourth Edition

A Comprehensive 10-Volume Set

Mehdi Khosrow-Pour, D.B.A. (Information Resources Management Association, USA)
ISBN: 978-1-5225-2255-3; © 2018; Pg: 8,104; Release Date: July 31, 2017

The **Encyclopedia of Information Science and Technology, Fourth Edition** is a 10-volume set which includes 705 original and previously unpublished research articles covering a full range of perspectives, applications, and techniques contributed by thousands of experts and researchers from around the globe. This authoritative encyclopedia is an all-encompassing, well-established reference source that is ideally designed to disseminate the most forward-thinking and diverse research findings. With critical perspectives on the impact of information science management and new technologies in modern settings, including but not limited to computer science, education, healthcare, government, engineering, business, and natural and physical sciences, it is a pivotal and relevant source of knowledge that will benefit every professional within the field of information science and technology and is an invaluable addition to every academic and corporate library.

Scan for Online Bookstore

Pricing Information

Hardcover: **$5,695** E-Book: **$5,695** Hardcover + E-Book: **$6,895**

Recommend this Title to Your Institution's Library: www.igi-global.com/books

www.ingramcontent.com/pod-product-compliance
Lightning Source LLC
LaVergne TN
LVHW081349300926
849161LV00008B/2104